CELLULAR AND MOLECULAR APPROACHES TO FISH IONIC REGULATION

This is Volume 14 in the

FISH PHYSIOLOGY series

Edited by W. S. Hoar, D. J. Randall, and A. P. Farrell

A complete list of books in this series appears at the end of the volume.

CELLULAR AND MOLECULAR APPROACHES TO FISH IONIC REGULATION

Edited by

CHRIS M. WOOD

Department of Biology
McMaster University
Hamilton, Ontario, Canada

TREVOR J. SHUTTLEWORTH

Department of Physiology
University of Rochester
School of Medicine and Dentistry
Rochester, New York

ACADEMIC PRESS

San Diego New York Boston London Sydney Tokyo Toronto

Cover photograph: Confocal optical cross-section (0.5 m) of gills of 3-day-old guppy, *Poecilia reticulata*, after *in vivo* colabeling with DASPMI (green), to visualize the contours of the mitochondria-rich chloride cells, and concanavalin-A-fluorescein (red), to stain the glycoproteins in the apical crypt. Filament tips point to the right; apical crypts are at the epithelium–water interface. Note that essentially all cells may be considered mature. Magnification 400x. Courtesy of Li Jie, Gert Flik, and James C. Fenwick from the Universities of Nijmegen and Ottawa.

This book is printed on acid-free paper. ♾

Academic Press, Inc.
A Division of Harcourt Brace & Company
525 B Street, Suite 1900, San Diego, California 92101-4495

United Kingdom Edition published by
Academic Press Limited
24-28 Oval Road, London NW1 7DX

Library of Congress Cataloging-in-Publication Data

Hoar, William Stewart, date.
Fish physiology.

Vols. 8-12 edited by W.S. Hoar [et al.].
Vol. 13 edited by Nancy M. Sherwood, Choy L. Hew.
Vol. 14 edited by Chris M. Wood, Trevor J. Shuttleworth.
Includes bibliographies and indexes.
Contents: v. 1. Excretion, ionic regulation, and metabolism. -- [etc.] -- v. 13. Molecular endocrinology of fish -- v. 14. Cellular and molecular approaches to fish ionic regulation.
1. Fishes--Physiology--Collected works. I. Randall, David J., date. II. Conte, Frank P., date. III. Title.
QL639.1.H6 597'.01 76-84233
ISBN 0-12-350438-4 (v. 14)
PRINTED IN THE UNITED STATES OF AMERICA
95 96 97 98 99 00 BB 9 8 7 6 5 4 3 2 1

CONTENTS

4. Secretory Electrolyte Transport in Renal Proximal Tubules of Fish

Klaus W. Beyenbach

5. Functional Morphology of the Elasmobranch Nephron and Retention of Urea

Eric R. Lacy and Enrico Reale

6. Solute Transport by Flounder Renal Cells in Primary Culture

J. Larry Renfro

7. Phenotypic Expression and Natriuretic Peptide-Activated Chloride Secretion in Cultured Shark *(Squalus acanthias)* Rectal Gland Epithelial Cells

John D. Valentich, Karl J. Karnaky, Jr., and William M. Moran

8. Primary Cultures of Teleost Branchial Epithelial Cells

Peter Pärt and Elisabeth Bergström

9. Proton Pumps in Fish Gills

Hong Lin and David Randall

10. Ultrastructural and Morphometric Studies on Ion and Acid–Base Transport Processes in Freshwater Fish

Greg Goss, Steve Perry, and Pierre Laurent

11. Hormonal Control of Gill Na^+,K^+-ATPase and Chloride Cell Function

Stephen D. McCormick

12. Calcium Transport Processes in Fishes

Gert Flik, Pieter M. Verbost, and Sjoerd E. Wendelaar Bonga

CONTRIBUTORS

Numbers in parentheses indicate the pages on which the authors' contributions begin.

PAUL M. ANDERSON *(57), Department of Biochemistry and Molecular Biology, University of Minnesota, Duluth, Duluth, Minnesota 55812*

ELISABETH BERGSTRÖM *(207), Department of Environmental Toxicology, Uppsala University, S-752 36 Uppsala, Sweden*

KLAUS W. BEYENBACH *(85), Section of Physiology, Cornell University, Ithaca, New York 14853*

GERT FLIK *(317), Department of Animal Physiology, Faculty of Science, University of Nijmegen, NL-6525 ED Nijmegen, The Netherlands*

GREG GOSS *(257), Division of Cell Biology, Hospital for Sick Children, Toronto, Ontario, Canada M5G 1X8*

KARL J. KARNAKY, JR. *(173), Department of Cell Biology and Anatomy and Marine Biomedical and Environmental Sciences Program, Medical University of South Carolina, Charleston, South Carolina 29425, and Mount Desert Island Biological Laboratory, Salsbury Cove, Maine 04672*

ERIC R. LACY *(107), Department of Cell Biology and Anatomy, Medical University of South Carolina, Charleston, South Carolina 29425*

PIERRE LAURENT *(257), Laboratoire de Morphologie Fonctionelle et Ultrastructurale des Adaptations, Centre d'Ecologie et de Physiologie Energetique, CNRS, F-67037 Strasbourg, France*

HONG LIN *(229), Department of Zoology, University of British Columbia, Vancouver, British Columbia, Canada V6T 1Z4*

CHRISTOPHER A. LORETZ *(25), Department of Biological Sciences, State University of New York at Buffalo, Buffalo, New York 14260*

WILLIAM S. MARSHALL *(1), Department of Biology, St. Francis Xavier University, Antigonish, Nova Scotia, Canada B2G 2W5*

STEPHEN D. MCCORMICK *(285), Anadromous Fish Research Center, National Biological Service, Turners Falls, Massachusetts 01376*

WILLIAM M. MORAN *(173), Department of Biology, University of Central Arkansas, Conway, Arkansas 72035*

PETER PÄRT *(207), Department of Environmental Toxicology, Uppsala University, S-752 36 Uppsala, Sweden*

STEVE PERRY *(257), Department of Biology, University of Ottawa, Ottawa, Ontario, Canada K1N 6N5*

DAVID RANDALL *(229), Department of Zoology, University of British Columbia, Vancouver, British Columbia, Canada V6T 1Z4*

ENRICO REALE *(107), Laboratory of Cell Biology and Electron Microscopy, Hannover Medical School, 30625 Hannover, Germany*

J. LARRY RENFRO *(147), Department of Physiology and Neurobiology, University of Connecticut, Storrs, Connecticut 06269*

JOHN D. VALENTICH *(173), Department of Internal Medicine, University of Texas Medical Branch, Galveston, Texas 77555, and Mount Desert Island Biological Laboratory, Salsbury Cove, Maine 04672*

PIETER M. VERBOST *(317), Department of Animal Physiology, Faculty of Science, University of Nijmegen, NL-6525 ED Nijmegen, The Netherlands*

SJOERD E. WENDELAAR BONGA *(317), Department of Animal Physiology, Faculty of Science, University of Nijmegen, NL-6525 ED Nijmegen, The Netherlands*

PREFACE

The debut volume of **Fish Physiology** focused largely on ion regulation and summarized the work of almost a century, most of it at the level of the whole animal. Even though over a quarter-century old, its utility has stood the test of time; many of its chapters remain fundamental references in the field. In 1984, another **Fish Physiology** volume moved the focus to a more mechanistic level with comprehensive descriptions of gill fine structure and transport processes. Since then, the past decade has seen an explosive development of new cellular, molecular, tissue culture, biochemical, and electrophysiological approaches for studying the physiology of transport and synthesis. Many of these powerful techniques are now being used to dissect the mechanisms of ion regulation and osmolyte metabolism, and the knowledge base is changing rapidly. When we conceived of this volume in 1993, it seemed a particularly opportune time for a book emphasizing these new approaches and their application to the ion-regulatory physiology of fish. Rather than trying to cover all areas of ion-regulatory physiology, our interest has been in the new approaches themselves, and we recruited individuals and groups who have been developing and exploiting these new approaches most effectively. Authors were encouraged to be insightful rather than exhaustive in their reviews, and to keep their focus at the cellular level. At the same time, they were encouraged to relate their chapters to a whole animal framework, but not to provide detail at that level.

We are very grateful to the authors for their cooperation with this difficult mandate, and for their enthusiasm and dedication to the project, and to the many reviewers for their constructive guidance. We thank Bill Hoar, Dave Randall, and Tony Farrell for the legacy of this series, and Dave in particular for encouraging us in this effort. The support of Chuck Crumly and the staff at Academic Press is also greatly appreciated. Finally, the editors thank DeLoach Vineyards for providing the medium in which the idea for this volume was initiated, and in which it differentiated into its final form.

CHRIS M. WOOD
TREVOR J. SHUTTLEWORTH

1

TRANSPORT PROCESSES IN ISOLATED TELEOST EPITHELIA: OPERCULAR EPITHELIUM AND URINARY BLADDER

WILLIAM S. MARSHALL

I. INTRODUCTION

The isolated opercular epithelium has provided many insights into the operation of seawater chloride cells. Although a major goal has been to study chloride cells in a model of the gill epithelium, what should not be ignored is that the skin and opercular epithelium are in themselves osmoregulatory organs that contribute significantly to ion balance of the whole animal. Because of the success of the seawater opercular epithelia of tilapia (*Oreochromis mossambicus*), killifish (*Fundulus heteroclitus*), and the longjawed mudsucker (*Gillichthys mirabilis*), there has been a long search for a skin epithelium from a *freshwater* animal that could help resolve some of the mechanisms of ion uptake in freshwater fish. From attempts to date, teleost chloride cells apparently do not proliferate in primary culture of rainbow trout (*Oncorhynchus mykiss*) gill (Pärt *et al.*, 1993) and opercular epithelia (Marshall *et al.*, 1995b), but the use of

CELLULAR AND MOLECULAR APPROACHES
TO FISH IONIC REGULATION

short-term primary cultures has allowed patch clamp experiments to be performed on chloride cells.

The urinary bladder epithelium of teleosts has been studied, mounted in Ussing-style membrane chambers, and recent studies have used patch clamp and microelectrode methodology. Although the urinary bladder does not model the secretory functions of the teleost nephron, the urinary bladder is an important location for ion re-uptake from urine in freshwater fish; also it serves to reabsorb ions and water from urine of seawater teleosts.

II. ION TRANSPORT BY CHLORIDE CELLS

A. Ion Secretion in Seawater Opercular Epithelia

1. The Current Model for NaCl Secretion

The current model for ion transport by chloride cells in seawater is a modification of that proposed initially by Silva *et al.* (1977). The model is depicted in Fig. 1A. More exhaustive reviews include Maetz (1970), Zadunaisky (1984), Karnaky (1986), Péqueux *et al.* (1988), and Wood and Marshall (1994). The Ca^{2+} transport pathway is not shown; for this see Flik *et al.* (Chapter 12, this volume). Cl^- secretory current varies with the density of chloride cells (Marshall and Nishioka, 1980; Karnaky *et al.*, 1984) and by the vibrating probe technique Cl^- current was localized specifically to these cells (Foskett and Scheffey, 1982), hence chloride cells are responsible for ion secretion in the gill and opercular epithelium. The ultimate driving force for Cl^- secretion is the Na^+ electrochemical gradient established by Na^+,K^+-ATPase; the enzyme is localized to the basolateral membrane, based on tritiated ouabain autoradiography (Karnaky *et al.*, 1976) and ultracytochemical localization of K^+-NPPase (K^+-dependent phosphatase activity) on the tubular system of chloride cells (Hootman and Philpott, 1979). In addition, ouabain on the basal side but not the apical side rapidly inhibits the ion transport (Marshall, 1977; see Section II,A,4). Cl^- enters the cell across the basolateral membrane in a Na^+-dependent (Marshall, 1981a; Degnan, 1984) cotransport that is inhibited by the "loop" diuretics furosemide (Degnan *et al.*, 1977) and bumetanide (Eriksson *et al.*, 1985; Péqueux *et al.*, 1988), but not by thiazide-type diuretics (Eriksson *et al.*, 1985; Eriksson and Wistrand, 1986). Cl^- accumulates intracellularly such that Cl^- exit occurs down its electrochemical gradient through anion channels in the apical membrane (Marshall *et al.*, 1995b; see Section II,A,3). The accumulation of K^+ intracellularly by Na^+,K^+-ATPase and presumably also by $Na^+,K^+,2Cl^-$

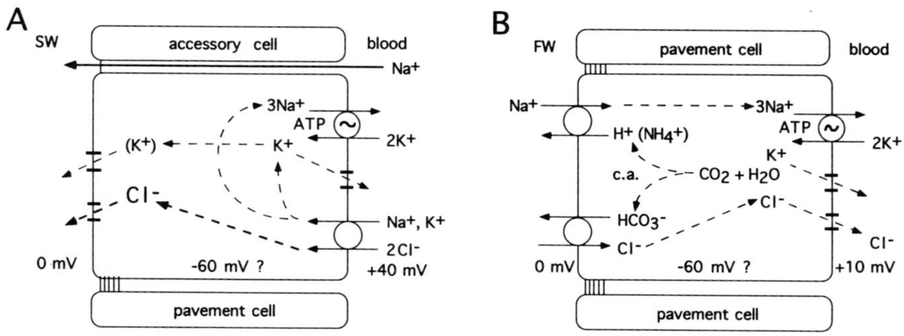

Fig. 1. (A) Diagrammatic summary of a model of marine teleost chloride cell function. Passive diffusion is indicated by dashed lines; active and cotransport by solid lines. Inhibitors are adjacent to transport entities. "Leaky" intercellular junctions are single strands; "tight" intercellular junctions are multistranded. The paracellular pathway is selective for cations and favors Na^+ efflux aided by the serosa positive transepithelial potential. (B) A model of freshwater mitochondria-rich teleost gill epithelium. Conventions as per part A.; c.a. is carbonic anhydrase. See text for details.

cotransport (see Section II,A,4) in turn is limited by basolateral K^+ conductance that is inhibited by Ba^{2+} (Degnan, 1985). Na^+ secretion occurs down its electrochemical gradient via a cation-selective paracellular pathway (Degnan and Zadunaisky, 1980; Marshall, 1981a) that is located between chloride cells and the smaller adjacent (or accessory) cells and is permeable to La^{3+} (Sardet *et al.*, 1979; Hootman and Philpott, 1980). There is also a smaller secondary active K^+ secretion (Marshall, 1981b). Although this is the current model, many uncertainties remain in the mechanisms themselves and in the regulation of the transport.

2. Regulation of Cl^- Secretion

Bern and Madsen (1992) provide a complete review of the endocrinology of teleost osmoregulation and the adaptive responses, particularly to prolactin in freshwater acclimation and to cortisol and growth hormone in seawater acclimation. Here only rapid-acting hormones and neurotransmitters will be considered. The effects and second messengers of several rapid-acting hormones are summarized in Table I. It is not clear whether the chloride cells would be physiologically exposed to urotensin I and II, eicosanoids, glucagon, or vasoactive intestinal polypeptide (VIP), but certainly there appears to be multihormonal control of Cl^- secretion by Cl^- cells. The second messenger for glucagon is cAMP, based on adenylate cyclase stimulation of rainbow trout gill epithelial cells (Guibbolini and Lahlou, 1987). The eicosanoids may themselves be second messengers in these responses (Van Praag *et al.*, 1987). Prostaglandin E_1, for instance,

Table I
Summary of Rapid Hormonal Effects on Cl^- Secretion by Marine Teleost Skin and Opercular Epithelium

Hormone or neurotransmitter	Effect	Intracellular messenger	Reference
Clonidine (α_2) and epinephrine	↓	Ca^{2+}	Marshall *et al.* (1993)
Urotensin II	↓	?	Marshall and Bern (1979)
Acetylcholine (musc)	↓	?	May and Degnan (1985)
Prostaglandin E_2	↓	?	Van Praag *et al.* (1987)
Isoproterenol (β)[a]	↑	cAMP	Mendelsohn *et al.* (1981)
Glucagon	↑	cAMP	Foskett *et al.* (1982a)
Urotensin I	↑	cAMP?	Marshall and Bern (1979)
VIP	↑	cAMP?	Foskett *et al.* (1982a)
Leukotrienes C_4,D_4,E_4	↑	?	Van Praag *et al.* (1987)

[a] Nonspecific agonists (epinephrine, norepinephrine) inhibit Cl^- secretion.

stimulates adenylate cyclase in trout gills (Guibbolini and Lahlou, 1987), but its effect on ion transport is unknown. The most physiologically relevant response is the inhibition of Cl^- secretion by epinephrine via α_2 adrenoceptors because the application of the endogenous nonspecific agonists (epinephrine and norepinephrine) consistently produces inhibitory effects. Also a catecholamine-mediated inhibition of Cl^- secretion is fully consistent with the stress-induced inhibition of Cl^- secretion in the "instantaneous" reduction in ion secretion when seawater-adapted killifish are transferred to fresh water (Potts and Evans, 1967; Maetz *et al.*, 1967). The intracellular mediator of this effect is a rise in intracellular Ca^{2+}, based on an inhibition of Cl^- secretion by ionomycin (a Ca^{2+} ionophore) and by thapsigargin (an inhibitor of intracellular Ca^{2+} ATPase), and a "blunting" of the clonidine effect by Ca^{2+}-deficient bathing solutions (Marshall *et al.*, 1993) and by the calmodulin blocker trifluoperazine (May and Degnan, 1985; Marshall *et al.*, 1993). Interestingly, the less efficient Ca^{2+} ionophore A23187 was repeatedly ineffective (May and Degnan, 1985; Marshall *et al.*, 1993), a result that diverted attention from Ca^{2+} as a suspected intracellular messenger for almost a decade. The teleost α_2 receptors appear to be unusual, because it is the α_1 receptors that are normally associated with mediation via Ca^{2+}. Of particular importance is the fact that while Ca^{2+} stimulates Cl^- transport in many epithelia, such as Cl^- secretion in airway (reviewed by Riordan, 1993) and Cl^- uptake by colonic epithelia (Tabcharani *et al.*, 1990), the effect is opposite in the teleost opercular epithelium. This teleostean system is unusual because in most systems it is α_1 receptors (not α_2 receptors) that are normally associated with mediation via Ca^{2+}.

3. Apical Membrane Events

A hormonally regulated apical membrane Cl^- conductance has been generally accepted to be the means of Cl^- exit in Cl^- secreting epithelia (Klyce and Wong, 1977). Strong inhibition of Cl^- secretion by α_2 adrenergic agonists reproducibly produces a small reduction in total epithelial conductance (Péqueux *et al.*, 1988; Marshall *et al.*, 1993), but because of the high conductance (a result of a relatively permeable paracellular shunt pathway in the epithelium; Sardet *et al.*, 1979; Hootman and Philpott, 1980), an accurate estimate of the conductance change of the apical membrane per se is not possible. The opercular epithelium has a relatively high transepithelial conductance, about 6.0–8.0 mS · cm^{-2} (Degnan *et al.*, 1977; Péqueux *et al.*, 1988), a result of the cation-selective paracellular shunt. For this reason, even large changes in the apical membrane resistance may produce only a small change in the transepithelial resistance. Foskett *et al.* (1982b) and Péqueux *et al.* (1988) have argued previously that this slight change in the tissue resistance is less than compelling evidence that clonidine acts exclusively to close apical membrane Cl^- channels and that a basolateral location may be as likely.

In some cases inhibition of Cl^- secretion in fish skin is also connected to decreases in conductance, as is true for inhibitions by mucosally added Cu^{2+} (Degnan, 1985, 1986) and diphenylamine-2-carboxylate (DPC) (Marshall *et al.*, 1995b) and serosally added ouabain and thiocyanate (SCN^-) (Marshall, 1981a). These results are consistent with reductions in apical membrane Cl^- conductance either by direct inhibition of the channels (by DPC and Cu^{2+}) or indirectly by reducing the availability of Cl^- intracellularly (following ouabain inhibition of Na^+,K^+-ATPase). Similarly, stimulation of Cl^- secretion, if apical membrane Cl^- channels are involved, should increase epithelial conductance. The only published intracellular recordings from opercular epithelial cells (Zadunaisky *et al.*, 1988) indicate an apical membrane potential of only -18 mV, which is too small to drive Cl^- outward across the apical membrane into seawater. Although the likely problem is shunting around the shank of the microelectrode when the tubular system of chloride cells is disrupted, this study showed that stimulation of Cl^- secretion by isoproterenol did reduce the apical membrane fractional resistance, suggestive of an increase in apical membrane conductance coincident with stimulation of Cl^- secretion (Zadunaisky *et al.*, 1988). Hence there is good reason to suspect a hormonally regulated apical membrane anion channel in chloride cells. Because of difficulties in obtaining acceptable microelectrode impalements of Cl^- cells, the preferred technique to resolve the issue is the patch clamp (Sakmann and Neher, 1984).

In patch clamp experiments it is necessary that the membrane surface be as clean as possible to enhance the probability of the formation of a high-resistance seal (on the order of several GΩ; Sakmann and Neher, 1984) between the micropipette glass and the membrane. Chloride cells *in situ* have a small apical membrane surface in an invaginated apical crypt, and the crypt is often filled with a "fuzzy coat" of (presumably) a polyanionic mucus (e.g., Marshall and Nishioka, 1980; Karnaky, 1986). Because GΩ seal formation *in situ* was unlikely to succeed, we adapted previous methods (Marshall and Hanrahan, 1991) for short-term primary culture of opercular epithelium (Marshall *et al.*, 1995b). In culture, cells with numerous mitochondria were present, as detected by fluorescent labeling with the mitochondrial stain dimethylaminostyrylethylpyridinium iodide (DASPEI; Marshall and Nishioka, 1980; Marshall *et al.*, 1995b). These large DASPEI-positive cells were present in young (< 48 hr) primary cultures and were absent from older cultures. We frequently (55 of 156 seals in control cells) identified a low-conductance channel averaging 8.8 pS in cell-attached patches of the apical membranes of these cells. The channel had slight inward rectification (Fig. 2), inactivated upon excision of the on-cell patch, and was not voltage sensitive (Marshall *et al.*, 1995b). The inactivation on excision and the low conductance are characteristics similar to those of the low-conductance channel of anion-transporting

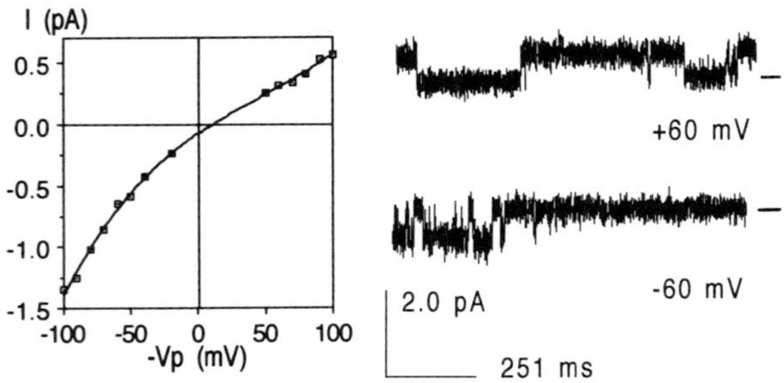

Fig. 2. Example electrophysiology of the low-conductance cAMP-activated anion channel of marine killifish (*Fundulus heteroclitus*) opercular epithelial cells in primary culture. Left: Current–voltage plot of a cell-attached patch with 150 m*M* NaCl plus 0.1 m*M* $CaCl_2$ in the pipet and 150 m*M* NaCl plus 1.0 m*M* $CaCl_2$ in the bath. Voltages are with respect to ground in the pipet ($-V_p$); current is in picoamps (pA). At physiological voltages ($-V_p$ = -60), current is 0.6 pA and conductance about 10 pS. The slight inward rectification may result from high intracellular [Cl^-]; the curve is fitted by second-order polynomial ($r > 0.99$). Right: Example traces of cell-attached patch low-pass filtered to 1.0 kHz. Closed state indicated by marginal lines. Openings at positive voltages are up; downward deflections represent inward (+) current.

epithelia known as Cystic Fibrosis Transmembrane Conductance Regulator, CFTR (Gray *et al.*, 1989; Tabcharani *et al.*, 1990; Riordan, 1993). In addition, when the cultures were stimulated with 0.1 m*M* isobutylmethylxanthine and 1.0 m*M* dibutyryl-cAMP, there was a significant increase in the incidence of the channel from 35.3% to 61.9% of patches and in the occurrence of multiple copies of the channel in the patches (from 2.2% to 38.5% of patches). Hence the channel is activated by cAMP, as is CFTR (Tabcharani *et al.*, 1990). The selectivity of the channel for Cl^- over gluconate was 1.0 : 0.07.

The opercular epithelium anion channel is unlike the "small" (Gögelein *et al.*, 1987) or the "larger" (Greger *et al.*, 1987) conductance anion channels from shark rectal gland that are associated with Cl^- secretion in that tissue. The larger conductance outward-rectifying anion channel that is associated with volume regulation in mammalian cells (Tabcharani *et al.*, 1990) was seen only rarely (2 of 156 seals) in the killifish opercular epithelial cells and therefore is apparently not important in Cl^- secretion by chloride cells. The addition of the disulfonic stilbene DIDS to the apical side of opercular membranes in Ussing chambers had no inhibitory effect on Cl^- secretion rate, but DPC (0.1–2.0 m*M*) rapidly inhibited Cl^- secretion and, more importantly, abolished the stimulation of Cl^- secretion by the β adrenergic agonist isoproterenol (Marshall *et al.*, 1995b) and by cAMP (Fig. 3). Insensitivity to mucosal DIDS and inhibition by DPC are characteristics in common with CFTR (Tabcharani *et al.*, 1990). Therefore it appears that a CFTR-like anion channel in the apical membrane of marine teleost Cl^- cells is responsible for the cAMP-activated Cl^- secretion. Studies of the differences in regulation of anion channels in fish may therefore be valuable to ongoing investigations in cystic fibrosis, in addition to understanding the rapid acclimation steps in fish ion balance.

4. Basolateral Membrane Events

The existence and operation of Na^+,K^+-ATPase is now established and there is a well-recognized Na^+-dependent, furosemide-sensitive cotransporter at the basolateral membrane (see the foregoing). Equally important for efficient transcellular Cl^- transport is a K^+ conductance that allows rapid recycling of K^+ from the pump. Such a mechanism is even more important if the cotransporter requires K^+, that is, the $Na^+,K^+,2Cl^-$ cotransporter that is common to many Cl^--transporting epithelia, including the marine teleost intestinal epithelium (Musch *et al.*, 1982) and elasmobranch rectal gland (Hannafin *et al.*, 1983).

To address the question of K^+ dependence of the cotransporter, we examined the effect of serosal K^+-free solutions on Cl^- transport rate by the killifish opercular epithelium. By comparing the speed of the inhibitions

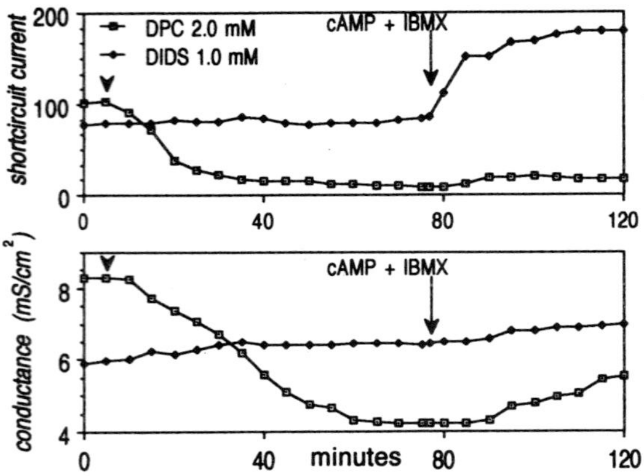

Fig. 3. In paired opercular membranes from a seawater killifish *in vitro* bathed in symmetrical saline, there was marked inhibition of Cl^- secretion (as short-circuit current in $\mu amp \cdot cm^{-2}$) and conductance (in $mS \cdot cm^{-2}$) by the Cl^- channel blocker diphenylamine-2-carboxylate (DPC, 2.0 m*M*, mucosal side). There was no effect of the disulfonic stilbene DIDS (1.0 m*M*, mucosal side). DPC and DIDS additions are indicated by the arrowheads. Note that the subsequent addition of the phosphodiesterase inhibitor 3-isobutyl-1-methylxanthine (IBMX 0.1 m*M*, serosal) and dibutyryl cyclic AMP (cAMP 0.5 m*M*, serosal) augmented Cl^- secretion after DIDS but not after DPC and that conductance was not restored to control levels.

by the Na^+,K^+-ATPase inhibitor ouabain to that by K^+-free solutions, we could distinguish between a K^+-free effect on the pump from that on the cotransporter. Briefly, the K^+-free solutions started to inhibit Cl^- secretion rapidly (0.66 ± 0.24 min, $N = 5$), whereas ouabain initiated inhibition only after 5–8 min (6.65 ± 0.67 min, $N = 5$). Differences in diffusion coefficients of K^+ versus ouabain cannot account fully for the longer delay with ouabain. We therefore conclude that K^+-free solutions block Cl^- transport directly at the cotransporter, thus the cotransporter apparently requires K^+ and is the Na^+,K^+,$2Cl^-$ cotransporter.

The following demonstrates the importance of the basolateral K^+ conductance. If the pump has the generally recognized stoichiometry of $3Na^+ : 2K^+$ and the cotransporter has the stoichiometry of $1Na^+ : 1K^+ : 2Cl^-$, then the approximate ratio of Cl^- transported to K^+ recycled across the basolateral membrane is $6Cl^- : 5K^+$. The opercular epithelium transports upward of 10 $\mu mol \cdot cm^{-2} \cdot h^{-1}$ of Cl^-, hence about 8 $\mu mol \cdot cm^{-2} \cdot h^{-1}$ of transmembrane K^+ recycling is likely. This is huge compared to the trans*epithelial* K^+ secretory rate of a few nmol $\cdot cm^{-2} \cdot h^{-1}$ (Marshall, 1981b). Degnan (1985) observed large, dose-dependent inhibitions of Cl^- secretion by the opercular epithelium with the addition of serosal Ba^{2+} (0.05–5.0 m*M*), strongly suggestive of a reliance of Cl^- secretion on basolateral Ba^{2+}-sensitive K^+ channels. These inhibitions did not significantly affect transepithelial conductance, consistent with the idea that the basolateral membrane has other conductive pathways in addition to K^+ channels. In spite of their importance, the characteristics and regulation of these K^+ channels are undescribed.

B. Ion Uptake in Freshwater Opercular Epithelia

1. Current Models

Almost all the information on ion uptake by freshwater fish is derived from whole-animal experiments with augmentation from short-lived *in vitro* perfused gill and head experiments. Ion uptake is inextricably linked to acid/base balance and the most elucidating experiments measure ion flux and acid/base parameters simultaneously. Because this generally requires repeated sampling of blood, large (>200g) animals have been used almost exclusively and the most important species has been the rainbow trout. Figure 1B summarizes a current model for the operation of the rainbow trout gill epithelium, fully acknowledging the fact that although a single cell is shown, the transport operations may well be partitioned into several cell types, including mitochondria-rich (MR) cells and pavement cells at least. Parenthetically, the term "mitochondria-rich" cells used by some (Wood and Marshall, 1994) or "ionocytes" (Conte and Lin, 1967; Flik *et al.*, Chapter 12, this volume) instead of "chloride cells" accurately reflects the uncertainty of which ions are transported by these cells in freshwater fish.

The model includes electrically silent anion and cation exchangers (e.g., Maetz, 1970) that allow uptake of NaCl in roughly equivalent amounts with concomitant excretion of acid or base equivalents depending on the relative speed of operation of the two exchangers. Classically, the Na^+–H^+ exchange may also accept NH_4^+ when ammonia is present (Maetz, 1970). The transport is dependent on carbonic anhydrase and on a basolaterally located Na^+,K^+-ATPase that is ouabain sensitive. The Ca^{2+} active uptake pathway is not shown; for details on this, see Flik *et al.* (Chapter 12, this volume). The "active" step for Cl^- transport almost certainly is at the apical membrane because of the large electrochemical gradient operating against Cl^- uptake, but this may be driven by a HCO_3^- (or pH) gradient and an obligatory electroneutral exchange with HCO_3^-. A second mechanism, similar to that proposed for amphibian skin (Ehrenfeld *et al.*, 1985), has been presented for freshwater fish (Lin and Randall, 1991; Chapter 9, this volume); the model involves apical membrane H^+-ATPase that would help create an electrochemical gradient to drive Na^+ uptake across the apical membrane via Na^+ channels. Pharmacologically both epithelial Na^+ channels and Na^+–H^+ exchange are known to be blocked by amiloride, so the definitive proof will require biochemical and electrophysiological identification of the H^+-ATPase and the putative Na^+channel. Both models involve as yet undemonstrated basolateral Cl^- and K^+ passive conductances. A likely candidate for the basolateral Cl^- channel may be that now identified in basolateral membrane of teleost urinary bladder (Chang and Loretz, 1991).

2. Freshwater Opercular Epithelium NaCl Transport

Although the need for a flat epithelial model to study the operation of freshwater MR cells has been recognized repeatedly (Karnaky and Kinter, 1977; Marshall, 1977; Kirschner, 1983; Karnaky, 1986; Péqueux *et al.*, 1988), until very recently there has not been a viable option. Early attempts involved adapting *Gillichthys* or killifish to dilute (5–10%) seawater and then mounting the epithelia in membrane chambers with saline on both surfaces (Degnan *et al.*, 1977; Marshall, 1977). This produced the anomalous result of frank *secretion* of ions by these epithelia and the conclusion that these euryhaline animals must retain hormonally inhibited seawater-type Cl^- cells even in relatively dilute, certainly hypotonic, media. Killifish gill chloride cells do not apparently change to the freshwater ultrastructural pattern unless fully acclimated to fresh water (salinity < 0.15 ppt; Philpott and Copeland, 1963). Among other species examined, brook trout opercular epithelium was disappointing because of the lack of MR cells (by DASPEI) and concomitant lack of NaCl transepithelial transport (Marshall, 1985). Rainbow trout were more promising in that a few MR cells were present (Marshall *et al.*, 1992) in the epithelium overlying the cleithrum, but again there was no evidence of NaCl transport, although a small Ca^{2+} active transport in the uptake direction was present. McCormick *et al.* (1992) found net Ca^{2+} uptake across freshwater tilapia opercular epithelium that was stimulated by acclimation to soft water, but again, only comparatively few MR cells were present and there was no detectable NaCl uptake (Foskett *et al.*, 1981). Attention refocused on killifish and acclimation of these animals to low-Ca^{2+} fresh water (1.0 m*M* NaCl, 0.1 m*M* Ca^{2+}) produced permanent changes in the epithelium from a Cl^--secreting tissue to one adapted to ion uptake (Wood and Marshall, 1994).

The chloride cells in killifish gill epithelium are essentially indistinguishable from those in the opercular membrane (Karnaky, 1986), and this is an essential component of the usefulness of the epithelium as a model of the gill. In the development of a corresponding freshwater model this similarity needs to be readdressed because freshwater MR cells are distinct from their seawater counterparts in a variety of ways. In both the gill (Hossler *et al.,* 1985) and in our preliminary work with the opercular epithelium the apical pit is present in freshwater MR cells and the apical membrane surface area is elaborated by folding and/or microvilli. The accessory cells are generally lacking and the tight junctions between MR cells and their neighboring pavement or MR cells are more well developed. This is true for gill epithelia (Pisam and Rambourg, 1991) and in the killifish opercular epithelium. Therefore, the MR cells of the freshwater opercular epithelium resemble closely those of the freshwater gill.

The freshwater killifish opercular epithelium at open circuit and with fresh water (1.0 m*M* NaCl and 0.1 m*M* Ca^{2+}) bathing the mucosal surface actively transports Cl^- in the absorptive direction, based on nonagreement of the observed flux ratio with that predicted by the Ussing flux ratio equation (Wood and Marshall, 1994). Na^+ fluxes were not significantly different from the predicted Ussing flux ratio, hence it appears that the cation-selective shunt that is normally present in seawater killifish opercular epithelium is not entirely eliminated in fresh water (Wood and Marshall, 1994). The influxes of Na^+ and Cl^- (about 120 nmol $\cdot$ cm^{-2} $\cdot$ h^{-1} for Na^+ and 90 nmol $\cdot$ cm^{-2} $\cdot$ h^{-1} for Cl^-) were usually smaller than the respective effluxes, hence the preparation was not in positive ion balance. Because the killifish opercular epithelium has a large *negative* inside transepithelial potential in fresh water of about -55–65 mV (unlike most freshwater teleost transgill potentials *in vivo;* Potts, 1984), it appears that a cation-selective shunt is present in the isolated epithelium that is not present *in vivo*. In symmetrical saline and short-circuited, the freshwater opercular epithelium often has net Cl^- uptake (W. S. Marshall and co-workers, unpublished results) and has no net NaCl secretion. The mechanisms of ion uptake therefore can be studied with the freshwater opercular epithelium and in preliminary work we have found Cl^- uptake inhibition by SCN^- (1.0 m*M*) but no apparent effect of SITS and DIDS or of amiloride (on Na^+ uptake).

A major component of the prolactin-mediated freshwater adaptation in euryhaline fish is the reduction in ion turnover and overall reduced permeability of the gill epithelium (Maetz, 1970; Bern and Madsen, 1992). Comparing the freshwater and seawater opercular epithelia, the apparent Na^+ permeability (Na^+ efflux divided by serosal Na^+ concentration) of the epithelium of freshwater-adapted killifish is approximately 2.3×10^{-6} cm $\cdot$ s^{-1} (Marshall *et al.,* 1995a), substantially lower than that for seawater opercular epithelium at 8.7×10^{-6} cm.s^{-1} (Degnan and Zadunaisky, 1980). The smaller number of MR cells and more elaborate tight junctions are consistent with the observed lower ionic permeability of the freshwater skin and gill epithelium.

3. Freshwater Opercular Epithelium Ca^{2+} Transport

The freshwater killifish opercular epithelium actively absorbs Ca^{2+} from soft (0.1 m*M* Ca^{2+}) fresh water with a net flux of about 20–40 nmol · cm^{-2} · h^{-1} (Burghardt, 1993). The Ca^{2+} uptake rate across the killifish membrane is much more rapid than that across tilapia opercular epithelium (McCormick *et al.*, 1992) or rainbow trout cleithrum skin (Marshall *et al.*, 1992). In all three preparations, Ca^{2+} uptake varies with the density of MR cells (McCormick *et al.*, 1992; Marshall *et al.*, 1992, 1995a; Burghardt, 1993), is partially inhibited by La^{3+} and Mg^{2+}, and is saturable (apparent $K_{1/2}$ = 0.35 m*M*; Marshall *et al.*, 1995a). The killifish opercular epithelium may also prove to be valuable for the resolution of mechanisms of transepithelial Ca^{2+} and NaCl transport and their regulation.

III. ION TRANSPORT BY URINARY BLADDER

The urinary bladder (urinary sinus, a derivative of the archinephric duct) of teleosts is an accessory osmoregulatory organ to the kidney in that urine is retained by the urinary bladder for some time during which the ionic composition is modified. For example, whereas ureteral urine of trout is about 10 m*M* NaCl, voluntarily released urine (Curtis and Wood, 1991) and urine extracted from the bladder postmortem (Marshall, 1988) have NaCl concentrations of 2–3 m*M*. In freshwater teleosts the urinary bladder reabsorbs NaCl with a minimum of accompanying water, that is, it has a low osmotic permeability. Prolactin decreases osmotic permeability and increases the rate of ion (Na^+) reabsorption (reviewed by Utida *et al.*, 1972; Bern and Madsen, 1992). In this way, the animal retains maximal amounts of monovalent ions and excretes a very dilute urine. In seawater teleosts the urinary bladder also reabsorbs NaCl but instead the osmotic permeability of the epithelium is higher and the resulting fluid reabsorption tends to concentrate the urine with respect to untransported ionic species, particularly Mg^{2+} and Ca^{2+} (Beyenbach and Kirschner, 1975; Howe and Gutknecht, 1978; Loretz and Bern, 1980).

A. Freshwater Teleosts

The freshwater teleost urinary bladder NaCl uptake may be via coupled transport, as in rainbow trout (Fossat and Lahlou, 1979), a partially coupled mechanism, as in starry flounder (*Platichthys stellatus*) (Demarest, 1984), or uncoupled paired anion and cation exchangers, for instance in brook trout (Marshall, 1986) and longjawed mudsucker (Loretz and Bern, 1980). In all cases most ion uptake is by electrically silent processes. There are varying amounts of apparent exchange diffusion components and the paracellular pathway permeability is highly variable such that some epithelia are typically "tight" and others are more "leaky" (i.e., the apical tight junctions are more permeable than the transcellular pathway).

In rainbow trout the $K_{1/2}$ for NaCl uptake was 8 mM and 30–50% of Cl^- transepithelial fluxes appear to be exchange diffusion (Fossat and Lahlou, 1979). The transepithelial resistance is exceptionally low among the teleost preparations examined thus far, 200 $\Omega \cdot cm^2$ (Fossat and Lahlou, 1979) when these animals are adapted to hard fresh water (composition approximately 1.6 mM Ca^{2+}, 0.16 mM NaCl as per Avella *et al.*, 1987). By cable analysis with microelectrodes the shunt resistance is also low, 205 $\Omega \cdot cm^2$, although the apical and basal membrane resistances are 2.8 and 2.0 $k\Omega \cdot cm^2$, respectively (Harvey and Lahlou, 1986). For this reason, the authors concluded that the rainbow trout urinary bladder is a "leaky" epithelium. In contrast, rainbow trout and brook trout urinary bladders have a very high transepithelial resistance (5.0–10.0 $k\Omega \cdot cm^2$, Marshall, 1988) when acclimated to soft fresh water (0.1 mM Ca^{2+}, 0.17 mM NaCl). Because prolactin is well known to decrease osmotic (and ionic) permeability of the urinary bladder (Utida *et al.*, 1972; Bern and Madsen, 1992) and because prolactin titers are lower in "hard" water, the difference in shunt permeability between these two studies may well be the higher prolactin levels and concomitant lower permeability of urinary bladders from animals adapted to soft fresh water. The hormonal control of urinary bladder transport characteristics, especially in connection with water hardness and pH, needs to be examined further. Harvey and Lahlou (1986) measured intracellular activities for Na^+, Cl^-, and K^+ and obtained 16, 21, and 87 mM, respectively, with the epithelium bathed in symmetrical saline. This indicates active accumulation of Cl^- and K^+ above electrochemical equilibrium. Entry of Cl^- across the apical membrane was dependent on mucosal Na^+, implying that apically located NaCl cotransport down the electrochemical gradient for Na^+ results in transport of Cl^- up its electrochemical gradient. Hence rainbow trout kept in hard fresh water have urinary bladder transport characteristics resembling the model in Fig. 4A.

Brook trout urinary bladder has independent Na^+ and Cl^- electrically neutral uptake mechanisms (Marshall, 1986; Fig. 4B). The epithelium has a high resistance and low hydraulic conductivity (1.6×10^{-7} $cm \cdot s^{-1} \cdot atm \cdot {}^{-1}$) and low permeability to mannitol (Marshall, 1988), indicative of a high-resistance shunt pathway. Passive loss of Na^+ and Cl^- as unidirectional efflux was very low and the reduction of mucosal Cl^- reduced Cl^- loss by 0.1 μequiv $\cdot cm^{-2} \cdot h^{-1}$, hence exchange diffu-

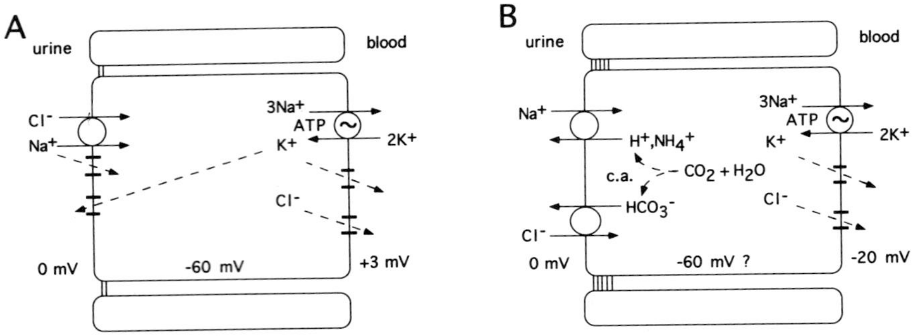

Fig. 4. (A) Diagrammatic summary of urinary bladder function for most seawater teleosts, flounder in dilute seawater and trout in hard fresh water. The degree of coupling of Na^+ to Cl^- uptake is variable, as is the relative tightness of the intercellular junctions. I_{sc} may be Na^+ uptake and/or K^+ secretion. Hydraulic conductivity is generally high. (B) Model of trout urinary bladder function in soft fresh water. Na^+ and Cl^- uptake are not tightly coupled to each other and the epithelium has high resistance and well-developed tight junctions. I_{sc} and hydraulic conductivity are low.

sion accounted for a very small (<10%) portion of Cl^- uptake (Marshall, 1988). With a mimic of dilute urine on the apical side (2.0 m*M* NaCl), the urinary bladder at open circuit had Cl^- forward and backfluxes that were not significantly different, indicating that the preparation *in vitro* can support the ion gradients seen *in vivo* (Marshall, 1986). Na^+,K^+,$2Cl^-$ cotransport was not indicated because the uptake was insensitive to apically added bumetanide (Marshall, 1986). Na^+ uptake was partially inhibited by 0.1 m*M* amiloride, and in Cl^- free saline, Na^+ uptake was accompanied by extrusion of an equal amount of acid equivalents, indicating the presence of Na^+/H^+ exchange (Marshall and Bryson, 1991). Because serosal NH_4^+ stimulated Na^+ uptake when the mucosal side was bathed in mock urine, the exchanger apparently can operate to exchange Na^+ for NH_4^+. When pH_i, measured by BCECF microspectrofluorometry, was decreased by hypercapnia, recovery to normal pH_i was blocked by apically added amiloride, indicating involvement of Na^+/H^+ exchange in pH_i regulation (Marshall and Bryson, 1991).

The *Gillichthys* urinary bladder has two zones, one of cuboidal cells that transport little (in fresh water or seawater) and a columnar region that transports NaCl at a high rate (Loretz and Bern, 1980; Fig. 4A). Like the freshwater rainbow trout and brook trout, *Gillichthys* adapted to 5% seawater have neutral NaCl uptake and the transepithelial resistance, averaging 1.8–2.6 kΩ · cm^2, is similar to that of the brook trout. The low passive permeability to ions and nonlinearity of I–V relations suggest that the epithelium is "tight." In starry flounder, active uptake of NaCl occurs by a partially coupled, electrically silent process (Demarest, 1984; Fig. 4A). The degree of coupling is approximately 60%, inasmuch as removal of Na (or Cl) inhibits the uptake of the other ion by about 60% (Demarest, 1984).

B. Seawater Teleosts

In seawater-adapted *Gillichthys* the urine is rich in Ca^{2+}, Mg^{2+}, and SO_4^{2-} and has a higher osmolality than freshwater urine, demonstrating a role of the bladder in reabsorbing NaCl and water while excreting a urine rich in divalent salts (Loretz and Bern, 1980). As a result, Ca/Mg phosphate commonly precipitates in the urinary bladder of seawater teleosts such as *Gillichthys* and the sculpin (*Myoxocephalus octodecimspinosus;* C. A. Loretz, unpublished data; Maren *et al.*, 1992). NaCl uptake is partially coupled (Fig. 4A) and has an additional amiloride-sensitive electrogenic component, the latter showing seasonal variation (peak in winter and nadir in summer). The electrogenic NaCl uptake is also reduced by hypophysectomy and is restored by cortisol injection (Loretz and Bern, 1983). The seawater bladder has lower resistance than the freshwater counterpart (0.3–0.8 $k\Omega \cdot cm^2$) (Loretz and Bern, 1980). Seawater urinary bladder cells in culture possess stretch-activated, cation-nonselective channels with conductances of 55 and 116 pS (Chang and Loretz, 1991). The channel is likely involved in cell volume regulation but could also contribute to Na^+ uptake (and K^+ secretion). In the seawater *Gillichthys* bladder there is a large (75 pS) anion channel that is voltage sensitive in the physiological range and that increases open probability with depolarizing voltages (Figs. 4A and 5; Chang and Loretz, 1993). This channel is inhibited by DPC, added on the cytoplasmic side of excised inside-out patches. In the intact epithelium, serosally added DPC blocks Cl^- uptake, hence this channel affords a means by which Cl^-, having entered across the apical membrane via electroneutral exchange (as in brook trout) or cotransport (as in *Gillichthys*), can exit across the basolateral membrane.

Urinary bladders from seawater-adapted starry flounder have a fivefold higher transepithelial resistance than do freshwater-acclimated animals, about 2.0 $k\Omega \cdot cm^2$, and the mannitol permeability is low, indicating a low-permeability shunt pathway, such that the majority of passive ion flux occurs via nonconductive (presumably transcellular) means (Demarest and Machen, 1984). The winter flounder (*Pseudopleuronectes americanus*) urinary bladder has a basolaterally located Na^+,K^+-ATPase, identified by tritiated ouabain autoradiography, that provides the driving force for active ion (and water) transport by the tissue (Renfro *et al.*, 1976). Consistent with this, the transport is dependent on Na^+,K^+-ATPase, as

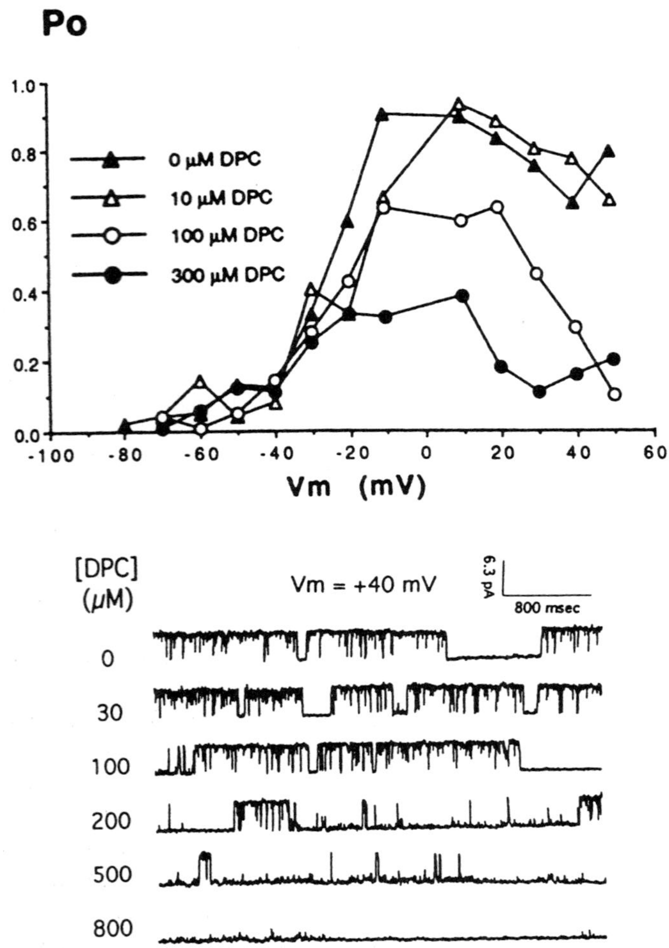

Fig. 5. DPC blockade of anion channel activity in *Gillichthys* urinary bladder. Bottom: Channel–current records from an inside-out membrane patch containing a single high-activity anion channel clamped at V_m = +40 mV and bathed in KCl-basic saline. DPC was cumulatively added to yield the stated concentrations. Downward deflections represent inward (+) current. For visual presentation, current records were low-pass filtered at 500 Hz. Top: Voltage dependence of DPC blockade. Data were collected from an inside-out patch containing a low-activity channel. (From Chang and Loretz, 1993, reprinted with permission of the American Physiological Society.)

serosal ouabain, with or without mucosal papaverine, inhibited Na^+ and Cl^- uptake (Stokes, 1984). Seawater-adapted winter flounder have a high resistance urinary bladder similar to the starry flounder and there is electrically neutral (Renfro, 1977; Stokes, 1984) amiloride- and SITS-insensitive NaCl uptake (Stokes, 1984). The uptake is inhibited by thiazide-type diuretics but not by bumetanide (Stokes, 1984). This NaCl cotransporter has now been cloned and gene expression includes a 3.7-kb mRNA localized only to the urinary bladder (Gamba *et al.*, 1993). There also is substantial Ba^{2+}-sensitive K^+ secretion by this epithelium (Dawson and Frizzell, 1989; Fig. 4A). Analysis of passive ion fluxes with winter flounder confirmed those with starry flounder and indicated that most passive flux occurs not via the low-permeability shunt but via a cellular pathway (Stokes, 1988). Using amphotericin B to permeabilize the apical membrane, Dawson and Frizzell (1989) demonstrated that the basolateral membrane is anion (Cl^-) selective, consistent with the existence of basolateral anion channels.

IV. FUTURE DIRECTIONS

A. Opercular Epithelium

The presence of the multidrug resistance protein, also known as P-glycoprotein, in winter flounder testis (Chan *et al.*, 1992), as well as the association of P-glycoprotein with expression of CFTR in mammalian systems (Riordan, 1993) make P-glycoprotein a very appealing candidate for future work with seawater Cl^- and freshwater MR cells. Molecular biological approaches should be used to elucidate the teleostean version of the CFTR gene and of MDR, the gene that encodes the P-glycoprotein. Teleostean systems may be helpful in advancing understanding in these areas. For instance, while P-glycoprotein is functionally present (Karnaky *et al.*, 1993) in opercular epithelium the large conductance outward rectifying anion channel seems to be absent (Marshall *et al.*, 1995). These data do not support the association in mammalian systems of P-glycoprotein with an outward rectifying anion channel (Valverde *et al.*, 1992) but do point to the side-by-side operation of MDR and CFTR in teleostean Cl^- cells. Virtually no work has been done on the renallike operation of the gills in secretion of drugs, toxins, and other xenobiotics, but promising preliminary results now suggest that P-glycoprotein operates in killifish opercular epithelium (Karnaky *et al.*, 1993) and killifish chloride cells take up and transport methylene blue dye (Fisher, 1989). Also, the characteristics and regulation of basolateral K^+ channels that are essential for Cl^- secretion should be examined.

Because so little is now known about the freshwater MR cells, at least from *in vitro* epithelia, direct molecular biological approaches to determining which transport entities are on these cells may be immediately productive and would help to focus physiological studies. In addition, the mechanisms of Ca^{2+} uptake and the regulation of this transport need to be elucidated and compared to acid/base and NaCl transport in the same preparations. The existence of ultrastructurally distinguishable subtypes of MR cells in freshwater fish (e.g., Pisam and Rambourg, 1991) is strongly suggestive of functional heterogeneity among MR cells. The killifish and tilapia opercular epithelia, now established in seawater and freshwater versions, should be used to examine in detail the hormonally influenced transitions between the seawater and freshwater ion-transporting cells to augment the foundation work by Foskett *et al.* (1981) and McCormick (1990). What redifferentiation or *de novo* development of freshwater and seawater chloride cells occurs during salinity adaptation? To what degree are cells retained and temporarily inhibited (by rapid-acting hormones, neurotransmitters, or physical covering by motile pavement cells; e,g., Goss *et al.,* 1992a,b; Chapter 10, this volume) while estuarine fish undertake sorties into waters of different salinity? What are the cues to changeover of one cell type to another? By addressing these questions, we may increase our understanding of the adaptive strategies for organisms facing environmental extremes.

B. Urinary Bladder

The large variation of results with the urinary bladder among diverse species of teleosts may reflect the different osmoregulatory strategies (some completely euryhaline and estuarine, whereas others are anadromous migrators or freshwater residents). In studies with freshwater animals, Ca^{2+} concentration in the water of acclimation should be carefully controlled (or at least reported). Resolution of the apical membrane events in the (almost universal) electrically neutral ion uptake would be most valuable and the use of mucosal thiazide-type diuretics (Stokes, 1984) and serosal DPC (Chang and Loretz, 1993) on the other preparations could help build a common experimental base. On the molecular level, the role of prolactin should be examined more closely, especially with respect to the widely accepted effect of reducing osmotic permeability of the epithelium. The regulation of expression of the NaCl cotransporter will be very important. Overall, the urinary bladder may provide definitive answers to some questions more easily than would similar studies on the gills and kidney.

ACKNOWLEDGMENTS

Thanks to C. A. Loretz, C. M. Wood, and S. E. Bryson for help in manuscript preparation. Work by the author is supported by research grants from the Natural Sciences and Engineering Research Council of Canada and the Department of Fisheries and Oceans Canada.

REFERENCES

Avella, M., Masoni, A., Bornancin, M., and Mayer-Gostan, N. (1987). Gill morphology and sodium influx in the rainbow trout (*Salmo gairdneri*) acclimated to artificial freshwater environments. *J. Exp. Zool.* **241,** 159–169.

Bern, H. A., and Madsen, S. S. (1992). A selective survey of the endocrine system of the rainbow trout (*Oncorhynchus mykiss*) with emphasis on the hormonal regulation of ion balance. *Aquaculture* **100,** 237–262.

Beyenbach, K. W., and Kirschner, L. B. (1975). Kidney and urinary bladder functions of the rainbow trout in Mg and Na excretion. *Am. J. Physiol.* **229,** 389–393.

Burghardt, J. (1993). Ca^{2+} transport *in vitro* in the opercular epithelium of freshwater adapted *Fundulus heteroclitus*. Honours Thesis, St. Francis Xavier University, Antigonish, Nova Scotia, Canada.

Chan, K. M., Davies, P. L., Childs, S., Vienot, L., and Ling, V. (1992). P-Glycoprotein genes in the winter flounder, *Pleuronectes* [*sic*] americanus: Isolation of two types of genomic clones carrying 3′ terminal exons. *Biochim. Biophys. Acta* **1171,** 65–72.

Chang, W., and Loretz, C. A. (1991). Identification of a stretch-activated monovalent cation channel from teleost urinary bladder cells. *J. Exp. Zool.* **259,** 304–315.

Chang, W., and Loretz, C. A. (1993). DPC blockade of transepithelial chloride absorption and single channels in teleost urinary bladder. *Am. J. Physiol.* **265,** R66–R75.

Conte, F. P., and Lin, D. H. Y. (1967). Kinetics of cellular morphogenesis in gill epithelium during sea water adaptation of *Oncorhynchus* (Walbaum). *Comp. Biochem. Physiol.* **23,** 945–957.

Curtis, B. J., and Wood, C. M. (1991). The function of the urinary bladder *in vivo* in the freshwater rainbow trout. *J. Exp. Biol.* **155,** 567–583.

Dawson, D. C., and Frizzell, R. A. (1989). Mechanism of active K^+ secretion by the flounder urinary bladder. *Pflüegers Arch.* **414,** 393–400.

Degnan, K. J. (1984). The sodium and chloride dependence of chloride secretion by the opercular epithelium. *J. Exp. Zool.* **231,** 11–17.

Degnan, K. J. (1985). The role of K^+ and Cl^- conductances in chloride secretion by the opercular epithelium. *J. Exp. Zool.* **236,** 19–25.

Degnan, K. J. (1986). Cyclic AMP stimulation of Cl^- secretion by the opercular epithelium: The apical membrane chloride conductance. *J. Exp. Zool.* **238,** 141–146.

Degnan, K. J., and Zadunaisky, J. A. (1980). Passive sodium movements across the opercular epithelium: The paracellular shunt pathway and ionic conductance. *J. Membr. Biol.* **55,** 175–185.

Degnan, K. J., Karnaky, K. J., Jr., and Zadunaisky, J. A. (1977). Active chloride transport in the *in vitro* opercular skin of a teleost (*Fundulus heteroclitus*), a gill-like epithelium rich in chloride cells. *J. Physiol.* (*London*) **271,** 155–191.

Demarest, J. R. (1984). Ion and water transport by the flounder urinary bladder: Salinity dependence. *Am. J. Physiol.* **246,** F395–F401.

Demarest, J. R., and Machen, T. E. (1984). Passive and active ion transport by the urinary bladder of a euryhaline flounder. *Am. J. Physiol.* **246,** F402–F408.

Ehrenfeld, J., Garcia-Romeu, F., and Harvey, B. J. (1985). Electrogenic active proton pump in *Rana esculenta* skin and its role in sodium ion transport. *J. Physiol.* (*London*) **359,** 331–355.

Eriksson, O., and Wistrand, P. J. (1986). Chloride transport inhibition by various types of 'loop' diuretics in fish opercular epithelium. *Acta Physiol. Scand.* **126,** 93–101.

Eriksson, O., Mayer-Gostan, N., and Wistrand, P. J. (1985). The use of fish opercular epithelium as a model tissue for studying intrinsic activities of loop diuretics. *Acta Physiol. Scand.* **125,** 55–66.

Fisher, R. J. (1989). Measurement of the intracellular pH of chloride cells in the opercular membrane of *Fundulus heteroclitus*. Honours Thesis, St. Francis Xavier University, Antigonish, Nova Scotia, Canada.

Foskett, J. K., Machen, T. E., and Bern, H. A. (1982b). Chloride secretion and conductance of teleost opercular membrane: Effects of prolactin. *Am. J. Physiol.* **242,** R380–R389,

Foskett, J. K., and Scheffey, C. (1982). The chloride cell: Definitive identification as the salt-secretory cell in teleosts. *Science* **215,** 164–166.

Foskett, J. K., Logsdon, D., Turner, T., Machen, T. E., and Bern, H. A. (1981). Differentiation of the chloride extrusion mechanism during seawater adaptation of a teleost fish, the cichlid *Sarotherodon mossambicus*. *J. Exp. Biol.* **93,** 209–224.

Foskett, J. K., Hubbard, G. M., Machen, T. E., and Bern, H. A. (1982a). Effects of epinephrine, glucagon and vasoactive intestinal polypeptide on chloride secretion by teleost opercular membrane. *J. Comp. Physiol.* **146,** 27–34.

Fossat, B., and Lahlou, B. (1979). The mechanism of coupled transport of sodium and chloride in isolated urinary bladder of the trout. *J. Physiol.* (*London*) **294,** 211–222.

Gamba, G., Saltzberg, S. N., Lombardi, M., Miyanoshita, A., Lytton, J., Hediger, M. A., Brenner, B. M., and Hebert, S. C. (1993). Primary structure and functional expression of a cDNA encoding the thiazide-sensitive, electroneutral sodium-chloride cotransporter. *Proc. Natl. Acad. Sci. U.S.A.* **90,** 2749–2753.

Garcia-Romeu, F., and Ehrenfeld, J. (1975). Chloride transport through the nonshort-circuited isolated skin of *Rana esculenta*. *Am. J. Physiol.* **228,** 845–849.

Gögelein, H., Schlatter, E., and Greger, R. (1987). The "small" conductance chloride channel in the luminal membrane of the rectal gland of the dogfish (*Squalus acanthias*). *Pflüegers Arch.* **409,** 122–125.

Goss, G. G., Perry, S. F., Wood, C. M., and Laurent, P. L. (1992a). Mechanisms of ion and acid–base regulation at the gills of freshwater fish. *J. Exp. Zool.* **263,** 143–159.

Goss, G. G., Laurent, P., and Perry, S. F. (1992b). Evidence for a morphological component in acid–base regulation during environmental hypercapnia in brown bullhead (*Ictalurus nebulosus*). *Cell Tissue Res.* **268,** 539–552.

Gray, M. A., Harris, A., Coleman, L., Greenwell, J. R., and Argent, B. E. (1989). Two types of chloride channel on duct cells cultured from human fetal pancreas. *Am. J. Physiol.* **257,** C240–C251.

Greger, R., Schlatter, E., and Gögelein, H. (1987). Chloride channels in the luminal membrane of the rectal gland of the dogfish (*Squalus acanthias*). *Pflüegers Arch.* **409,** 114–121.

Guibbolini, M., and Lahlou, B. (1987). Basal and stimulated adenylate cyclase activity in the gill epithelium of the rainbow trout. *Gen. Comp. Endocrinol.* **67,** 178–188.

Hannafin, J., Kinne-Saffran, E., Friedman, D., and Kinne, R. (1983). Presence of a sodium–potassium chloride cotransport system in the rectal gland of *Squalus acanthias*. *J. Membr. Biol.* **75,** 73–83.

Harvey, B. J., and Lahlou, B. (1986). Ion-selective micro-electrode studies of the electrochemical potentials in trout urinary bladder. *J. Physiol.* (*London*) **370,** 467–488.

Hootman, S. R., and Philpott, C. W. (1979). Ultracytochemical localization of Na^+,K^+-activated ATPase in chloride cells from the gills of a euryhaline teleost. *Anat. Rec.* **193,** 99–130.

Hootman, S. R., and Philpott, C. W. (1980). Accessory cells in teleost branchial epithelium. *Am. J. Physiol.* **238,** R185–R198.

Hossler, F. E., Musil, G., Karnaky, K. J., Jr., and Epstein, F. H. (1985). Surface ultrastructure of the gill arch of the killifish, *Fundulus heteroclitus,* from seawater and freshwater, with special reference to the morphology of the apical crypts of chloride cells. *J. Morphol.* **185,** 377–386.

Howe, D., and Gutknecht, J. (1978). Role of urinary bladder in osmoregulation in marine teleost, *Opsanus tau*. *Am. J. Physiol.* **235,** R48–R54.

Karnaky, K. J., Jr. (1986). Structure and function of the chloride cells of *Fundulus heteroclitus* and other teleosts. *Am. Zool.* **26,** 209–224.

Karnaky, K. J., Jr., and Kinter, W. B. (1977). Killifish opercular skin: A flat epithelium with a high density of chloride cells. *J. Exp. Zool.* **199,** 355–364.

Karnaky, K. J., Jr., Kinter, L. B., Kinter, W. B., and Stirling, C. E. (1976). Teleost chloride cell. II. Autoradiographic localization of gill Na^+,K^+-ATPase in killifish (*Fundulus heteroclitus*) adapted to low and high salinity environments. *J. Cell Biol.* **70,** 157–177.

Karnaky, K. J., Jr., Degnan, K. J., Garretson, L. T., and Zadunaisky, J. A. (1984). Identification and quantification of mitochondria-rich cells in transporting epithelia. *Am. J. Physiol.* **246,** R770–R775.

Karnaky, K. J., Jr., Suggs, K., French, S., and Willingham, M. C. (1993). Evidence for the multidrug transporter, P-glycoprotein, in the killifish, *Fundulus heteroclitus*. *Bull. Mt. Desert Island Biol. Lab.* **32,** 61–62.

Kirschner, L. B. (1983). Sodium chloride absorption across the body surface: Frog skins and other epithelia. *Am. J. Physiol.* **244,** R429–R443.

Klyce, S. D., and Wong, R. K. S. (1977). Site and mode of adrenaline action on chloride transport across rabbit corneal epithelium. *J. Physiol.* (*London*) **266,** 777–799.

Lin, H., and Randall, D. (1991). Evidence for the presence of an electrogenic proton pump on the trout gill epithelium. *J. Exp. Biol.* **161,** 119–134.

Loretz, C. A., and Bern, H. A. (1980). Ion transport by the urinary bladder of the gobiid teleost, *Gillichthys mirabilis*. *Am. J. Physiol.* **239,** R415–R423.

Loretz, C. A., and Bern, H. A. (1983). Control of ion transport by *Gillichthys mirabilis* urinary bladder. *Am. J. Physiol.* **245,** R45–R52.

Maetz, J. (1970). Mechanisms of salt and water transfer across membranes in teleosts in relation to the aquatic environment. *Mem. Soc. Endocrinol.* **18,** 3–29.

Maetz, J., Sawyer, W. H., Pickford, G. E., and Mayer, N. (1967). Evolution de la balance minerále du sodium chez *Fundulus heteroclitus* au cours du transfert d'eau de mer d'eau douce: effets du hypophysectomie et de la prolactine. *Gen. Comp. Endocrinol.* **8,** 163–176.

Maren, T. H., Fine, A., Swenson, E. R., and Rothman, D. (1992). Renal acid–base physiology in marine teleost, the long-horned sculpin (*Myoxocephalus octodecimspinosus*). *Am. J. Physiol.* **263,** F49–F55.

Marshall, W. S. (1977). Transepithelial potential and short-circuit current across the isolated skin of *Gillichthys mirabilis* (Teleostei: Gobiidae) acclimated to 5% and 100% seawater. *J. Comp. Physiol.* **114,** 157–165.

Marshall, W. S. (1981a). Sodium dependency of active chloride transport across isolated fish skin (*Gillichthys mirabilis*). *J. Physiol.* (*London*) **319,** 165–178.

Marshall, W. S. (1981b). Active transport of Rb^+ across skin of the teleost *Gillichthys mirabilis*. *Am. J. Physiol.* **241,** F482–F486.

Marshall, W. S. (1985). Paracellular ion transport in trout opercular epithelium models osmoregulatory effects of acid precipitation. *Can. J. Zool.* **63,** 1816–1822.

Marshall, W. S. (1986). Independent Na^+ and Cl^- active transport by urinary bladder epithelium in brook trout. *Am. J. Physiol.* **250,** R227–R234.

Marshall, W. S. (1988). Passive solute and fluid transport in brook trout (*Salvelinus fontinalis*) urinary bladder epithelium. *Can. J. Zool.* **66,** 912–918.

Marshall, W. S., and Bern, H. A. (1979). Teleostean urophysis: Urotensin II and ion transport across the isolated skin of a marine teleost. *Science* **204,** 519–521.

Marshall, W. S., and Bryson, S. E. (1991). Intracellular pH regulation in trout urinary bladder epithelium: Na^+–H^+ (NH_4^+) exchange. *Am. J. Physiol.* **261,** R652–R658.

Marshall, W. S., and Hanrahan, J. W. (1991). Anion channels in the apical membrane of mammalian corneal epithelium primary cultures. *Invest. Ophthalmol. Visual Sci.* **32,** 1562–1568.

Marshall, W. S., Nishioka, R. S. (1980). Relation of mitochondria rich chloride cells to active chloride transport in the skin of a marine teleost. *J. Exp. Zool.* **214,** 147–156.

Marshall, W. S., Bryson, S. E., and Wood, C. M. (1992). Calcium transport by isolated skin of rainbow trout. *J. Exp. Biol.* **166,** 297–316.

Marshall, W. S., Bryson, S. E., and Garg, D. (1993). α_2-Adrenergic inhibition of Cl^- transport by opercular epithelium is mediated by intracellular Ca^{2+}. *Proc. Natl. Acad. Sci. U.S.A.* **90,** 5504–5508.

Marshall, W. S., Bryson, S. E., Burghardt, J. S., and Verbost, P. M. (1995a). Ca^{2+} transport by opercular epithelium of the freshwater adapted euryhaline teleost, *Fundulus heteroclitus*. *J. Comp. Physiol.* (in press).

Marshall, W. S., Bryson, S. E., Midelfart, A., and Hamilton, F. W. (1995b). Low conductance anion channel activated by cAMP in teleost Cl^- secreting cells. *Am. J. Physiol.* **268.**

May, S. A., and Degnan, K. J. (1985). Converging adrenergic and cholinergic mechanisms in the inhibition of Cl^- secretion in fish opercular epithelium. *J. Comp. Physiol., B* **156,** 183–189.

McCormick, S. D. (1990). Cortisol directly stimulates differentiation of chloride cells in tilapia opercular membrane. *Am. J. Physiol.* **259,** R857–R863.

McCormick, S. D., Hasegawa, S., and Hirano, T. (1992). Calcium uptake in the skin of a freshwater teleost. *Proc. Natl. Acad. Sci. U.S.A.* **89,** 3635–3638.

Mendelsohn, S. A., Cherksey, B. D., and Degnan, K. J. (1981). Adrenergic regulation of chloride secretion across the opercular epithelium: The role of cyclic AMP. *J. Comp. Physiol.* **145,** 29–35.

Musch, M. W., Orellana, S. A., Kimberg, L. S., Field, M., Halm, D. R., Krasny, E. J., Jr., and Frizzell, R. A. (1982). Na^+–K^+-Cl^- cotransport in the intestine of a marine teleost. *Nature* (*London*) **300,** 351–353.

Pärt, P., Norrgren, L., Bergström, E., and Sjöberg, P. (1993). Primary cultures of epithelial cells from rainbow trout gills. *J. Exp. Biol.* **175,** 219–232.

Péqueux, A., Gilles, R., and Marshall, W. S. (1988). NaCl transport in gills and related structures. *Adv. Comp. Environ. Physiol.* **1,** 1–73.

Philpott, C. W., and Copeland, D. E. (1963). Fine structure of chloride cells from three species of *Fundulus*. *J. Cell Biol.* **18,** 389–404.

Pisam, M., and Rambourg, A. (1991). Mitochondria-rich cells in the gill epithelium of teleosts fishes: An ultrastructural approach. *Int. Rev. Cytol.* **130,** 191–232.

Potts, W. T. W. (1984). Transepithelial potentials in fish gills. *In* "Fish Physiology" (W. S. Hoar, and D. J. Randall, eds.), Vol. 10B, pp. 105–128. Academic Press, New York.

Potts, W. T. W., and Evans, D. H. (1967). Sodium and chloride balance in the killifish *Fundulus heteroclitus*. *Biol. Bull.* (*Woods Hole, Mass.*) **133,** 411–425.

Renfro, J. L. (1977). Interdependence of active Na^+ and Cl^- transport in the isolated urinary bladder of the teleost, *Pseudopleuronectes americanus*. *J. Exp. Zool.* **199,** 383–390.

Renfro, J. L., Miller, D. S., Karnaky, K. J., Jr., and Kinter, W. B. (1976). Na-K-ATPase localization in teleost urinary bladder by [^{3}H]ouabain autoradiography. *Am. J. Physiol.* **231,** 1735–1743.

Riordan, J. R. (1993). The cystic fibrosis transmembrane conductance regulator. *Annu. Rev. Physiol.* **55,** 609–630.

Sakmann, B., and Neher, E. (1984). Patch clamp techniques for studying ionic channels in excitable membranes. *Annu. Rev. Physiol.* **46,** 455–472.

Sardet, C., Pisam, M., and Maetz, J. (1979). The surface epithelium of teleostean fish gills: Cellular and junctional adaptations of the chloride cell in relation to salt adaptation. *J. Cell Biol.* **80,** 96–117.

Silva, P., Solomon, R., Spokes, K., and Epstein, F. H. (1977). Ouabain inhibition of gill Na-K-ATPase: Relationship to active chloride transport. *J. Exp. Zool.* **199,** 419–426.

Stokes, J. B. (1984). Sodium chloride absorption by the urinary bladder of the winter flounder: A thiazide-sensitive, electrically neutral transport system. *J. Clin. Invest.* **74,** 7–16.

Stokes, J. B. (1988). Passive NaCl transport in the flounder urinary bladder: Predominance of a cellular pathway. *Am. J. Physiol.* **255,** F229–F236.

Tabcharani, J. A., Low, W., Elie, D., and Hanrahan, J. W. (1990). Low conductance chloride channel activated by cAMP in the epithelial cell line T_{84}. *FEBS Lett.* **270,** 157–164.

Utida, S., Hirano, T., Oide, H., Ando, M., Johnson, D. W., and Bern, H. A. (1972). Hormonal control of the intestine and urinary bladder in teleost osmoregulation. *Gen. Comp. Endocrinol., Suppl.* **3,** 317–327.

Valverde, M. A., Díaz, M., Sepúlveda, F. V., Gill, D. R., Hyde, S. C., and Higgins, C. F. (1992). Volume-regulated chloride channels associated with human multidrug-resistance P-glycoprotein. *Nature* (*London*) **355,** 830–833.

Van Praag, D., Farber, S. J., Minkin, E., and Primor, N. (1987). Production of eicosanoids by the killifish gills and opercular epithelia and their effect on active transport of ions. *Gen. Comp. Endocrinol.* **67,** 50–57.

Wood, C. M., and Marshall, W. S. (1994). Ion balance, acid–base regulation and chloride cell function in the common killifish, *Fundulus heteroclitus*—a freely euryhaline estuarine teleost. *Estuaries* **17:1A,** 34–52.

Zadunaisky, J. A. (1984). The chloride cell: The active transport of chloride and the paracellular pathways. *In* "Fish Physiology" (W. S. Hoar and D. J. Randall, eds.), pp. 129–176. Vol. 10B. Academic Press, Orlando, FL.

Zadunaisky, J. A., Curci, S., Schettino, T., and Scheide, J. I. (1988). Intracellular recordings in the opercular epithelium of *Fundus* [*sic*] *heteroclitus*. *J. Exp. Zool.* **247,** 126–130.

2

ELECTROPHYSIOLOGY OF ION TRANSPORT IN TELEOST INTESTINAL CELLS

CHRISTOPHER A. LORETZ

I. OVERVIEW OF TELEOST INTESTINAL ION TRANSPORT PROCESSES

The teleost intestine performs several functions, including the absorptive transport of nutrients essential for metabolism and growth, and the active transport of salts and water as part of hydromineral homeostasis. The absorption of dietary Na^+ and Cl^- from the intestinal lumen in freshwater fishes replaces salt lost by diffusion to the dilute external medium, whereas in seawater fishes, salt absorption from the intestine following the ingestion of seawater drives the absorption of water to replace that lost to the hyperosmotic external environment; the additional salt load is actively secreted across the gills and other surface epithelia (Smith, 1930, 1932; and see reviews elsewhere in this volume). Nutrient absorption from the intestinal lumen is coupled to Na^+ transport driven by the inwardly directed electrochemical gradient for Na^+; as a result, there will be competition for the Na^+ electrochemical gradient and possible interdependence of electrolyte and nonelectrolyte transport. The emphasis in this chapter will be on systems that transport electrolytes exclusively and contribute to teleost iono- and osmoregulation. For a discussion of nutrient-absorptive systems, the reader is referred to published reviews (e.g., Ferraris and Ahearn, 1983; Karasov and Diamond, 1983; Buddington *et al.*, 1987; Collie and Hirano, 1987).

The electrophysiological properties of the teleost intestine reflect the nature and magnitude of ion transport across the epithelium. The ion transport processes giving rise to these properties are summarized in this chapter, with particular attention to recent technical advances that allow description of the individual membrane conductances that contribute to the transport and electrical properties.

A. Regional Specialization of Alimentary Tract Function

The teleost intestinal tract is divided into morphologically and histologically recognizable segments, each characterized by distinct ion transport properties relating to a particular phase of the intestine's role in osmoregulation. The teleost intestine is variably divided into the following segments: esophagus, stomach, anterior/middle intestine, and posterior intestine (Fig. 1A). Whereas comprehensive studies of the processing of fluid along the alimentary tract are few (Smith, 1930; Hickman, 1968; Shehadeh and Gordon, 1969; Skadhauge, 1973, 1974; Kristensen and Skadhauge, 1974; Kirsch and Meister, 1982; Sleet and Weber, 1982; Parmalee and Renfro, 1983), each of these intestinal segments has received attention individually, providing understanding of their respective roles in osmoregulation. As initially set forth decades ago (Smith, 1930, 1932), the role of drinking-coupled ion and water absorption in seawater fishes is the retrieval of environmental water to replace that lost by diffusion to the hyperosmotic environment. Primary regulation appears to be at the level of drinking (thereby controlling delivery of transportable substrate), which increases with seawater adaptation and is under the control of hormones such as angiotensin II (Nishimura, 1987). Not surprisingly, most of the available data on intestinal electrolyte transport are from stenohaline seawater and seawater-adapted euryhaline fishes, where intestinal transport is most active.

1. ESOPHAGUS

In seawater fishes, where drinking occurs as the first step in the compensation for water lost to the hyperosmotic external medium, processing begins with the desalinization of ingested seawater across the esophageal epithelium. This desalinization occurs through both the passive and active transport of NaCl from the lumen; in the Japanese and European eels (*Anguilla japonica* and *Anguilla anguilla,* respectively; Hirano and Mayer-Gostan, 1976; Kirsch, 1978), the winter flounder (*Pseudopleuronectes americanus;* Parmalee and Renfro, 1983), and the cod (*Gadus morhua*), sole (*Solea solea*), European flounder (*Platichthys flesus*), and plaice (*Pleuronectes platessa;* Kirsch and Meister, 1982), the water permeability of the esophagus is substantially less than that of the intestine, thereby limiting the energetically expensive recycling of water. In the several species examined, the process of desalinization is complete by the time the ingested fluid is delivered to the anterior intestine (Hickman, 1968; Shehadeh and Gordon, 1969; Kirsch and Meister, 1982; Parmalee and Renfro, 1983). This is in contrast to the mechanism initially proposed by Smith (1930) of gastric dilution through fluid secretion, followed by intestinal reabsorption.

2. STOMACH

Although the transport characteristics of the morphologically and functionally complex gastric epithelium have not been thoroughly studied, Parmalee and Renfro (1983) used $^{28}Mg^{2+}$ as a nontransported marker to demonstrate that some body water contributes to the dilution of ingested seawater in the stomach. It might be expected that electrical characteristics of the stomach are dominated by the transport processes associated with gastric acid secretion.

3. INTESTINE

In many species, the intestine is morphologically and functionally divided into narrower anterior and middle segments and a wider-bore posterior segment (several centimeters in length) separated by a muscular sphincter that exhibits sufficient tone to prevent the reflux forward of fluid from the typically distended posterior segment (Smith, 1930; Ferraris and Ahearn, 1983; Loretz, 1983, 1987a; Fig. 1A). Posterior intestinal distention may reflect fluid retention for extended processing. Once desalinization is complete, the anterior, middle, and posterior intestinal segments are responsible for the active transport of NaCl, which, through the establishment of a water potential gradient, drives water uptake from the lumen.

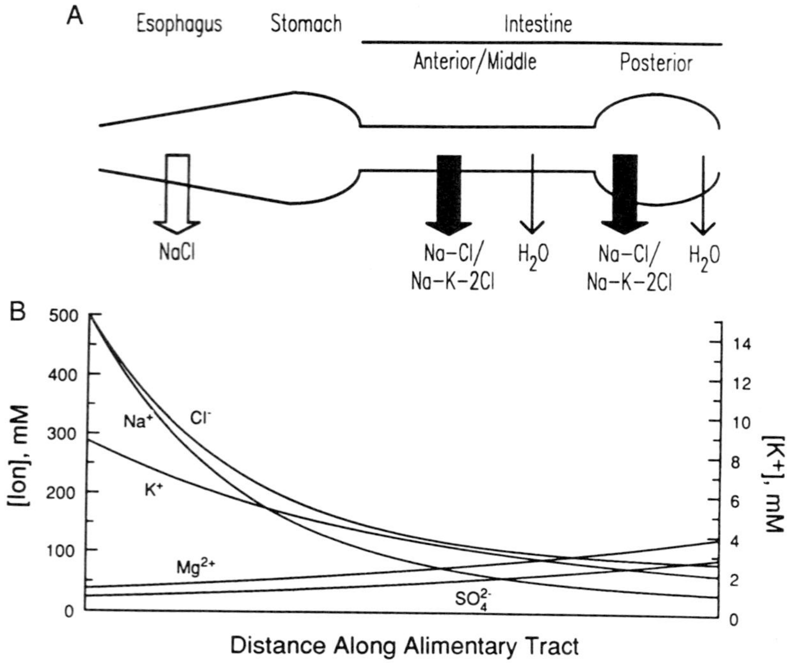

Fig. 1. Major ion transport activities across the alimentary tract of a seawater teleost are illustrated in the upper panel (A): open arrows, passive NaCl absorption; solid arrows, active NaCl transport via Na^+–Cl^- and/or Na^+–K^+–$2Cl^-$ cotransport; thin arrow, water transport. Nutrient transporters (also present) are not shown. The change in ionic composition of ingested seawater along its passage through the alimentary tract is presented graphically in the lower panel (B); the data are representative of literature values for a number of seawater teleosts (Smith, 1930; Hickman, 1968; Shehadeh and Gordon, 1969; Skadhauge, 1973, 1974; Kirsch and Meister, 1982; Sleet and Weber, 1982; Parmalee and Renfro, 1983; C. A. Loretz, unpublished data).

The luminal fluid remains nearly isosmotic with body fluids throughout its passage through the intestine; the relative contributions of Mg^{2+} and SO_4^{2-} to overall fluid composition increase as isosmotic NaCl and water absorption reduces luminal volume. The nutrient-absorptive functions of the more anterior and posterior segments differ, with the anterior intestine and its diverticula, the pyloric ceca (where present), being more active (Ferraris and Ahearn, 1983; Karasov and Diamond, 1983; Buddington *et al.*, 1987; Collie and Hirano, 1987). The ion-transporting function is present along the length of the intestine, judging from the continuous modification of luminal fluid (Fig. 1B; although the electrical characteristics differ somewhat as described in the following). Along the length of the gut about

10–15% of the ingested Mg^{2+} and SO_4^{2-} is absorbed across the intestinal wall and appears in the excreted urine (Smith, 1930; Hickman, 1968; Parmalee and Renfro, 1983). As a result, since Mg^{2+} and SO_4^{2-} are absorbed only very little, they can function as volume markers reflecting the relative absorption of fluid along the gut. About 20–30% of the ingested fluid volume is ultimately excreted in the rectal fluid (Smith, 1930; Hickman, 1968; Shehadeh and Gordon, 1969; Sleet and Weber, 1982; Parmalee and Renfro, 1983).

The key histological feature to be noted from the standpoint of ion transport is the absence of intramural glands (crypts of Lieberkühn and Brunner's glands) in the teleost intestine. The crypt tissue in the intestines of tetrapods is represented functionally as well as evolutionarily in the elasmobranchs as the rectal gland (see Valentich *et al.*, Chapter 7, this volume) and in some other fishes as extramural secretory tissues of various morphologies (Loretz, 1987a). Consequently, and consistent with what has been proposed with respect to the crypts as the site of active Cl^- secretion in mammals (Welsh *et al.*, 1982), fish intestine is generally accepted to be exclusively a NaCl-absorbing tissue (Field *et al.*, 1978; Frizzell *et al.*, 1979a; Loretz, 1983, 1987a; Field, 1993), although an apparent Cl^- secretory pathway has been proposed for the flounder (O'Grady and Wolters, 1990). The teleost intestine also possesses K^+ and HCO_3^- secretory capabilities (Stewart *et al.*, 1980; Dixon and Loretz, 1986; Musch *et al.*, 1990). There is likely some adaptive benefit to the separation of absorptive and secretory components (cf. Loretz, 1987a), namely, delivery in elasmobranchs and some other fishes of the concentrated secretory fluid from the rectal gland to the terminal intestine, or directly to the exterior by the dendritic organ of some marine ploticid catfish (van Lennep and Lanzing, 1967; Kowarsky, 1973). The advantage of this mode of secretion is to reduce the recycling of salt and water.

B. Intestinal Electrophysiology

1. Transepithelial Electrical Characteristics

Teleost intestine, in general and quite unlike its serosa-positive mammalian counterpart, exhibits a serosa-negative transepithelial potential (V_t) of a few mV when mounted in Ussing-type chambers and bathed in symmetrical saline solutions. Table I summarizes literature values for transepithelial electrophysiological characteristics for several well-studied teleost species. In the winter flounder, V_t increases in magnitude when the mucosal bathing solution is replaced by an artificial intestinal fluid with reduced Na^+, K^+, and Cl^- concentrations representative of those *in vivo* (Halm *et al.*, 1985b). As presented in the table, the transepithelial resistance (R_t) varies along the length of the alimentary tract. The anterior and middle segments of intestine have a very low electrical resistance (typically <100 Ω cm²) and can be characterized as "leaky" epithelia, where R_t is largely a function of the resistance (or, more appropriately, the conductance or "leakiness") of the paracellular shunt pathway (terminal junction, belt desmosomes, and lateral intercellular space) between epithelial cells (Powell, 1981; Kottra and Frömter, 1983). Posterior intestinal segments of several species (winter flounder, goby, coho salmon) typically exhibit higher resistance than anterior or middle intestinal segments, similar to the pattern seen in mammalian small intestine and colon (Powell, 1981). The relative contribution of muscularis mucosa and muscularis externa to the total intestinal wall thickness varies among species (Loretz, 1987a); the outer muscle layers contribute little to the electrical resistance of the tissue (Ando and Kobayashi, 1978). Feeding and nutritional state influence electrophysiological characteristics. Collie (1985) reported increased R_t across both anterior/middle and posterior intestinal segments of fed freshwater-adapted coho salmon compared with those of unfed animals. In seawater-adapted winter flounder, although the effects of feeding on R_t were not reported, the measured short-circuit current (I_{SC}, the current required to reduce V_t to zero and a measure of net ionic transport across the tissue) was greater in unfed animals (Musch *et al.*, 1987).

Table I
Summary of Transepithelial Potential (V_t) and Transepithelial Resistance (R_t) Values

Species	V_t[a]	R_t	Reference[b]
		Esophagus	
Winter flounder, *Pseudopleuronectes americanus*			
Seawater-adapted	0 mV	90 Ω cm²	1
With seawater on mucosal surface	−5 mV	92 Ω cm²	1
		Anterior/Middle Intestine	
Winter flounder, *Pseudopleuronectes americanus*			
Seawater-adapted	−1 to −6 mV	35–80 Ω cm²	2–15
Seawater-adapted with AIF[c]	−20 mV	67 Ω cm²	6
15% seawater-adapted	NR[d]	46 Ω cm²	3

(*continues*)

Table I (*continued*)

Species	V_t[a]	R_t	Reference[b]
European flounder, *Platichthys flesus*			
Seawater-adapted	−2 mV	32–71 Ω cm^2	16, 17
Freshwater-adapted	−1 mV	62 Ω cm^2	18
Japanese eel, *Anguilla japonica*			
Seawater-adapted	−1 to −10 mV	83–84 Ω cm^2	19–22
Freshwater-adapted	0 to −3 mV	94–105 Ω cm^2	19
Goby, *Gillichthys mirabilis*			
Seawater-adapted	−1 mV	52 Ω cm^2	23
Coho salmon, *Oncorhynchus kisutch*			
Seawater-adapted	−1 mV	76 Ω cm^2	24
Freshwater-adapted, unfed	+1 to +2 mV	41 Ω cm^2	24
Freshwater-adapted, fed	+1 to +2 mV	102 Ω cm^2	24
		Posterior Intestine	
Winter flounder, *Pseudopleuronectes americanus*			
Seawater-adapted	−6 mV	91 Ω cm^2	25
Japanese eel, *Anguilla japonica*			
Seawater-adapted	−3 to −10 mV	36–61 Ω cm^2	20, 26, 27
Freshwater-adapted	0 to −3 mV	76–88 Ω cm^2	26, 27
Goby, *Gillichthys mirabilis*			
Seawater-adapted	−5 to −9 mV	140–400 Ω cm^2	23, 28–33
5% seawater-adapted	−10 mV	160–267 Ω cm^2	23, 31
Coho salmon, *Oncorhynchus kisutch*			
Seawater-adapted	−1 mV	113 Ω cm^2	24
Freshwater-adapted, unfed	+1 to +2 mV	83 Ω cm^2	24
Freshwater-adapted, fed	+1 to +2 mV	181 Ω cm^2	24

[a] All V_t are reported as the serosal side with respect to mucosal side ground.

[b] Key to references: (1) Parmalee and Renfro (1983); (2) Field *et al.* (1978); (3) Field *et al.* (1980); (4) Frizzell *et al.* (1979b); (5) Halm *et al.* (1985a); (6) Halm *et al.* (1985b); (7) Helman and Beyenbach (1978); (8) Krasny and Frizzell (1984); (9) Musch *et al.* (1982); (10) Musch *et al.* (1987); (11) O'Grady (1989); (12) O'Grady *et al.* (1988); (13) Rao and Nash (1988); (14) Rao *et al.* (1984); (15) Smith *et al.* (1980); (16) Mackay and Lahlou (1980); (17) Smith *et al.* (1975); (18) Gibson *et al.* (1987); (19) Ando (1975); (20) Ando (1980); (21) Ando (1981); (22) Ando (1983); (23) Loretz (1983); (24) Collie (1985); (25) C. A. Loretz, unpublished data; (26) Ando and Kobayashi (1978); (27) Ando *et al.* (1975); (28) Dixon and Loretz (1986); (29) Loretz (1987b); (30) Loretz (1990); (31) Loretz *et al.* (1983); (32) Loretz *et al.* (1985); (33) Mooney and Loretz (1987).

[c] AIF, artificial intestinal fluid.

[d] NR, not reported.

2. Ionic Basis for the Transepithelial Potential

Ion transport across the teleost intestine, summarized in Fig. 2, depends ultimately on the basolateral membrane Na^+-K^+-ATPase, which establishes a large inwardly directed electrochemical gradient for Na^+. Transepithelial salt absorption begins with the coupled, electrically neutral entry of Na^+ and Cl^- across the apical membrane, driven by the Na^+ electrochemical gradient. Na^+ that enters the cell is subsequently extruded by the Na^+-K^+-ATPase and Cl^- exits passively down its electrochemical gradient into the serosal solution; Cl^- efflux across the basolateral membrane may occur via conductive (anion channel: Loretz and Fourtner, 1988) and nonconductive (K^+–Cl^- symport: Stewart *et al.*, 1980; and Cl^-/HCO_3^- antiport: Dixon and Loretz, 1986) pathways. In an early model for intestinal transport in the winter flounder (Field *et al.*, 1978; Frizzell *et al.*, 1979b), the observed serosa-negative V_t was proposed to result from the active transport of Na^+ across both the basal and lateral cell membranes; in this model, Na^+ delivered to the lateral intercellular space (LIS) would diffuse either through the long and tortuous LIS to the serosal solution or, preferentially, across the cation-selective terminal junction into the mucosal solution, thereby establishing serosal negativity. This model is supported by the measured excess of net Cl^- absorption over Na^+ absorption under short-circuit conditions in both the winter flounder and goby intestines (Field *et al.*, 1978; Loretz, 1983). Further refinement to this model included a cotransporter stoichiometry of 1 Na^+ : 1K^+ : 2Cl^- (Musch *et al.*, 1982; O'Grady *et al.*, 1986), although, as discussed in Section III,A, this cotransporter may operate in more than one stoichiometric mode.

The I_{sc} across teleost intestine reflects the overall net ion transport activity and, in addition to Na^+ and Cl^- absorption, includes K^+ and HCO_3^- secretion as well, based on $^{86}Rb^+$ flux and pH-stat measurements, respectively (Stewart *et al.*, 1980; Dixon and Loretz, 1986). K^+ secretion across the apical membrane occurs via passive transport; blockade of apical membrane K^+ conductance with Ba^{2+} reverses K^+ secretion to become absorption (Musch *et al.*, 1982).

II. EQUIVALENT CIRCUIT ANALYSIS OF INTESTINAL EPITHELIUM

A. Fine Structure in Relation to Electrophysiology

The teleost intestine is lined by a simple columnar epithelium composed of two cell types, the predominant absorptive cells and mucus cells (Field

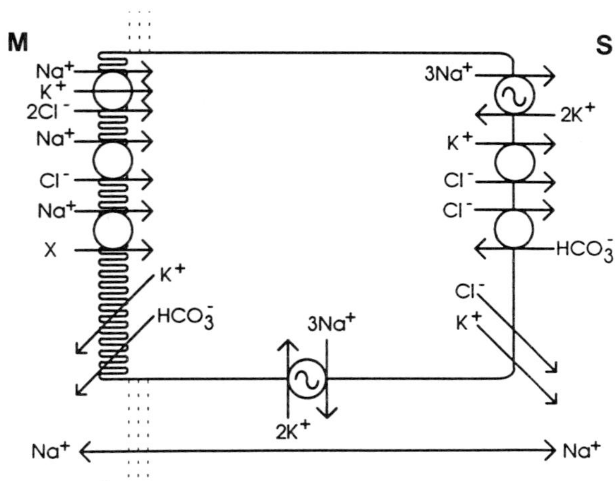

Fig. 2. Schematic cellular model of ion transport pathways across the teleost intestinal epithelium (M, mucosal solution; S, serosal solution). Transepithelial transport ultimately depends on the basolateral membrane Na^+-K^+-ATPase to establish a transmembrane electrochemical gradient for Na^+. Coupled transmembrane ion movements are driven by the appropriate electrochemical gradients, whereas channel-mediated ion flows occur down electrochemical gradients established by either active or coupled transport. X in the figure represents nonelectrolytes (glucose or amino acids) transported through apical membrane Na^+-coupled mechanisms. Diffusion of Na^+ through a cation-selective terminal junction (dashed line) and lateral intercellular space is depicted at the bottom of the figure.

et al., 1978; Loretz, 1983; Loretz *et al.*, 1985). The epithelial cells are joined by terminal junctions at their apical ends; beneath and near that zone, lateral cell membranes are joined by numerous spot desmosomes and gap junctions (Fig. 3; Field *et al.*, 1978; Curtis *et al.*, 1984; Loretz *et al.*, 1985). The gap junctional connections are the presumed basis for the electrical coupling evidenced by the small variation among cells in measured apical membrane potential difference (V_a; see Fig. 5) compared with that seen in the less well-coupled rabbit small intestinal epithelium (Curtis *et al.*, 1984). The specific functions of gap junctional connections in teleost intestine, beyond the presumed electrical coupling and transfer of small-molecular-weight solutes among cells, remain undefined. The LIS is narrow and can be highly convoluted, forming interdigitations of lateral cell membrane; the conformation of the LIS may be a determinant of paracellular shunt conductance (Field *et al.*, 1978; Loretz *et al.*, 1985).

B. Equivalent Electrical Circuit for Teleost Intestine

Epithelial can be schematically represented using Thévenin equivalent electrical circuits in which the apical, basolateral, and terminal junction/

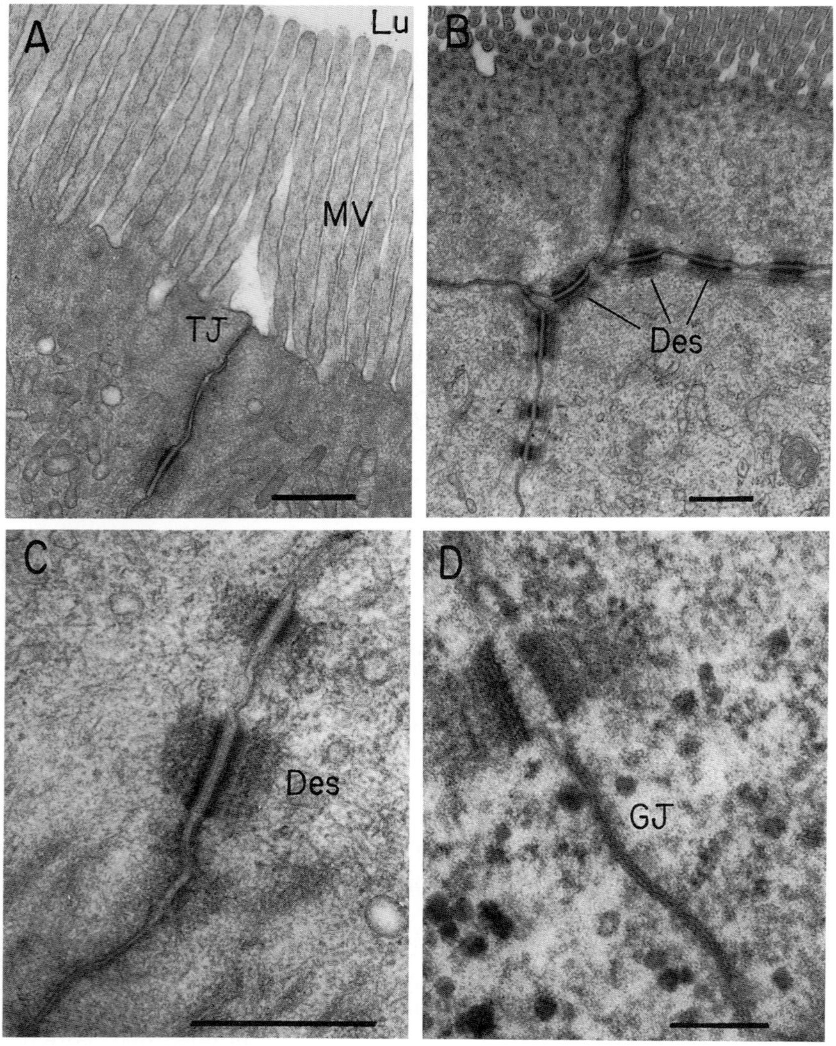

Fig. 3. Teleost intestinal epithelial cells are joined on their lateral surfaces to form a permeability barrier and to provide for electrical coupling and intercellular communication. Junctional types in goby intestinal epithelium are illustrated in these electron micrographs from the author's laboratory. (A) Terminal junctions (TJ) join cells at their apical ends and confer cation selectivity on the paracellular shunt pathway (Lu, lumen; MV, microvilli). (B) In the region below the terminal junction, cells are joined by abundant desmosomes (Des) as seen in this oblique section. (C) At higher magnification, the fine structure of spot desmosomes is seen; characteristic of spot desmosomes, a 30-nm gap with central stratum separates 20-nm-thick cell membrane plaques with their associated intermediate filament arrays extending into the cytoplasm. (D) Gap junctions (GJ) are visualized as parallel flat arrangements of cell membrane that extend over a distance of several hundred nanometers and where the intermembrane space is reduced to 2–3 nm. Calibration bars: (A–C), 0.5 μm; (D), 0.1 μm.

paracellular shunt barriers between the mucosal and serosal solutions are each modeled as an electromotive force for transmembrane current flow and a resistance representing, respectively, the relevant electrochemical gradient and conductance (Reuss and Finn, 1974; Schultz, 1980; Helman and Thompson, 1982). An equivalent electrical circuit for fish intestinal epithelium is diagrammed in Fig. 4A; in this circuit, the measured transepithelial potential, V_t, results, from both the paracellular shunt and series-combined apical and basolateral membrane electromotive forces (Halm *et al.*, 1985a,b). The relative contributions of these two components in detemining V_t are related to the ratio of the resistances in the two transepithelial branches of the network. The dependence of V_t on E_a, E_b, and E_p is expressed mathematically through a voltage divider ratio approach as

$$V_t = -\frac{R_a + R_b}{R_{sum}}E_p - \frac{R_p}{R_{sum}}(E_a - E_b) \qquad (1)$$

where E_a, E_b, and E_p are the equivalent electromotive forces across the apical membrane, basolateral membrane, and the paracellular shunt pathway, respectively; R_a, R_b, and R_p are the corresponding resistances across the apical membrane, basolateral membrane, and paracellular shunt pathway; and $R_{sum} = R_a + R_b + R_p$. This analysis redefines the origin of V_t relative to transcellular and transjunctional components compared with the earlier model of Field *et al.* (1978). In a tight epithelium, where the paracellular shunt resistance, R_p, is very high [i.e., when $R_p \gg (R_a + R_b)$], V_t will approach in magnitude the difference between the inwardly directed electromotive forces for the apical and basolateral membranes, E_a and E_b, respectively. In a leaky epithelium, where R_p is low, V_t will approach E_p; when bathed *in vitro* with symmetrical saline solutions, E_p will be near zero, even when salt transport into the LIS and the consequent transjunctional diffusion potential are taken into account (Halm *et al.*, 1985a), and V_t will consequently be small. Halm *et al.*, (1985a) have calculated from paracellular permeability ratios that the contribution of transjunctional diffusion to V_t will be about 1 mV per 10 m*M* elevation above ambient of NaCl concentration in the LIS. As the authors note, evaluation of the actual sustainable gradient is difficult.

The measured apical and basolateral cell membrane electrical potentials, V_t and V_b, can differ from E_a and E_b, the respective electromotive driving forces across these membranes, through interaction via the paracellular shunt pathway in accordance with the relative conductances of the pathways for current flow. Again, applying a voltage divider ratio approach to the circuit in Fig. 4A, V_a and V_b are represented mathematically by the equations

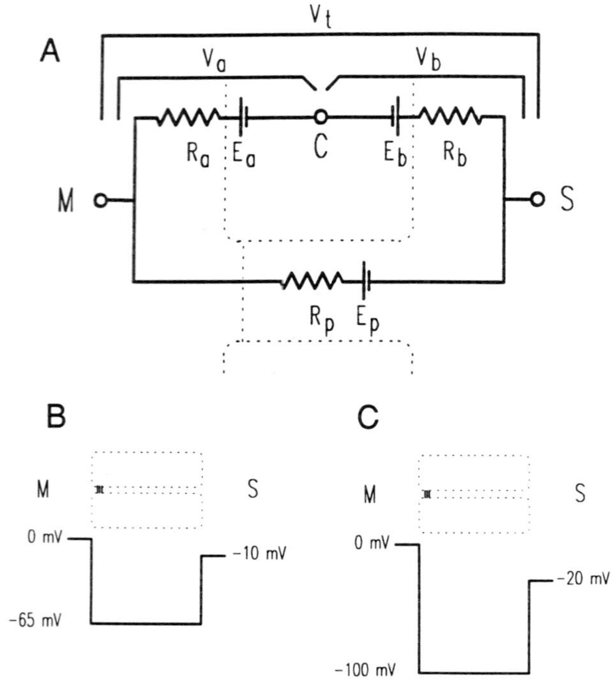

Fig. 4. (A) Equivalent electrical circuit schematic diagram of teleost intestinal epithelium. Dashed lines represent the simple epithelium with paracellular shunt pathway separating mucosal (M) and serosal (S) solutions and the intracellular compartment (C). Symbols as in text. (B) The electrical potential profile across the teleost intestinal cell is well shaped with a serosal-directed V_t of about -10 mV in symmetrical saline bathing solutions. (C) In asymmetrical solutions resembling *in vivo* conditions, cellular and transepithelial hyperpolarization occurs.

$$V_a = -\frac{R_b + R_p}{R_{sum}}E_a - \frac{R_a}{R_{sum}}(E_b + E_p) \tag{2}$$

and

$$V_b = -\frac{R_a + R_p}{R_{sum}}E_b - \frac{R_b}{R_{sum}}(E_a - E_p) \tag{3}$$

Measurements of V_a and V_b are accomplished using conventional glass intracellular microelectrodes. In the winter flounder and goby intestines bathed symmetrically with standard fish saline, V_a is about -65 mV (cell interior negative), resulting from a dominant apical membrane K^+ conductance, and V_b is about -55 mV (cell interior negative), reflecting a dominant basolateral membrane Cl^- conductance; ion substitution studies and direct pharmacological blockade of membrane conductances confirm the dependence of V_a and V_b on K^+ and Cl^- conductances, respectively (Halm *et al.*, 1985a,b; Loretz *et al.*, 1985). The electrical potential profile across the epithelium appears as a well (Fig. 4B).

In asymmetrical bathing solutions (as would occur *in vivo*), and as a result of NaCl transport into the lateral intercellular space, the contribution of E_p to V_t will be greater. When the mucosal surface of the epithelium is bathed in artificial intestinal fluid, hyperpolarization of V_a and V_t is observed, consistent with a K^+-permeable apical membrane and cation-selective terminal junction (Table I and Fig. 4C; Halm *et al.*, 1985b; C. A. Loretz, unpublished data).

Quantification of the equivalent circuit elements is accomplished through experimental manipulation during microelectrode recording. Briefly, with change to a single electromotive force or resistance, and with the assumption that the remaining electromotive forces and resistances of the circuit remain unchanged (as assumption more likely to be met for small perturbations to the system), the solution of a series of simultaneous equations yields the values of individual network components (Reuss and Finn, 1974; Halm *et al.*, 1985a). Changes in electromotive forces are achieved by altering the concentration of permeable ionic species, whereas membrane conductances can be modified by pharmacological agents. For example, mucosal Ba^{2+} treatment to block apical K^+ conductance causes concomitant changes in V_t, V_a, R_t, and the fractional resistance, f_R (the ratio of apical membrane resistance, R_a, to the total cellular resistance, $R_a + R_b$). The effect of Ba^{2+} on V_a and f_R is illustrated in Fig. 5A; the increased size of deflections in the V_a trace, representing the response of V_a to constant transepithelial current pulses, reflects the elevation of f_R resulting from Ba^{2+} blockade. Network analysis of this type performed on winter flounder middle intestine and goby posterior intestine (Halm *et al.*, 1985a,b; Loretz *et al.*, 1985) generated the data presented in Table II.

The degree to which the paracellular shunt influences V_a and V_b (and, therefore, V_t) is a function of the relative contribution of R_p to R_t. In flounder middle intestine, R_p is low; the shunting effect results in the similarity of V_a and V_b and, consequently, a modest V_t. In the goby posterior intestine, where R_p is higher, the shunting effect is smaller and consequently V_a and V_b will be closer approximations of E_a and E_b, and V_t will be greater. In these two teleost systems, the only ones for which data are available, the absolute cellular membrane resistances are similar between the two tissues and the observed difference in R_t relates predominantly to differences in paracellular conductance. Overall, with their relatively high paracellular conductances, these are classified as "leaky"

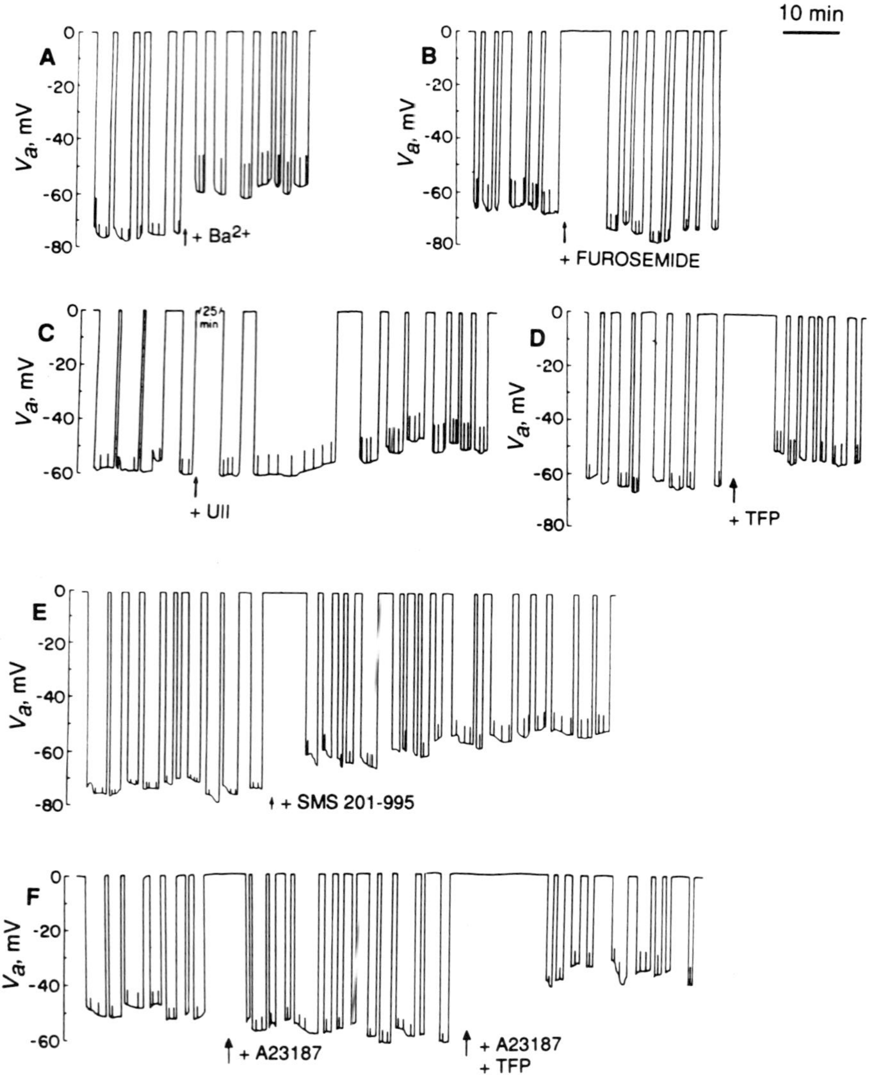

Fig. 5. Conventional glass intracellular microelectrode recordings of apical membrane potential (V_a) in goby posterior intestinal epithelial cells *in situ* in the epithelium. The deflections in the individual traces are the change in V_a in response to constant transepithelial current pulses (10–50 μA cm^{-2}). Individual impalements are separated by a return to zero of the V_a trace. (A) Mucosal addition of 2.0 m*M* Ba^{2+} causes depolarization of V_a by reducing apical membrane K^+ conductance. (B) Mucosal addition of 10 μM furosemide inhibits the Na^+–Cl^- and/or Na^+–K^+–$2Cl^-$ cotransporter and results in hyperpolarization of V_a via a reduction of intracellular Cl^- activity. (C, E) Urotensin II (100 n*M*, serosal) and the structurally similar somatostatin analog SMS 201-995 (100 n*M*, serosal) both increase NaCl absorption via stimulation of the Na^+–Cl^- and/or Na^+–K^+–$2Cl^-$ cotransporter resulting in an increase in intracellular Cl^- activity. (D) Trifluoperazine (TFP, 100 μM, serosal) inhibits the Ca^{2+}–calmodulin complex and subsequently Ca^{2+}–calmodulin-dependent protein kinase, thereby

Table II
Equivalent Circuit Component Values for Winter Flounder and Goby Intestine[a]

Tissue	f_R	R_a	R_b	R_p	R_t	G_p/G_t[b]
Middle intestine (winter flounder)	0.24	192 Ω cm²	667 Ω cm²	37 Ω cm²	36 Ω cm²	0.96
Posterior intestine (goby)	0.48	386 Ω cm²	417 Ω cm²	206 Ω cm²	164 Ω cm²	0.80

[a] Data taken from Halm *et al.* (1985a,b) and Loretz *et al.* (1985).
[b] G_p/G_t, conductance of the paracellular shunt pathway relative to the total transepithelial conductance; other abbreviations as in text.

epithelia. The calculated values for the fraction of tissue conductance attributable to the paracellular shunt are within the range reported for a variety of mammalian and other vertebrate intestinal epithelia (Powell, 1981).

III. TRANSCELLULAR ION MOVEMENTS

A. Thermodynamic Considerations

Despite the continuing development of mechanistic models of transepithelial ion transport for the teleost intestine, there has been no rigorous examination of the electrochemical driving forces for the proposed ion movements across membranes of the epithelium either *in vivo* or *in vitro*. In its general form, the electrochemical potential ($\Delta\bar{\mu}$) for ion transport across a membrane from side 1 to side 2 includes contributions from the chemical and electrical gradients:

$$\Delta\bar{\mu} = RT \ln C_2/C_1 + zF\Delta E \tag{4}$$

where C_1 and C_2 are the concentrations (or activities) of the ion on the two sides of the membrane, ΔE is the electrical potential difference ($E_2 - E_1$)

mimicking the apical membrane depolarization by urotensin II, which exerts its effects through a reduction in intracellular Ca^{2+} activity and subsequent Ca^{2+}-dependent processes. (F) Calcium ionophore A23187 (1 μM, serosal) inhibits NaCl absorption and causes hyperpolarization of V_a. Both effects can be overridden by TFP (100 μM). (Panels A–C from Loretz *et al.* (1985). Copyright © 1985, The American Physiological Society. Panels D and F from Loretz, C. A. (1987b). Regulation of goby, intestinal ion absorption, by the calcium messenger system. *J. Exp. Zool.* **244,** 67–78. Copyright © 1987, Wiley–Liss, a division of John Wiley & Sons, Inc. Panel E from Loretz, C. A. (1990). Recognition by goby intestine of a somatostatin analog. *J. Exp. Zool. Suppl.* **4,** 31–36. Copyright © 1990, Wiley–Liss, a division of John Wiley & Sons, Inc. Used with permission.)

across the membrane, z is the valence, and R, T, and F have their usual meanings. For thermodynamically favorable transport reactions, $\Delta\bar{\mu} < 0$. Transepithelial electrical profiles for open-circuited tissues are available from intracellular microelectrode studies (Fig. 4 and 5). For the winter flounder and the goby, the cell membrane potentials are similar in magnitude to those seen in other vertebrates despite the observed serosa-negative transepithelial potential; for tissues bathed bilaterally in normal saline solutions resembling plasma, V_a of -60 to -70 mV and V_t of -5 to -15 mV are typical.

Table III presents representative ionic compositions for mucosal and serosal solutions used in Ussing-type chamber studies of ion transport. Symmetrical solutions are used in voltage clamp and radiotracer flux experiments to abolish transepithelial electrochemical gradients in the identification of active transport processes across short-circuited tissues. Whereas the composition of fluids bathing the mucosal and serosal surfaces of epithelial tissues *in vitro* can be set, determination of the intracellular activities is less easily achieved but some measurements are available. Data on ionic activities in the cytoplasmic compartment are limited; intracellular Cl^- and K^+ activities have been determined in winter flounder middle (Duffey *et al.*, 1979; Smith *et al.*, 1980) and posterior (C. A. Loretz and M. E. Duffey, unpublished data) intestinal cells using intracellular ion-specific microelectrodes. Reported activities of about 30 mM for Cl^- and of about 80 mM for K^+ are both greater than those expected for electrochemical equilibrium at a physiological membrane potential (V_m) of about -60 mV. Intracellular Na^+ activity for goldfish intestinal epithelial cells is about 15 mM (Zuidema *et al.*, 1986), a value less than that expected for electrochemical equilibrium. Using these representative values for transmembrane electrical gradients and fluid compartment compositions, $\Delta\bar{\mu}$ can be calculated. As summarized in Table III, the electrochemical potential for Na^+ ($\Delta\bar{\mu}^{Na}$) is directed inward across the apical membrane, whereas those for K^+ ($\Delta\bar{\mu}^{K}$) and Cl^- ($\Delta\bar{\mu}^{Cl}$) are directed outward. The electrochemical gradients will be similarly directed across the basolateral membrane with respect to the cell interior.

The thermodynamic feasibility of passive transmembrane movement *in vivo* under actual physiological conditions will depend on the composition of the intestinal fluid. The ionic composition of luminal fluid throughout the length of the intestine was presented earlier (Fig. 1); values for a region near the junction of the middle and posterior intestine are entered in Table III. Recalculation of electrochemical gradients using these more realistic values shows that although $\Delta\bar{\mu}^{Na}$ is altered little, $\Delta\bar{\mu}^{K}$ is reversed to favor K^+ entry and $\Delta\bar{\mu}^{Cl}$ opposes apical entry even more strongly as a result of apical membrane hyperpolarization and reduced luminal Cl^-

Table III
Transmembrane Electrochemical Gradients ($\Delta\bar{\mu}$) Calculated for Symmetrical Saline Bathing Solutions Assuming $V_{ms} = -10$ mV and $V_{mc} = -65$ mV and for Mucosal Intestinal Fluid at the Junction of Middle and Posterior Intestine Assuming $V_{ms} = -20$ mV and $V_{mc} = -100$ mV

Ion	Mucosal (mM)	Cytoplasmic (mM)	Serosal (mM)	$\Delta\bar{\mu}_{mc}^{ion}$ [a] (kJ mol^{-1})	$\Delta\bar{\mu}_{cs}^{ion}$ (kJ mol^{-1})
			Symmetrical Saline		
Na^+	160	15	160	−11.95 (m→c)	10.99 (s→c)
K^+	2.5	87	2.5	2.25 (c→m)	−3.21 (c→s)
Cl^{-1}	160	25	160	1.81 (c→m)	−0.85 (c→s)
			Mucosal Intestinal Fluid		
Na^+	48	15	160	−12.44 (m→c)	10.08 (s→c)
K^+	2.7	87	2.5	−1.32 (m→c)	−0.08 (c→s)
Cl^{-1}	99	25	160	6.35 (c→m)	−3.26 (c→s)

[a] $\Delta\bar{\mu}_{12}^{ion}$, electrochemical gradient for ion movement from side 1 to side 2, that is, in the absorptive direction (lumen→cell→blood side). m, mucosal side; c, cytoplasmic space; s, serosal side. The direction of the net electrochemical gradient for each ion is noted within parentheses.

concentration. Based on this analysis, the likelihood of K^+ secretion under physiological conditions is questionable.

Thermodynamic evaluation of the transport model in Fig. 2 can be made using the calculated values for $\Delta\bar{\mu}$ in Table III. Apical membrane uptake has been variably attributed to neutral Na^+–K^+–$2Cl^-$ or to Na^+–Cl^- cotransport driven by the inwardly directed Na^+ electrochemical gradient. Either of these stoichiometries is thermodynamically permissable for tissues *in vitro* with symmetrical normal saline bathing solutions as evidenced by the large negative $\Delta\bar{\mu}$ for cotransport calculated as the sum of $\Delta\bar{\mu}$ for the coupled ions:

$$\Delta\bar{\mu}^{Na\text{-}K\text{-}2Cl}{}_{mc} = -\ 6.08\ \text{kJ mol}^{-1}$$

and

$$\Delta\bar{\mu}^{Na\text{-}Cl}{}_{mc} = -\ 10.14\ \text{kJ mol}^{-1}$$

Recalculation of $\Delta\bar{\mu}$ for apical membrane cotransport in tissues with mucosal intestinal fluid and with the assumption of stable intracellular ionic activities (Schultz, 1981) yields diminished electrochemical gradients:

$$\Delta\bar{\mu}^{\text{Na-K-2Cl}}_{\text{mc}} = -\ 1.06\ \text{kJ mol}^{-1}$$

and

$$\Delta\bar{\mu}^{\text{Na-K-2Cl}}_{\text{mc}} = -\ 6.09\ \text{kJ mol}^{-1}$$

The dramatic reduction in $\Delta\bar{\mu}^{\text{Na-K-Cl}}_{\text{mc}}$ suggests limited capability for Na^+–K^+–$2Cl^-$ coupled salt uptake in this segment of intestine but, indeed, may reflect that it has been operating a great deal in anterior segments. This analysis is extended in Fig. 6 by calculation of $\Delta\bar{\mu}$ for cotransport along the length of the intestinal tract using the concentration profiles in Fig. 1. The fluid absorption that continues in the posterior intestine cannot be supported *in vivo* by Na^+–K^+–$2Cl^-$ coupled uptake although uptake via a $1Na^+ : 1Cl^-$ stoichiometry would be thermodynamically quite feasible. Two possibilities emerge to explain continued salt and fluid absorption along the entire length of the intestine: first, there may exist two cotransporter proteins exhibiting different stoichiometries and, second, a single contransporter may operate in two stoichiometrically different modes. Multiple cotransporters or operational modes may explain the apparent confusion regarding stoichiometry and continued NaCl uptake in the absence of luminal K^+ (Frizzell *et al.*, 1979b; Musch *et al.*, 1982; Halm *et al.*, 1985b). Interesting questions arise regarding regional segregation of mechanism along the intestine with, perhaps, $1Na^+$–$1K^+$–$2Cl^-$ stoichiometry expressed in the more anterior segments and $1\ Na^+$–$1Cl^-$ expressed in the more posterior segments. Clearly, more study is required. The involvement of parallel apical membrane antiporters (Na^+/H^+ and Cl^-/HCO_3^-) in mucosal uptake is not indicated. Removal of either Na^+ or Cl^- from the mucosal solution *in vitro* abolishes the net absorption of Cl^- or Na^+, respectively, by intact tissues; in these preparations, however, net absorption is inhibited only 40–70% by the specific inhibitors furosemide, bumetanide, and piretanide (Field *et al.*, 1978; Zeuthen *et al.*, 1978; Frizzell *et al.*, 1979b; C. A. Loretz, unpublished observation). Further supporting mucosal uptake by symport, Cl^-removal and furosemide addition produced equivalent reductions in Na^+ transport by intestinal brush border membrane vesicles (Eveloff *et al.*, 1980). Dissipation of the Na^+ electrochemical gradient across the apical membrane is not complete despite the approach of $\Delta\bar{\mu}$ for cotransport to zero; consequently, $\Delta\bar{\mu}^{\text{Na}}$ can still be used to drive nutrient and nonelectrolyte absorption from the lumen (Collie and Hirano, 1987).

The basolateral membrane mechanisms included in Fig. 2 are also thermodynamically justified. As is typical in transporting epithelia, basolateral Na^+ efflux and K^+ influx are active. The electrochemical gradient favoring basolateral membrane K^+ efflux is small under physiological

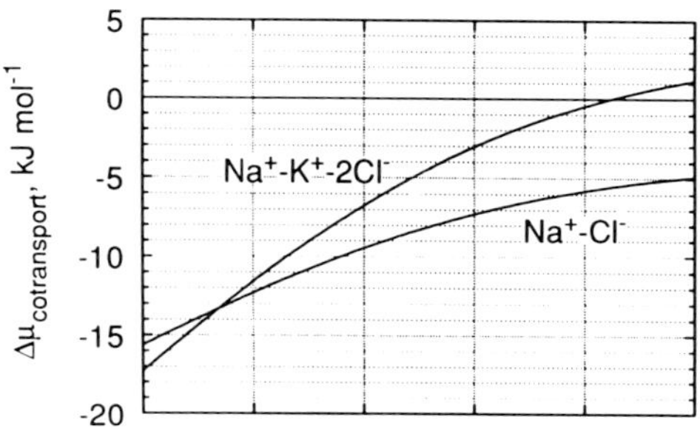

Fig. 6. Electrochemical gradients for Na^+–Cl^- and Na^+–K^+–$2Cl^-$ cotransport across the apical cell membrane (as $\Delta\bar{\mu}_{mc}^{Na-Cl}$) and $\Delta\bar{\mu}_{mc}^{Na-K-2Cl}$) are graphed as a function of alimentary tract length. Along most of the alimentary tract, both transport mechanisms are thermodynamically feasible, but in the terminal intestine, only Na^+–K^+–$2Cl^-$ cotransport (with $\Delta\bar{\mu} < 0$) is thermodynamically possible. The $\Delta\bar{\mu}$ for cotransport were calculated using the luminal ion concentrations presented in Fig. 1 and intracellular ion activities estimated from literature values (Na^+, 15 m*M;* K^+, 87 m*M;* Cl^-, 25 m*M*). See text for details.

conditions. K^+ exit across the basolateral membrane may be enhanced through its proposed coupling to Cl^- efflux, for which there is a large electrochemical gradient (Fig. 2; Stewart *et al.*, 1980).

B. Electrophysiological Correlates of Altered Ion Transport

Direct pharmacological inhibition of the apical membrane cotransporter by furosemide or bumetanide causes cellular hyperpolarization of V_a and reduction of f_R (Fig. 5B and Table IV; Musch *et al.*, 1982; Halm *et al.*, 1985b; Loretz *et al.*, 1985). Cotransporter activity is also inhibited physiologically through a Ca^{2+}–calmodulin-dependent cellular mechanism; elevation of intracellular Ca^{2+} by the ionophore A23187 brings about cellular hyperpolarization similar to that seen with direct inhibition (Fig. 5F and Table IV; Loretz and Assad, 1986; Loretz, 1987b).

Hormonal stimulation of transepithelial NaCl absorption by the caudal neurosecretory peptide urotensin II (UII) or the somatostatin analog SMS 201-995, which act through reductions in both hormone-sensitive adenylate cyclase activity and intracellular free Ca^{2+}, is accompanied by cellular depolarization and elevation of f_R, effects opposite to those of inhibitors of the cotransport mechanism (Figs. 5C and 5E and Table IV; Loretz *et al.*, 1985; Loretz, 1990). Inhibitors of calmodulin-mediated responses such as trifluoperazine (TFP; and other phenothiazines) and calmidazolium (R24571), which block the interaction between the Ca^{2+}–calmodulin complex and the enzymes it regulates (Weiss *et al.*, 1980), mimic hormonal stimulation by UII and SMS 201-995 (Fig. 5D and Table IV; Loretz, 1987b). Moreover, since TFP and R24571 will override the response to A23187, the inhibitory effect of Ca^{2+} is mediated through calmodulin and not via a direct action on the cotransporter (Figure 5F; Loretz, 1987b).

Halm *et al.* (1985a,b) proposed two mechanisms to account for the electrophysiological consequences of cotransporter inhibition. In the first, cotransporter inhibition in the face of continuing basolateral Cl^- efflux through both passive and K^+-coupled mechanisms leads to a fall in intracellular Cl^- activity. The reduction in intracellular Cl^- activity increases E_b and thus results in basolateral membrane hyperpolarization [and also apical membrane hyperpolarization, Eq. (2)]. In the second mechanism, the decrease in cell volume concomitant with Cl^- loss increases intracellular K^+ activity and thereby hyperpolarizes E_a. Conversely, and in accordance with Halm *et al.* (1985a,b), hormonal stimulation of cotransport would increase intracellular Cl^- activity and cell volume to cause cellular depolarization. The transport stimulation by UII and SMS 201-995 is not the result of cellular depolarization since Ba^{2+}, despite its marked depolarizing effect, does not influence active NaCl absorption (Loretz *et al.*, 1985).

IV. MEMBRANE ION CHANNELS IN TELEOST INTESTINAL EPITHELIUM

A. Isolation of Transmembrane Ionic Currents

Greater resolution of transmembrane (cytoplasmic–extracellular) ionic currents of enterocytes can be achieved by examination at both the single-cell and single-channel levels. Electrophysiological studies of individual dissociated enterocytes using the whole-cell voltage clamp technique removes the complications of paracellular conductances and of gap junction-mediated lateral current spread inherent in studies of cells in the intact epithelium. There are drawbacks in the interpretation of results from these isolated-cell studies, however, owing to the loss of electrical and functional polarity. Specifically, in an isolated cell, electrical polarity is compromised with $V_m = V_a = V_b$; that is, V_a is depolarized and V_b is hyperpolarized relative to *in situ* levels. Another drawback of whole-cell studies is the inability to distinguish between apical and basolateral membrane conductances that are examined in their ensemble behavior. Additionally, there can be reduced cytoplasmic influences in whole-cell recording following the exchange of pipette fluid for cytoplasm; this effect can be minimized with the perforated-patch technique, wherein nystatin in the recording pipette induces high electrical conductivity between the pipette and the

Table IV

Transport and Electrophysiological Effects on Teleost Intestinal Epithelium of Hormonal and Pharmacological Agents[a]

	Treatment					
	Urotensin II SMS201-995	TFP R24571	Furosemide Bumetanide	IBMX cGMP	A23187 Ionomycin	Ba^{2+} Quinidine Carbachol
Effect on NaCl absorption	↑[b]	↑	↓	↓	↓	↔
Effect on Na^+–Cl^- and/or Na^+–K^+–$2Cl^-$ cotransporter	↑ (via ↓ $[Ca^{2+}]_i$)	↑ (via Ca-CaM inhibition)	↓ (via direct inhibition)	↓ (via PK-G activation)	↓ (via ↑ $[Ca^{2+}]_i$)	↔
V_a[c]	Depolarization	Depolarization	Hyperpolarization	Hyperpolarization	Hyperpolarization	Depolarization
f_R[d]	↑	↔	↓	↓	↔	↑

[a] Data from Musch *et al.* (1982), Krasny and Frizzell (1984), Rao *et al.* (1984), Halm *et al.* (1985b), Loretz *et al.* (1985), O'Grady *et al.* (1986, 1988), Loretz (1987b, 1990, unpublished data), Rao and Nash (1988), and O'Grady (1989).

[b] ↑, increase; ↓, decrease; ↔, no change.

[c] V_a, apical membrane potential.

[d] f_R, fractional resistance of apical membrane.

cell interior but does not allow leakage of cytoplasmic macromolecules (Korn *et al.*, 1991). Similarly, with loss of the terminal junction as a barrier between apical and basolateral membrane domains in isolated cells, lateral diffusion in the membrane bilayer could permit intermingling of membrane proteins normally segregated to either the apical or basolateral domain.

Since its recent development, the patch clamp technique has provided a valuable perspective on membrane conductances and their physiological basis in membrane ion channels (Neher and Sakmann, 1976; Neher *et al.*, 1978; Hamill *et al.*, 1981). Application to intestinal and other epithelial cells of teleosts has allowed biophysical characterization of single ion channels, confirmation of conductive elements in cellular transport models, and demonstration of ion channels heretofore unknown based on electrophysiological analysis. In excised patches, there is fluid access to the cytoplasmic face of the membrane and voltage clamp control of the membrane potential. Limitations with this technique include uncertainty with respect to the *in situ* location of single ion channels due to lateral diffusion and, for excised patches, the loss of cytoplasmic regulatory factors. Despite their limitations, these subepithelial approaches are useful for examining ionic currents across cellular membranes.

B. Ensemble Channel Activity

Measurements of ensemble channel activity have been performed in dissociated enterocytes of the winter flounder using whole-cell and perforated-patch voltage clamp; O'Grady *et al.* (1991) described a voltage-activated K^+ current that could be blocked externally by Ba^{2+}, Cs^+, and charybdotoxin. V_m measured using this technique (-23 to -39 mV) was substantially lower than that typically recorded in this same tissue with conventional glass intracellular microelectrodes and may reflect the greater difficulty in achieving good electrical seals to the membrane with the larger-diameter recording pipettes used in these techniques. Consistent with its effects on the intact epithelium to inhibit K^+ secretion, cyclic GMP diminished the whole-cell K^+ current (Rao *et al.*, 1984; O'Grady *et al.*, 1991). Although cyclic GMP inhibition of transepithelial transport can be blocked by the cyclic nucleotide-dependent protein kinase inhibitor H-8, a direct effect of cyclic GMP on the channel cannot be excluded (Rao *et al.*, 1984; O'Grady *et al.*, 1988, 1991).

C. Single Ion Channels

1. Anion (Cl^-) Channel

Consistent with the general model for teleost intestinal cell transport, an anion channel from the basolateral cell membrane has been characterized using patch clamp (Loretz and Fourtner, 1988). This anion channel with a single-channel conductance of 20–90 pS is voltage dependent, with depolarization increasing its open probability (P_o, the fraction of time spent in the open state; Fig. 7). The observed steep dependence on voltage in the physiological range of V_m may address the issue of intermembrane crosstalk raised by Halm *et la.* (1985b), namely, how at steady state the basolateral membrane Cl^- efflux is matched to apical membrane influx. Simply, the cellular depolarization caused by Cl^- accumulation would increase P_o and thereby stimulate Cl^- efflux. Although depolarization would decrease the electrical driving force for Cl^- efflux, this is more than compensated by increased P_o such that the average channel current, calculated as $I_c \times P_o$, is increased (cf. Table 2 in Loretz and Fourtner, 1988). Additionally, the rise in intracellular Cl^- activity with cotransporter stimulation will further augment the chemical driving force for Cl^- exit.

2. Calcium-Activated Potassium Channel

Earlier studies using conventional glass intracellular microelectrodes demonstrated in enterocytes from both winter flounder and goby a substantial apical membrane K^+ conductance that could be inhibited by Ba^{2+}, quinidine, and carbachol (Musch *et al.*, 1982; Halm *et al.*, 1985a,b; Loretz *et al.*, 1985). Using patch clamp technique, a voltage-dependent, calcium-activated K^+ channel [K(Ca) channel] with a single-channel conductance of about 150 pS was observed; this channel was also inhibited by Ba^{2+}, quinidine, tetraethylammonium (TEA^+), and Cs^+ (Fig. 8; Loretz and

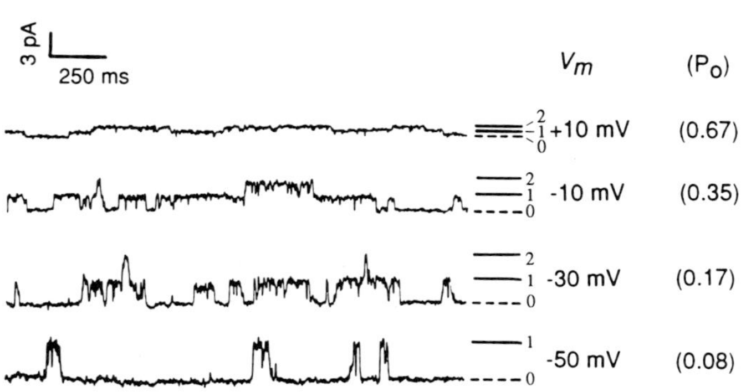

Fig. 7. Single-channel current record from an excised basolateral membrane patch of goby enterocyte demonstrating the voltage dependence of anion channel activity. In this record, upward deflections in the current record represent channel openings. The open probability of the channel (P_o) increases with depolarization of the membrane potential (V_m). This membrane patch containing two channels was bathed on the extracellular side with 144.5 m*M* and on the cytoplasmic side with 70 m*M* Cl^-. (From Loretz and Fourtner (1988). Copyright © 1988, The Company of Biologists, Ltd. Used with permission.)

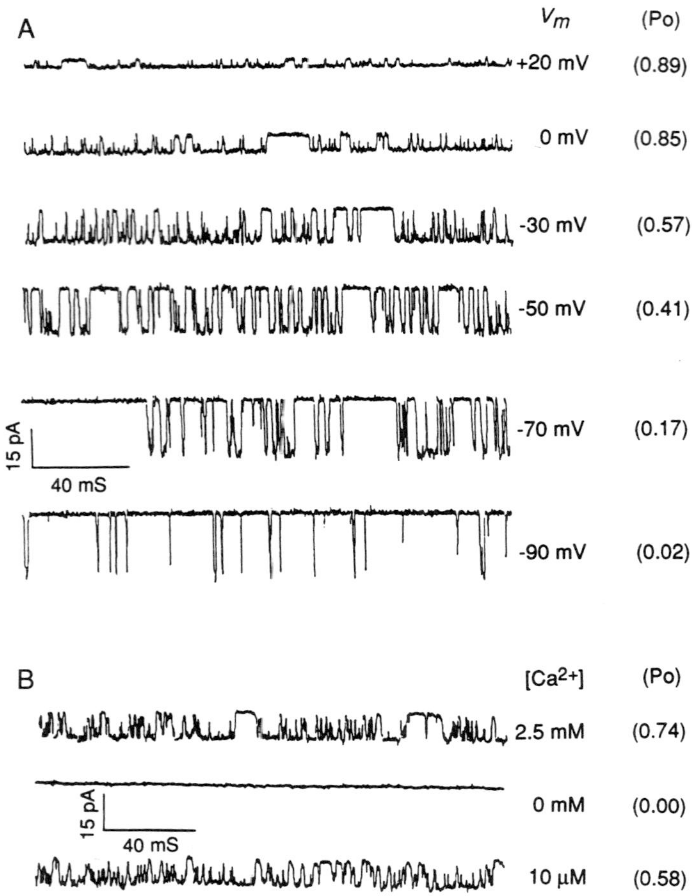

Fig. 8. Single-channel current records from basolateral membrane voltage-dependent, calcium-activated K^+ channels [K(Ca) channels] from goby enterocyte. Downward deflections represent channel openings. (A) The open probability (P_0) of the K(Ca) channel increases with depolarization of the membrane potential (V_m). This membrane patch was bathed on the extracellular side with 140 m*M* K^+ and on the cytoplasmic side with 35 m*M* K^+. (B) Calcium removal from the cytoplasmic face reversibly abolished channel activity in this membrane patch maintained at $V_m = -60$ mV. (From Loretz and Fourtner (1991). Copyright © 1991, The Company of Biologists, Ltd. Used with permission.)

Fourtner, 1990, 1991), suggesting that the several classes of K^+ conductance (based on the blocker sensitivity of cells in intact epithelia) exist in a single membrane ion channel. Similar pharmacological sensitivity is displayed by the expression product of a flounder intestinal K^+ channel mRNA in *Xenopus* oocytes, although this channel did not exhibit the voltage dependence characteristic of flounder and goby K^+ channels *in situ* (Fig. 9; Loretz and Fourtner, 1990, 1991; Sullivan *et al.*, 1990; O'Grady *et al.*, 1991). Sensitivity to Ba^{2+} and quinidine (or its dextrorotary stereoisomer, quinine) is similar, the difference in sensitivity to TEA^+ between the two systems resembles that described for neuromuscular K^+ channels where TEA^+ is more potent in blockade at the internal side than at the external side (Hille, 1992). Expression of the mRNA from winter flounder intestinal tissue offers no clue as to the cellular location of this K^+ channel *in vivo*. Patch clamp studies suggest a basolateral locus yet the actual cellular location(s) remain unresolved in light of the strong electrophysiological evidence that places the channel in the apical membrane (cf. Loretz and Fourtner, 1991). An apical membrane location would be consistent with the dependence of V_a on mucosal fluid K^+ concentration and the known K^+ secretory capability of this tissue (Stewart *et al.*, 1980; Musch *et al.*, 1982; Halm *et al.*, 1985a,b; Loretz *et al.*, 1985), whereas a basolateral membrane site might argue for a role in K^+ recycling across the basolateral membrane in parallel with Na^+-K^+-ATPase. K^+ absorption does occur across the intestine (Fig. 2) and a basolateral membrane K^+ channel would provide an exit pathway in addition to K^+–Cl^- cotransport. Independent regulation of apical and basolateral membrane K^+ channels through separate second messenger pathways, perhaps via protein phosphorylation, would afford flexible control over K^+ absorption and secretion (Sullivan *et al.*, 1990; Toskulkao *et al.*, 1990).

3. Mechanosensitive Cation Channel

The finding of a mechanosensitive (MS) cation channel was unpredicted based on electrophysiological data but not surprising based on the widespread tissue and phylogenetic occurrence of this channel type (Morris, 1990). The goby intestinal MS channel has a single-channel conductance of 67 pS and discriminates poorly between Na^+ and K^+ ($P_{Na}/P_K \approx 0.83$), suggesting that *in vivo* it may operate largely as a Na^+ channel owing to the large $\Delta\bar{\mu}^{Na}$ across the cell membrane (Table III and Fig. 10; Chang and Loretz, 1992a). Although a cellular location for this channel has not yet been determined (cf. Chang and Loretz, 1992a), it may be involved in mediating the changes in V_a following stimulation or inhibition of the apical membrane cotransporter. As referenced earlier, changes in cotransporter activity alter intracellular Cl^- activity to produce changes not only in E_b but also in cell volume. MS channel activation through transport-dependent increases in cell volume and membrane stretch will produce membrane depolarization through elevated cation conductance (specifically to Na^+, as noted in the foregoing). With speculative regard to cellular location, activation of *basolateral membrane* MS channels would increase f_R, consistent with the experimental data. Simi-

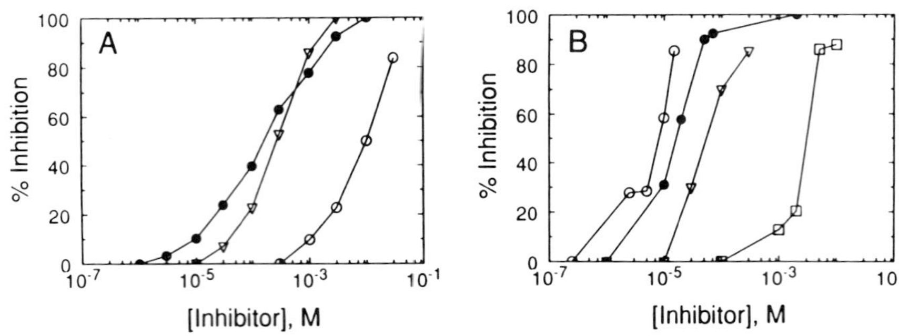

Fig. 9. (A) A winter flounder intestinal epithelial K^+ channel expressed in *Xenopus* oocyte is sensitive to pharmacological blockade by Ba^{2+} (solid circle), quinine (triangle), and TEA^+ (open circle). Calculated ID_{50} values are: Ba^{2+}, 200 μM; quinine (Q), 300 μM; TEA^+, 10 mM. (Redrawn from Sullivan *et al.* (1990). Copyright © 1990, The National Academy of Sciences. Used with permission). (B) The voltage-activated, calcium-dependent K^+ channel from goby posterior intestine studied using patch clamp technique is sensitive to pharmacological blockade by TEA^+, Ba^{2+}, quinidine (Q), and Cs^+ (square). The graph shows the inhibition of average single-channel current ($=I_c \times P_o$) as a function of blocker concentration. Calculated ID_{50} values are: Ba^{2+}, 16 μM; TEA^+, 7 μM; quinidine, 60 μM; Cs^+, 3 mM (data from the author's laboratory).

larly, inhibition of cotransporter activity would reduce cell volume and membrane stretch, closing these MS cation channels to produce hyperpolarization and reduction of f_R.

V. FUTURE DIRECTIONS

Continued study of single ion channels will contribute to mechanistic refinements of the cellular model for teleost intestinal transport and will provide important clues with respect to functional regulation. For example, mechanical stretch delivered to goby intestine increases active transepithelial Na^+ and Cl^- absorption, suggesting a mechanism for local control of intestinal function through either peristaltic contraction or luminal distention by food (Chang and Loretz, 1992b). It is inviting to speculate on the role of MS channels in coupling transport activity to food ingestion and to seawater-induced drinking behaviors. Since Na^+-coupled nutrient transport draws upon the Na^+ electrochemical gradient and has electrochemical consequences to the cell, coordination of hydromineral and nutrient uptake systems to meet most effectively as possible the osmoregulatory and nutritional needs of the organism is an important consideration given the finite potential of the Na^+ electrochemical gradient to drive

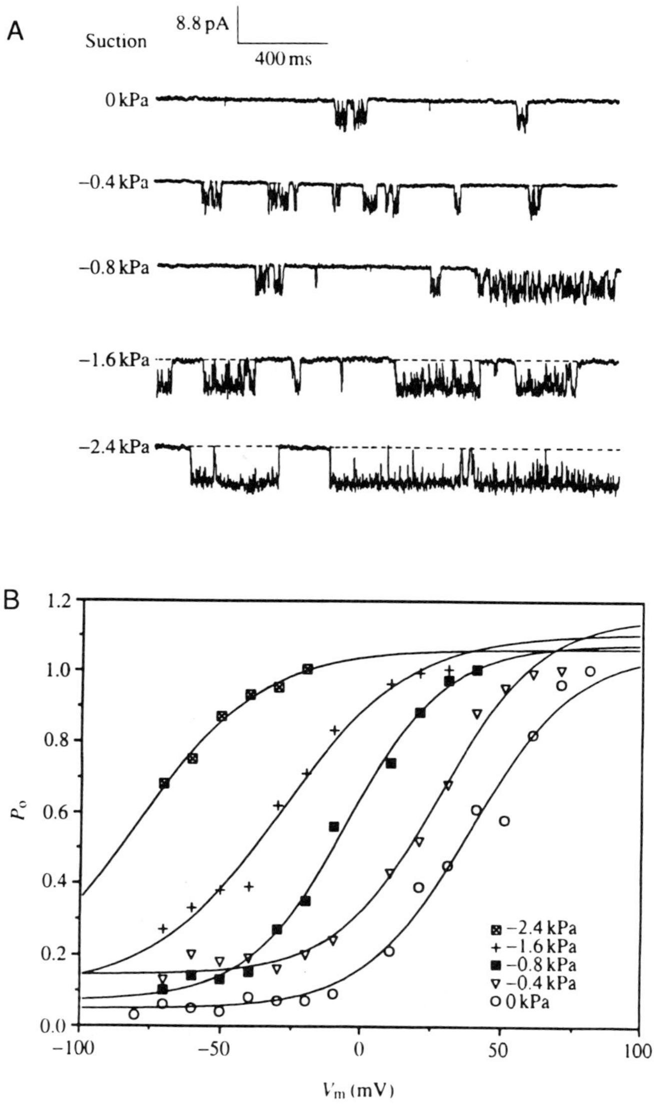

Fig. 10. (A) Mechanical stretch (applied as suction to the patch pipette) increased the open probability (P_o) of this mechanosensitive cation channel from goby enterocyte. Downward deflections represent channel openings. This membrane patch was bathed on the extracellular side with 161.4 m*M* Na^+/2.5 m*M* K^+ and on the cytoplasmic side with 140 m*M* K^+. V_m was voltage clamped at −60 mV. (B) Both membrane stretch and depolarization increase channel activity (expressed as P_o). (From Chang and Loretz (1992a). Copyright © 1992, The Company of Biologists, Ltd. Used with permission.)

transport. Biochemical (e.g., Toskulkao *et al.*, 1990) and molecular biological (e.g., Sullivan *et al.*, 1990) approaches will also contribute to the overall reconstruction of cellular events in the intact functional tissue and to an understanding of the role of physiological effectors, both intracellular second messengers in proximate control and extracellular hormonal agents, in coordinating organismal responses at the various osmoregulatory organs.

ACKNOWLEDGMENTS

Research in the author's laboratory has been supported by grants from the National Science Foundation (DCB-8718633 and DCB-9105874).

REFERENCES

Ando, M. (1975). Intestinal water transport and chloride pump in relation to sea-water adaptation of the eel, *Anguilla japonica*. *Comp. Biochem. Physiol., A* **52,** 229–233.

Ando, M. (1980). Chloride-dependent sodium and water transport in the seawater eel intestine. *J. Comp. Physiol., B* **138,** 87–91.

Ando, M. (1981). Effects of ouabain on chloride movements across the seawater eel intestine. *J. Comp. Physiol., B* **145,** 73–79.

Ando, M. (1983). Potassium-dependent chloride and water transport across the seawater eel intestine. *J. Membr. Biol.* **73,** 125–130.

Ando, M., and Kobayashi, M. (1978). Effects of stripping of the outer layers of the eel intestine on salt and water transport. *Comp. Biochem. Physiol., A* **61,** 497–501.

Ando, M., Utida, S., and Nagahama, H. (1975). Active transport of chloride in eel intestine with special reference to sea water adaptation. *Comp. Biochem. Physiol., A* **51,** 27–32.

Buddington, R. K., Chen, J. W., and Diamond, J. (1987). Genetic and phenotypic adaptation of intestinal nutrient transport to diet in fish. *J. Physiol.* (*London*) **393,** 261–281.

Chang, W., and Loretz, C. A. (1992a). Activation by membrane stretch and depolarization of an epithelial monovalent cation channel from teleost intestine. *J. Exp. Biol.* **169,** 87–104.

Chang, W., and Loretz, C. A. (1992b). Mechanical effects on Na^+ and Cl^- transport across goby posterior intestine. *Am. Zool.* **32,** 49A.

Collie, N. L. (1985). Intestinal nutrient transport in coho salmon (*Oncorhynchus kisutch*) and the effects of development, starvation, and seawater adaptation. *J. Comp. Physiol., B* **156,** 163–174.

Collie, N. L., and Hirano, T. (1987). Mechanisms of hormone actions on intestinal transport. *In* "Vertebrate Endocrinology: Fundamentals and Biomedical Implications" (P. K. T. Pang and M. P. Schreibman, eds.), pp. 239–270. Academic Press, Orlando, FL.

Curtis, R. L., Trier, J. S., Frizzell, R. A., Lindem, N. M., and Madara, J. L. (1984).

Flounder intestinal absorptive cells have abundant gap junctions and may be coupled. *Am. J. Physiol.* **246,** C77–C83.

Dixon, J. M., and Loretz, C. A. (1986). Luminal alkalinization in the intestine of the goby. *J. Comp. Physiol., B* **156,** 803–811.

Duffey, M. E., Thompson, S. M., Frizzell, R. A., and Schultz, S. G. (1979). Intracellular chloride activities and active chloride absorption in the intestinal epithelium of the winter flounder. *J. Membr. Biol.* **50,** 331–341.

Eveloff, J., Field, M., Kinne, R., and Murer, H. (1980). Sodium cotransport systems in intestine and kidney of the winter flounder. *J. Comp. Physiol.* **135,** 175–182.

Ferraris, R. P., and Ahearn, G. A. (1983). Intestinal glucose transport in carnivorous and herbivorous marine fishes. *J. Comp. Physiol., B* **152,** 79–90.

Field, M. (1993). Intestinal electrolyte secretion: History of a paradigm. *Arch. Surg. (Chicago)* **128,** 273–278.

Field, M., Karnaky, K. J., Jr., Smith, P. L., Bolton, J. E., and Kinter, W. B. (1978). Ion transport across the isolated intestinal mucosa of the winter flounder, *Pseudopleuronectes americanus:* I. Functional and structural properties of cellular and paracellular pathways for Na and Cl. *J. Membr. Biol.* **41,** 265–293.

Field, M., Smith, P. L., and Bolton, J. E. (1980). Ion transport across the isolated intestinal mucosa of the winter flounder, *Pseudopleuronectes americanus:* II. Effects of cyclic AMP. *J. Membr. Biol.* **55,** 157–163.

Frizzell, R. A., Field, M., and Schultz, S. G. (1979a). Sodium-coupled chloride transport by epithelial tissues. *Am. J. Physiol.* **236,** F1–F8.

Frizzell, R. A., Smith, P. L., Vosburgh, E., and Field, M. (1979b). Coupled sodium–chloride influx across brush border of flounder intestine. *J. Membr. Biol.* **46,** 27–39.

Gibson, J. S., Ellory, J. C., and Lahlou, B. (1987). Salinity acclimation and intestinal salt transport in the flounder: The role of the basolateral cell membrane. *J. Exp. Biol.* **128,** 371–382.

Halm, D. R., Krasny, E. J., Jr., and Frizzell, R. A. (1985a). Electrophysiology of flounder intestinal mucosa. I. Conductance properties of the cellular and paracellular pathways. *J. Gen. Physiol.* **85,** 843–864.

Halm, D. R., Krasny, E. J., Jr., and Frizzell, R. A. (1985b). Electrophysiology of flounder intestinal mucosa. II. Relation of the electrical potential profile to coupled NaCl absorption. *J. Gen. Physiol.* **85,** 865–883.

Hamill, O. P., Marty, A., Neher, E., Sakmann, B., and Sigworth, F. J. (1981). Improved patch-clamp techniques for high-resolution current recording from cells and cell-free membrane patches. *Pflüegers Arch.* **391,** 85–100.

Helman, S. I., and Beyenbach, K. (1978). Electrophysiological study of the apical barrier of the intestinal epithelium of winter flounder (*Pseudopleuronectes americanus*). *Bull. Mt. Desert Island Biol. Lab.* **18,** 51–54.

Helman, S. I., and Thompson, S. M. (1982). Interpretation and use of electrical equivalent circuits in studies of epithelial tissues. *Am. J. Physiol.* **243,** F519–F531.

Hickman, C. P., Jr. (1968). Ingestion, intestinal absorption, and elimination of seawater and salts in the southern flounder, *Paralichthys lethostigma. Can. J. Zool.* **46,** 457–466.

Hille, B. (1992). "Ionic Channels of Excitable Membranes," 2nd ed. Sinauer Assoc., Sunderland, MA.

Hirano, T., and Mayer-Gostan, N. (1976). Eel esophagus as an osmoregulatory organ. *Proc. Natl. Acad. Sci. U.S.A.* **73,** 1348–1350.

Karasov, W. H., and Diamond, J. M. (1983). Adaptive regulation of sugar and amino acid transport by vertebrate intestine. *Am. J. Physiol.* **245,** G443–G462.

Kirsch, R. (1978). Role of the esophagus in osmoregulation in teleost fishes. *In* "Osmotic and Volume Regulation" (C. B. Jørgensen and E. Skadhauge, eds.), pp. 138–150. Academic Press, New York.

Kirsch, R., and Meister, M. F. (1982). Progressive processing of ingested water in the gut of sea-water teleosts. *J. Exp. Biol.* **98,** 67–81.

Korn, S. J., Marty, A., Conner, J. A., and Horn, R. (1991). Perforated patch recording. *Methods Neurosci.* **4,** 264–373.

Kottra, G., and Frömter, E. (1983). Functional properties of the paracellular pathway in some leaky epithelia. *J. Exp. Biol.* **106,** 217–229.

Kowarsky, J. (1973). Extra-branchial pathways of salt exchange in a teleost fish. *Comp. Biochem. Physiol., A* **46,** 477–486.

Krasny, E. J., Jr., and Frizzell, R. A. (1984). Intestinal ion transport in marine teleosts. *In* "Chloride Transport Coupling in Biological Membranes and Epithelia" (G. A. Gerencser, ed.), pp. 205–220. Elsevier, New York.

Kristensen, K., and Skadhauge, E. (1974). Flow along the gut and intestinal absorption of salt and water in euryhaline teleosts: A theoretical analysis. *J. Exp. Biol.* **60,** 557–566.

Loretz, C. A. (1983). Ion transport by the intestine of the goby, *Gillichthys mirabilis*. *Comp. Biochem. Physiol., A* **75,** 205–210.

Loretz, C. A. (1987a). Rectal gland and crypts of Lieberkühn: Is there a phylogenetic basis for functional similarity? *Zool. Sci.* **4,** 933–944.

Loretz, C. A. (1987b). Regulation of goby intestinal ion absorption by the calcium messenger system. *J. Exp. Zool.* **244,** 67–78.

Loretz, C. A. (1990). Recognition by goby intestine of a somatostatin analog, SMS 201-995. *J. Exp. Zool., Suppl.* **4,** 31–36.

Loretz, C. A., and Assad, J. A. (1986). Urotensin II lowers cytoplasmic free calcium concentration in goby enterocytes: Measurements using quin 2. *Gen. Comp. Endocrinol.* **64,** 355–361.

Loretz, C. A., and Fourtner, C. R. (1988). Functional characterization of a voltage-gated anion channel from teleost fish intestinal epithelium. *J. Exp. Biol.* **136,** 383–403.

Loretz, C. A., and Fourtner, C. R. (1990). Pharmacological blockade of single K(Ca) channels from fish intestinal epithelium. *Am. Zool.* **30,** 66A.

Loretz, C. A., and Fourtner, C. R. (1991). Identification of a basolateral membrane potassium channel from teleost intestinal epithelial cells. *J. Exp. Biol.* **159,** 45–64.

Loretz, C. A., Freel, R. W., and Bern, H. A. (1983). Specificity of response of intestinal ion transport systems to a pair of natural peptide hormone analogs: Somatostatin and urotensin II. *Gen. Comp. Endocrinol.* **52,** 198–206.

Loretz, C. A., Howard, M. E., and Siegel, A. J. (1985). Ion transport in goby intestine: Cellular mechanism of urotensin II stimulation. *Am. J. Physiol.* **249,** G284–G293.

Mackay, W. C., and Lahlou, B. (1980). Relationships between Na^+ and Cl^- fluxes in the intestine of the European flounder, *Platichthys flesus*. *In* "Epithelial Transport in the Lower Vertebrates" (B. Lahlou, ed.), pp. 151–162. Cambridge Univ. Press, Cambridge, UK.

Mooney, S. M., and Loretz, C. A. (1987). Differential effects of two bile salts on ion transport characteristics of teleost intestine. *Comp. Biochem. Physiol., A* **86,** 367–372.

Morris, C. E. (1990). Mechanosensitive ion channels. *J. Membr. Biol.* **113,** 93–107.

Musch, M. W., Orellana, S. A., Kimberg, L. S., Field, M., Halm, D. R., Krasny, E. J., Jr., and Frizzell, R. A. (1982). Na^+–K^+–Cl^- co-transport in the intestine of a marine teleost. *Nature (London)* **300,** 351–353.

Musch, M. W., McConnell, F. M., Goldstein, L., and Field, M. (1987). Tyrosine transport in winter flounder intestine: Interaction with Na^+–K^+–$2Cl^-$ cotransport. *Am. J. Physiol.* **253,** R264–R269.

Musch, M. W., O'Grady, S. M., and Field, M. (1990). Ion transport of marine teleost intestine. *In* "Methods in Enzymology" (C. Guthrie and G. R. Fink, eds.), Vol. 194, pp. 746–753. Academic Press, San Diego, CA.

Neher, E., and Sakmann, B. (1976). Single channel currents recorded from membrane of denervated frog muscle fibres. *Nature (London)* **260,** 799–802.

Neher, E., Sakmann, B., and Steinbach, J. H. (1978). The extracellular patch clamp: A method for resolving currents through individual open channels in biological membranes. *Pflüegers Arch.* **375,** 219–228.

Nishimura, H. (1987). Role of renin–angiotensin system in osmoregulation. *In* "Vertebrate Endocrinology: Fundamentals and Biomedical Implications" (P. K. T. Pang and M. P. Schreibman, eds.), pp. 157–187. Academic Press, Orlando, FL.

O'Grady, S. M. (1989). Cyclic nucleotide-mediated effects of ANF and VIP on flounder intestinal ion transport. *Am. J. Physiol.* **256,** C142–C146.

O'Grady, S. M., and Wolters, P. J. (1990). Evidence for chloride secretion in the intestine of the winter flounder. *Am. J. Physiol.* **258,** C243–C247.

O'Grady, S. M., Musch, M. M., and Field, M. (1986). Stiochiometry and ion affinities of the Na–K–Cl cotransport system in the intestine of the winter flounder (*Pseudopleuronectes americanus*). *J. Membr. Biol.* **91,** 33–41.

O'Grady, S. M., DeJonge, H. R., Vaandrager, A. B., and Field, M. (1988). Cyclic nucleotide-dependent protein kinase inhibition by H-8: Effects on ion transport. *Am. J. Physiol.* **254,** C115–C121.

O'Grady, S. M., Cooper, K. E., and Rae, J. L. (1991). Cyclic GMP regulation of a voltage-activated K channel in dissociated enterocytes. *J. Membr. Biol.* **124,** 159–167.

Parmalee, J. T., and Renfro, J. L. (1983). Esophageal desalinization of seawater in flounder: Role of active sodium transport. *Am. J. Physiol.* **245,** R888–R893.

Powell, D. W. (1981). Barrier function of epithelia. *Am. J. Physiol.* **241,** G275–G288.

Rao, M. C., and Nash, N. T. (1988). 8-BrcAMP does not affect Na–K–2Cl cotransport in winter flounder intestine. *Am. J. Physiol.* **255,** C246–C251.

Rao, M. C., Nash, N. T., and Field, M. (1984). Differing effects of cGMP and cAMP on ion transport across flounder intestine. *Am. J. Physiol.* **246,** C167–C171.

Reuss, L., and Finn, A. L. (1974). Passive electrical properties of toad urinary bladder: Intracellular electrical coupling and transepithelial cellular and shunt conductances. *J. Gen. Physiol.* **64,** 1–25.

Schultz, S. G. (1980). "Basic Principles of Membrane Transport." Cambridge Univ. Press, Cambridge, UK.

Schultz, S. G. (1981). Homocellular regulatory mechanisms in sodium-transporting epithelia: Avoidance of extinction by "flush-through." *Am. J. Physiol.* **241,** F579–F590.

Shehadeh, Z. H., and Gordon, M. S. (1969). The role of the intestine in salinity adaptation of the rainbow trout, *Salmo gairdneri*. *Comp. Biochem. Physiol.* **30,** 397–418.

Skadhauge, E. (1973). Coupling of salt and water flow across and along intestinal epithelia. *In* "Transport Mechanisms in Epithelia" (H. Ussing and N. A. Thorn, eds.), pp. 315–324. Academic Press, New York.

Skadhauge, E. (1974). Coupling of transmural flows of NaCl and water in the intestine of the eel (*Anguilla anguilla*). *J. Exp. Biol.* **60,** 535–546.

Sleet, R. B., and Weber, L. J. (1982). The rate and manner of seawater ingestion by a marine teleost and corresponding seawater modification by the gut. *Comp. Biochem. Physiol., A.* **72,** 469–475.

Smith, H. W. (1930). The absorption and excretion of water and salts by marine teleosts. *Am. J. Physiol.* **93,** 480–505.

Smith, H. W. (1932). Water regulation and its evolution in the fishes. *Q. Rev. Biol.* **7,** 1–26.

Smith, M. W., Ellory, J. C., and Lahlou, B. (1975). Sodium and chloride transport by the intestine of the European flounder *Platichthys flesus* adapted to fresh or sea water. *Pflüegers Arch.* **357,** 303–312.

Smith, P. L., Welsh, M. J., Stewart, C. P., Frizzell, R. A., Orellana, S. A., and Field, M. (1980). Chloride absorption by the intestine of the winter flounder *Pseudopleuronectes americanus:* Mechanism of inhibition by reduced pH. *Bull. Mt. Desert Island Biol. Lab.* **20,** 96–101.

Stewart, C. P., Smith, P. L., Welsh, M. J., Frizzell, R. A., Musch, M. W., and Field, M. (1980). Potassium transport by the intestine of the winter flounder, *Pseudopleuronectes americanus:* Evidence for KCl co-transport. *Bull. Mt. Desert Island Biol. Lab.* **20,** 92–96.

Sullivan, S. K., Swamy, K., Greenspan, N. R., and Field, M. (1990). Epithelial K channel expressed in *Xenopus* oocytes is inactivated by protein kinase C. *Proc. Natl. Acad. Sci. U.S.A.* **87,** 4553–4556.

Toskulkao, C., Nash, N. T., Leach, K., and Rao, M. C. (1990). Second messenger-specific protein kinases in a salt-absorbing intestinal epithelium. *Am. J. Physiol.* **258,** C879–C888.

van Lennep, E. W., and Lanzing, W. J. R. (1967). The ultrastructure of glandular cells in the external dendritic organ of some marine catfish. *J. Ultrastruct. Res.* **18,** 333–344.

Weiss, B., Prozialeck, W., Cimino, M., Barnette, M. S., and Wallace, T. L. (1980). Pharmacological regulation of calmodulin. *Ann. N. Y. Acad. Sci.* **356,** 319–345.

Welsh, M. J., Smith, P. L., Fromm, M., and Frizzell, R. A. (1982). Crypts are the site of intestinal fluid and electrolyte secretion. *Science* **218,** 1219–1221.

Zeuthen, T., Ramos, M., and Ellory, J. C. (1978). Inhibition of active chloride transport by piretanide. *Nature (London)* **273,** 678–680.

Zuidema, T., Kamermans, M., and Siegenbeek von Heukelom, J. (1986). Influence of glucose absorption on ion activities in cells and submucosal space in goldfish intestine. *Pflüegers Arch.* **407,** 292–298.

3

UREA CYCLE IN FISH: MOLECULAR AND MITOCHONDRIAL STUDIES

PAUL M. ANDERSON

I. INTRODUCTION

A significant proportion of energy production in fish involves catabolism and oxidation of proteins and amino acids. Consistent with their water habitat, the major end product of nitrogen metabolism in most fish is ammonia. Excretion of ammonia across the gills into the surrounding water environment avoids the need to expend energy to detoxify and store ammonia urea (e.g., as in mammalian ureotoelic species) or as uric acid (e.g., as in uricotelic birds, crocodilians, and reptiles). The general view is that ammonia is primarily formed in liver mitochondria (by the coupled transaminase–glutamate dehydrogenase pathway) and is then transported in blood and excreted through the gills as NH_3. For a detailed discussion of ammonia metabolism and excretion in fish and the related topics of ionoregulation and acid–base balance, the reader is referred to several reviews (Randall and Wright, 1987; Campbell, 1991; Mommsen and Walsh, 1992; Wood, 1993).

Although the majority of fish species are ammonotelic, most fish excrete variable amounts of urea and in some species ureotelism predominates (for tables citing literature values of urea levels in blood and urea and ammonia excretion rates in various species, see Campbell and Anderson, 1991; Wood, 1993; see also Griffith, 1981). A high rate of urea synthesis and excretion can be induced in some fish experimentally, and occurs in some fish naturally, as the result of elevated water ammonia levels and/or alkalinity (e.g, pH 9.5–10), which reduces or eliminates the partial pressure for NH_3 across the gills, resulting in increased tissue ammonia levels and decreased ammonia excretion (Wright, 1993). In ureoosmotic elasmobranch species, blood and tissue urea concentrations as high as 0.4 *M* are maintained for the purpose of osmoregulation by low branchial permeability and active reabsorption in the kidney (Perlman and Goldstein, 1988; Wood, 1993). Recent observations indicate that active reabsorption of urea by kidney also occurs in a number of freshwater and marine teleosts. This may account for the observation that in primarily ammonotelic species of teleosts, plasma and tissue levels of urea are relatively high (1–10 m*M*). Thus, it has been suggested that urea retention for osmoregulatory or other purposes may be a characteristic feature of teleosts as well as of elasmobranchs (Mommsen and Walsh, 1992; Wood, 1993).

The two known metabolic pathways for biosynthesis of urea are uricolysis (uric acid derived from purine nucleotides is converted to urea) and the ornithine–urea cycle (hereafter referred to as the urea cycle). Although the three uricolytic pathway enzymes are apparently present in liver of most fish species, active uricolysis as the major source of urea has been documented in only a few species (Goldstein and Forster, 1965; Brown *et al.*, 1966; Cvancara, 1969; Hayashi *et al.*, 1989). Nevertheless, predominantly ammonotelic fish in which one or more of the urea cycle enzymes are not present are thought to form urea by the uricolytic pathway (Olson and Fromm, 1971; Vellas and Serfaty, 1974; Randall *et al.*, 1989; Campbell, 1991; Danulat and Kempe, 1992; Wright, 1993; Wright *et al.*, 1993). There has been little discussion or investigation of the metabolic pathway leading to uric acid in fish, however. A characteristic feature of uricotelic species is detoxification of intramitochondrially generated ammonia by reaction with glutamate to give glutamine, which exits the mitochondria as a neutral species and serves as a precursor for purine nucleotide biosynthesis and, subsequently, uric acid formation (Campbell, 1991). Thus, if the function of urea formation via uricolysis in fish was related to ammonia detoxification, one might expect to find significant levels of glutamine synthetase activity in liver localized in the mitochondria.

The existence of a very active and functional urea cycle in ureoosmotic elasmobranchs and chimaeras has been well established for some years (Brown, 1964; Schooler *et al.,* 1966; Watts and Watts, 1966; Goldstein, 1967, 1970; Read, 1967, 1970; Goldstein and Forster, 1971a,b; Casey and Anderson, 1982; Perlman and Goldstein, 1988). Although earlier studies suggested that the genes for urea cycle enzymes were absent or not expressed in all teleost fishes (Brown and Cohen, 1960), the full complement of enzymes or key enzymes of the urea cycle and/or urea synthesis have been shown to exist in a number of "typical" teleost species (Huggins *et al.,* 1969; Read, 1971; Wilson, 1973; Cvancara, 1974; Dragojevíc and Dida, 1977; Dépêche *et al.,* 1979; Chiu *et al.,* 1986; Mommsen and Walsh, 1989; Cao *et al.,* 1991). The levels of activities of urea cycle enzymes in these teleosts are normally very low. However, very active ureagenesis and expression of significant levels of urea cycle enzymes in some teleosts and other nonureoosmotic fishes have been demonstrated and appear to be correlated with unique environmental habitats, life cycle variations, or environmental adaptations (see Section III).

Salient features of the urea cycle in mammalian ureotelic species are illustrated in Fig. 1. For reviews of the important properties of the urea cycle in these species, see Meijer *et al.* (1990), Watford (1991), Atkinson (1992), and Morris (1992). The wealth of available information about mammalian urea cycle properties and function has been very useful for assessing the significance and functional implications of comparative differences observed in fish.

A number of reviews have been previously published that include or directly address the topic of the urea cycle in fish: Goldstein and Forster (1970), Thomson (1971), Goldstein (1967, 1970, 1972), Watts and Watts (1973), Cohen (1976), Perlman and Goldstein (1988), Anderson (1991), Campbell (1991), Campbell and Anderson (1991), Mommsen and Walsh (1991, 1992), and Wood (1993).

II. CARBAMOYL PHOSPHATE SYNTHESIS IN FISH

A. Carbamoyl-Phosphate Synthetases

Carbamoyl phosphate is the precursor for two major metabolic pathways, the urea cycle (and/or arginine biosynthesis) and pyrimidine nucleotide biosynthesis. As noted in Fig. 1, the first step of the urea cycle (ammonia fixation) in mammalian and amphibian ureotelic species is catalyzed by carbamoyl-phosphate synthetase I (CPSase I). As noted in Section II,B, in fish a different type of CPSase apparently catalyzes formation

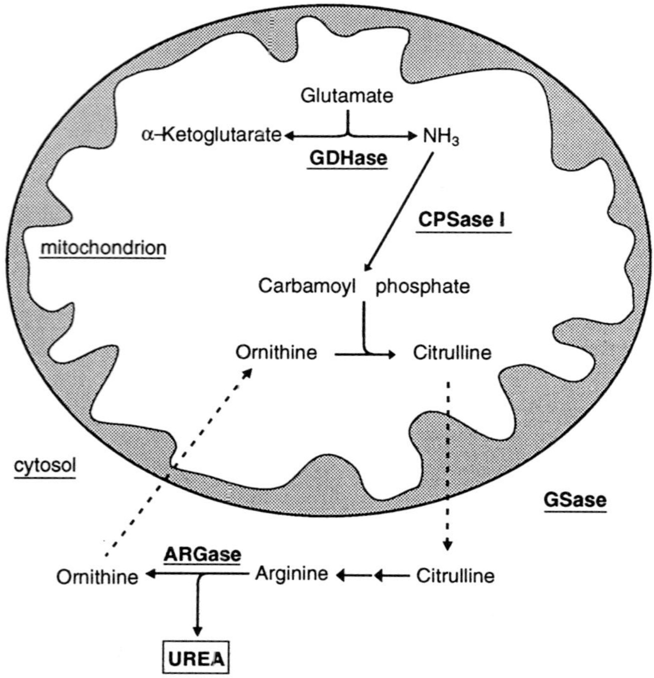

Fig. 1. Illustration of ammonia-dependent urea cycle. GDHase, glutamate dehydrogenase; GSase, glutamine synthetase; ARGase, arginase; CPSase, carbamoyl-phosphate synthetase. (Adapted from Anderson, 1991, with permission).

of carbamoyl phosphate utilized for urea synthesis. In most species yet another type of CPSase catalyzes carbamoyl phosphate formation utilized for pyrimidine nucleotide biosynthesis.

1. Higher Eukaryotes (except Fish)

In mammalian and amphibian ureotelic species, formation of carbamoyl phosphate for each pathway is catalyzed by two different CPSases (Jones, 1980; Evans, 1986; Campbell and Anderson, 1991). CPSase I is localized in the matrix of liver (and also small intestine) mitochondria and catalyzes incorporation of ammonia into carbamoyl phosphate as the first step of the urea cycle (Campbell, 1991) (Fig. 1):

$$NH_3 + 2ATP + HCO_3^- \xrightarrow{\text{NAcGlu, } Mg^{2+}} 2\ ADP + P_i + NH_2CO_2PO_3^{2-} \quad (1)$$

N-Acetyl-L-glutamate (NAcGlu) is a positive allosteric effector required for activity; synthesis of NAcGlu also occurs in the mitochondrial matrix and the levels of NAcGlu, and thus CPSase I activity, are affected by the extent of protein catabolism (Marshall, 1976; Cheung and Raijman, 1980; Morita *et al.*, 1982; Meijer *et al.*, 1990; Campbell, 1991). The molecular weight of the enzyme is 160,000 (single polypeptide chain). NH_3 rather than NH_4^+ is the substrate; the K_m for NH_3 in intact mitochondria is $\approx$13 μM (Cohen *et al.*, 1985). Since the concentration of CPSase I in the mitochondrial matrix is very high (0.5–1.5 m*M*!) and the concentration of NH_3 in the mitochondria has been estimated as 7 μM at pH 7.0, the rate of carbamoyl phosphate formation in normal liver is probably directly proportional to NH_3 concentration (Clarke, 1976; Cohen *et al.*, 1985). CPSase I is the only CPSase that does not utilize glutamine as the nitrogen-donating substrate.

The function of CPSase II is to catalyze carbamoyl phosphate formation as the first step in pyrimidine nucleotide biosynthesis (Jones, 1980; Evans, 1986). In contrast to CPSase I, this synthetase (1) is localized in the cytosol of liver and most tissues; (2) does not require NAcGlu for activity; (3) is part of a multifunctional complex that also includes the activities of the next two enzymes of the pyrimidine pathway, aspartate transcarbamylase and dihydro-orotase, on a single polypeptide chain (commonly referred to as CAD) with a molecular weight of $\approx$240,000; (4) is subject to end-product inhibition by UTP and to allosteric activation by 5-phosphoribosyl-1-pyrophosphate, and (5) utilizes glutamine as the physiologically significant nitrogen-donating substrate:

$$\text{glutamine} + H_2O + 2\text{ATP} + HCO_3^- \xrightarrow{Mg^{2+}} 2\text{ADP} + P_i + NH_2CO_2PO_3^{2-} + \text{glutamate} \quad (2)$$

2. PROKARYOTES AND LOWER EUKARYOTES

Enteric bacteria have a single CPSase that provides carbamoyl phosphate for both arginine and pyrimidine nucleotide biosynthesis (Evans, 1986; Meister, 1989). The enzyme of *E. coli* is allosterically inhibited by UMP and activated by ornithine, reflecting its dual role in these pathways. Like all CPSases except CPSase I, the *E. coli* enzyme utilizes glutamine as the physiologically significant nitrogen-donating substrate. The *E. coli* CPSase consists of two polypeptide chains coded for by two linked genes, *carA* and *carB*, respectively. The product of *carB* is fully functional in all respects with ammonia as nitrogen-donating substrate, but it cannot utilize glutamine as the nitrogen-donating substrate. The product of *carA* includes a "glutamine amide transfer" domain that functions as a glutaminase, providing ammonia that is transferred to the synthetase subunit (product of *carB*).

In *Bacillus subtilus*, two glutamine-dependent CPSases are expressed. The pyrimidine-related CPSase has properties similar to those of the *E.*

coli enzyme, that is, it is composed of two different subunits. The arginine-related enzyme appears to be a single polypeptide chain.

In fungi, separate CPSases catalyze carbamoyl phosphate formation for the two respective pathways. The arginine-related CPSases are composed of two different subunits, analogous to the *E. coli* enzyme. The pyrimidine-related CPSases are part of a bifunctional polypeptide with both aspartate transcarbamylase and glutamine-dependent CPSase activities.

3. Utilization of Glutamine

Utilization of the amide group of glutamine for biosynthesis of carbamoyl phosphate in the glutamine-dependent CPSases involves reaction of glutamine with a cysteine SH group on the enzyme to form a γ-glutamyl thioester intermediate, releasing ammonia, which reacts with an activated intermediate common to all CPSases; the γ-glutamyl thioester intermediate is subsequently hydrolyzed (Meister, 1989; Zalkin, 1993).

4. Structural Similarities

The amino acid sequences for rat, human, and tadpole (*Rana catesbeiana*) CPSase I, CPSase II from *Drosophila melanogaster* and hamster, *E. coli* CPSase, the yeast CPSases related to arginine biosynthesis and to pyrimidine biosynthesis, and the pyrimidine-related CPSases from *Dictyostelium discoideum* and *Bacillus subtilus* have been determined on the basis of the corresponding cDNA sequences (Simmer *et al.,* 1990; Bein *et al.,* 1991; Haraguchi *et al.,* 1991; Quinn *et al.,* 1991; Helbing and Atkinson, 1994). These and other studies have revealed a high degree of similarity in the deduced amino acid sequences and apparent structural domains in CPSases from a phylogenetically broad range of organisms (Simmer *et al.,* 1990; Evans *et al.,* 1993). As illustrated in Fig. 2, CPSases are composed of two basic domains, an N-terminal glutaminase domain and a C-terminal synthetase domain. CPSase I has an N-terminal signal sequence that is removed in the process of transfer into the mitochondria. Where the CPSase is part of a multifunctional complex, the additional polypeptide sequences constituting the other enzyme activities are extensions from the C-terminal end of the CPSase.

The C-terminal half of the glutaminase domain is homologous to the "glutamine amide transfer" (GAT) domain present in all known trp G-type amidotransferases; this domain contains a highly conserved sequence of 10 amino acids that includes the cysteine residue essential for glutamine amide transfer in the glutamine-dependent CPSases. As illustrated in Fig. 2, in human and rat CPSase I, the essential cysteine is replaced by a serine residue, which is consistent with the fact that glutamine cannot be used as a nitrogen-donating substrate. Interestingly, this residue has been found to be a cysteine in the "CPSase I" in the amphibian tadpole, suggesting that this CPSase may be an evolutionary intermediate (Helbing and Atkinson, 1994). In *E. coli* and certain other CPSases as described earlier, the glutaminase domain exists as a separate polypeptide chain coded for by a different gene.

The C-terminal synthetase domain is composed of two homologous halves, suggesting that this part of the gene arose from gene duplication and fusion (Nyunoya and Lusty, 1983). The binding sites for the different allosteric effectors for the various CPSases all appear to be located in the C-terminal end of the synthetase C-terminal subdomain (Evans *et al.*, 1993).

B. CPSases in Fish

1. CPSase III in *Squalus acanthias* and *Micropterus salmoides*

Earlier studies aimed at identifying CPSase activity in fish reasonably assumed that activity related to the urea cycle would be a CPSase I.

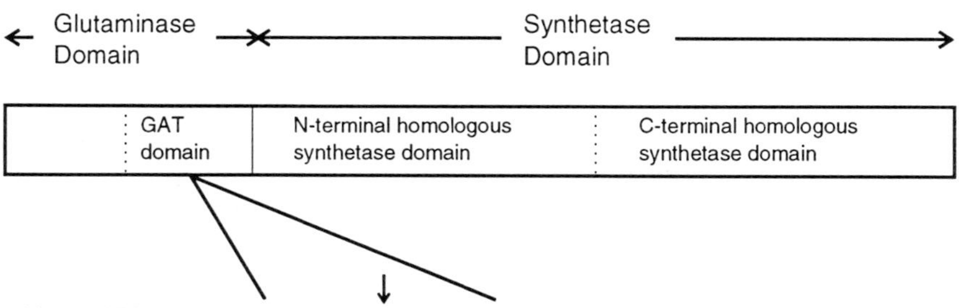

Human CPSase I	P L F G I S T G N L I
Rat CPSase I	P L F G I S T G N I I
Frog CPSase I	P I F G I C K G N E I
Shark CPSase III	P V F G I C M G N Q L
CAD CPSase II	P V F G I C L G H Q L
Yeast URA-2 CPSase	P V F G I C L G H Q L
Dict. PYR1-3 CPSase	A V F G V C M G N Q L
E. coli CarA CPSase	P V F G I C L G H Q L
Yeast CarA CPSase	P I F G I C L G H Q L

Fig. 2. Illustration of the domain structure of CPSases and comparative alignment of a highly conserved amino acid sequence in the glutaminase domain of glutamine-dependent CPSases. Arrow indicates position in the sequence of the critical cysteine residue in the glutamine binding site of glutamine-dependent CPSases. CAD and Dict.PYR1-3 refer to hamster and *D. discoideum* pyrimidine-specific CPSase II's, respectively; URA2 refers to the pyrimidine-specific CPSase; CarA refers to the glutaminase subunits of *E. coli* CPSase and the arginine-specific CPSase in yeast; GAT refers to "glutamine amide transfer." References are given in the text.

However, Anderson (1976) reported the presence of CPSase III activity in liver of *Micropterus salmoides* (largemouth bass) and, at much higher levels, in liver of ureoosmotic marine elasmobranchs and a holocephalan (Anderson, 1980). The latter study also established the presence of CPSase III activity in the freshwater elasmobranch *Potamotrygon circularis* (freshwater stingray) and the marine teleost *Porichthys notatus* (plainfin midshipman). CPSase III activity was first reported by Trammel and Campbell (1970, 1971) in invertebrates. Like CPSase I, CPSase III is a mitochondrial enzyme, requires NAcGlu for activity, and is not affected by allosteric effectors common to CPSase II. However, like CPSase II, CPSase III utilizes glutamine as the nitrogen-donating substrate. Watts and Watts (1966) had observed higher activity with glutamine than with ammonia for carbamoyl phosphate synthesis by elasmobranch liver extracts, but were unable at that time to establish the significance of this observation.

CPSase IIIs from the spiny dogfish (*Squalus acanthias*) and largemouth bass, representatives of marine ureoosmotic elasmobranchs and freshwater ammonotelic teleosts, respectively, have been isolated and characterized (Anderson, 1981; Casey and Anderson, 1983). The properties of both are very similar to CPSase I, except that glutamine serves as the nitrogen-donating substrate. Ammonia can replace glutamine as the nitrogen-donating substrate, but the K_m is quite high; unlike CPSase II and most other amidotransferases, however, the V_{max} with ammonia as substrate is less than one-fourth that attained with glutamine.

The discovery and characterization of NAcGlu- and glutamine-dependent CPSase III established that the NAcGlu binding site of CPSase I probably did not evolve from the glutamine binding site of glutamine-dependent CPSases. There is a significant synergistic relationship between these two sites in CPSase III; the apparent K_m's for NAcGlu and glutamine decrease significantly as the concentration of the other increases. At higher concentrations of Mg^{2+} and glutamine, reaction proceeds in the absence of NAcGlu at a rate that is about 25% of the rate with NAcGlu present. Ammonia-dependent activity has an absolute requirement for NAcGlu, however, and the K_m for NAcGlu is considerably higher than for the glutamine-dependent reaction. This synergistic interaction between the glutamine and NAcGlu binding sites may have physiological significance since both compounds are derived directly from glutamate in the mitochondria and CPSase III activity results in regeneration of glutamate. CPSase III catalyzes the same partial reactions as observed for other CPSases, indicating that the catalytic mechanism is essentially the same.

The gene for the dogfish CPSase III has been cloned and sequenced (Hong *et al.*, 1994). CPSase III has a high degree of sequence similarity with CPSase I and with other CPSases and likely has a domain structure similar to that of other CPSases. Of particular interest is the finding that the essential active site cysteine residue in the glutamine binding site of glutamine-dependent CPSases is preserved in CPSase III (Fig. 2).

2. Distribution and Detection of CPSases in Fish

CPSase III activity has been reported in several other teleost species and in the coelacanth (Mommsen and Walsh, 1989; Randall *et al.*, 1989). In ureoosmotic elasmobranchs the function of CPSase III is clearly related to the urea cycle (see Section II,C,1), and it is certainly reasonable to assume from this and from its structural similarity to CPSase I that its function in other fish species (where it is present) is also related to the urea cycle. Thus, establishing the presence of CPSase III activity is of considerable significance for understanding the nature and function of urea cycle activity in fish. The limited data available in the literature suggest that the levels of CPSase activity in most fish are very low or undetectable. The few reports documenting the presence of CPSase III activity have mostly been in relationship to circumstances related to higher than usual rates of urea synthesis. It is possible that in some circumstances the gene may be expressed only during certain portions of a life cycle or only during certain stressful environmental situations, and there may also be considerable individual variation within a given species.

A question related to the foregoing observations is whether CPSase III (as opposed to CPSase I) is a functional evolutionary trait of all fish. Mommsen and Walsh (1989) reported the presence of CPSase III activity in all fish species they examined, which represented a broad range of fish systematics. The levels of activity were not reported, however, except for two toadfish species (*Opsanus tau* and *Opsanus beta*); identification of CPSase activity as CPSase III was based on the observation of higher activity with glutamine than with ammonia as nitrogen-donating substrate. This definitive study, together with the work cited in Section II,B,1 and the demonstration of CPSase III activity in a tilapia fish (*Oreochromis alcalicus grahami*) adapted to an alkaline environment (Randall *et al.*, 1989), has led to the current assumption in the literature that CPSase III (as opposed to CPSase I) activity is, in fact, an evolutionary trait of all fish (Mommsen and Walsh, 1989, 1991, 1992; Wood, 1993). Current speculation is that CPSase I evolved from CPSase III (Mommsen and Walsh, 1989; Campbell and Anderson, 1991).

Until relatively recently ammonia was the only nitrogen-donating substrate normally utilized in assays for CPSase activity in fish. CPSase III is active with ammonia, but the V_{max} appears to be considerably lower than that obtained with glutamine. If the CPSase activities for the "typical'' teleosts referred to in Section I were due to the presence of CPSase III, the levels of CPSase activities could be higher than reported. Identification of a low level of CPSase activity in fish tissue extracts as a CPSase III rather than a CPSase II requires caution, since CPSase II activity with ammonia is considerably lower than that obtained with glutamine unless the NH_3/NH_4^+ concentration is quite high (e.g., 50–100 m*M*). Dependence on NAcGlu for activity and finding that glutamine-dependent activity is significantly greater than ammonia-dependent activity serve as the basic criteria for identifying activity as that of a CPSase III. Demonstration that the activity is not inhibited by UTP and/or stimulated by 5-phosphoribosyl-1-pyrophosphate provides confirming evidence that the activity is not a CPSase II. Using this kind of criteria, it was established that although CPSase III activity is present in liver of largemouth bass, a member of the sunfish family, the only activity present in liver of several closely related species (crappies and bluegills) and the bullhead (*Ameiurus* family) is CPSase II (Cao *et al.,* 1991). Similarly, only CPSase II activity was present in liver extracts of several species of marine teleosts (Anderson, 1980) and in the freshwater tilapia *Oreochromis niloticus* (Wright, 1993). Other criteria useful for distinguishing between CPSase III and CPSase II include subcellular localization (mitochondrial CPS III, cytosolic CPS II) and separation by size on gel filtration columns (Cao *et al.,* 1991).

The levels of CPSase II in fish tissue appear to be very low, but rather uniform, ranging from 0.001 to 0.002 μmoles/min/g tissue (Anderson, 1980, 1989; Cao *et al.,* 1991), although the reported levels in the freshwater tilapia (Wright, 1993) and Lake Magadi tilapia (Randall *et al.,* 1989) are higher (0.006 and 0.014 μmoles/min/g tissue, respectively). This is approximately the same level of activity as the CPSase activities reported by Huggins *et al.* (1969), who did not establish a dependence of the CPSase activities on NAcGlu; thus, these values could reflect primarily or only CPSase II activity. On the other hand, the reported apparent levels of ammonia-dependent CPSase activity (in the presence of NAcGlu) in rainbow trout (*Salmo gairdneri*) (Chiu *et al.,* 1986), several primitive freshwater bony fishes (shovel-nose sturgeon, *Scaphirhynctius platorynchus;* paddlefish, *Lepisosteus platostomus;* short-nose gar, *Lepisosteus platostomus;* bowfin, *Amia calva*) (Cvancara, 1974), and channel catfish (*Ictalurus punctatus*) (Wilson, 1973) are much higher (0.03–0.06 μmoles/min/g tissue); glutamine was not tested as a potential glutamine-dependent substrate and dependence on NAcGlu for activity was not demonstrated in these studies. These reported levels of CPSase activity are higher than that reported in liver of largemouth bass (≈0.01 μmoles/min/g tissue, of which 60–80% is CPSase III) and approximate the levels of CPSase III in liver extracts of ureoosmotic elasmobranchs (Anderson, 1980; Casey and Anderson, 1983; Cao *et al.*, 1991). These appear to be remarkably high levels of CPSase activity and would presumably be even higher if the activities are the result of CPSase III and glutamine was utilized as the nitrogen-donating substrate.

C. Physiological Significance of Glutamine-Dependent Urea Cycle and CPSase III: Studies with Isolated Mitochondria

1. Elasmobranchs

The urea cycle in the representative marine elasmobranch *Squalus acanthias* is similar to the urea cycle in mammalian and amphibian species, but there are significant differences (Fig. 3). The high level of glutamine-dependent CPSase III in liver of marine elasmobranchs (Anderson, 1980) is accompanied by uniquely high levels of glutamine synthetase (Webb and Brown, 1980). Both enzymes are localized in the mitochondrial matrix of *Squalus acanthias,* along with two other enzymes of the urea cycle, arginase and ornithine transcarbamylase (Casey and Anderson, 1982, 1985). In mammalian ureotelic species, glutamine synthetase in liver is

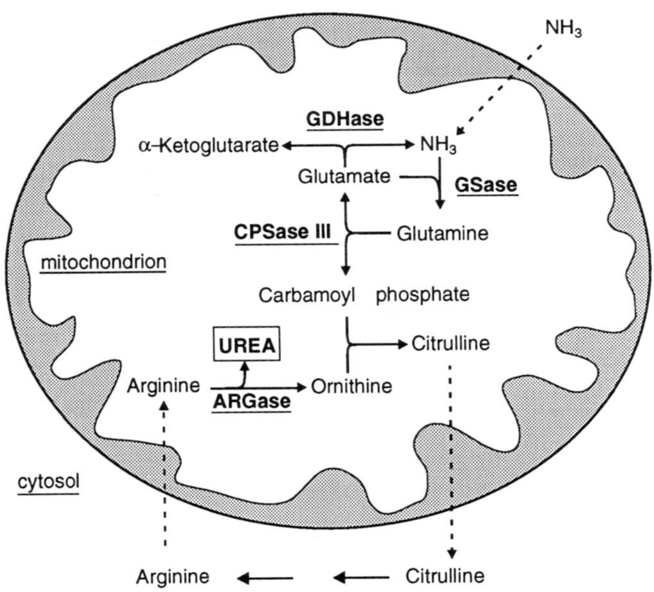

Fig. 3. Illustration of glutamine-dependent urea cycle in elasmobranchs. See Fig. 1 for abbreviations. (Adapted from Anderson, 1991, with permission.)

localized in the cytosol; arginase in mammalian ureotelic species is also localized in the cytosol, although about 10% of the arginase activity appears to be associated with the outer mitochondrial membrane (Campbell, 1991; Campbell and Anderson, 1991). When other tissues and species are considered, however, the subcellular compartmentation of arginase appears to be quite flexible (Cheung and Raijman, 1981; Campbell and Anderson, 1991). The mitochondrial localization of glutamine synthetase is well established for uricotelic species, where its function is analogous to that of CPSase I and ornithine transcarbamylase in mammalian ureotelic species, that is, assimilation of ammonia into a form that can be transported out of the mitochondria to the cytosol for further processing and excretion (Vorhaben and Campbell, 1972; Campbell, 1991).

Using isolated, respiring mitochondria, Anderson and Casey (1984) established that ammonia assimilation for citrulline synthesis (and, therefore, urea synthesis) involves mandatory intermediate formation of glutamine. With succinate as the energy source, isolated mitochondria synthesized citrulline (derived from reaction of the synthesized carbamoyl phosphate with added ornithine catalyzed by endogenous ornithine transcarbamylase) from glutamine or glutamate plus ammonia at a high rate. The observed higher rate of citrulline synthesis from glutamine compared to the rate from glutamate plus ammonia may simply reflect the fact that the observed rate appears to be near the limit of the mitochondria to generate ATP and synthesis from glutamine requires one less ATP. Selective inhibition of the glutamine-dependent CPSase III activity in the isolated mitochondria completely inhibited citrulline synthesis from glutamine or glutamate plus ammonia, whereas selective inhibition of glutamine synthetase in the isolated mitochondria inhibited citrulline synthesis from glutamate plus ammonia, but not from glutamine.

The properties of the mitochondrial glutamine synthetase are similar to those of mammalian glutamine synthetases, except that the K_m for (NH_3/NH_4^+) is very low (15 μM, compared to 300 μM for most mammalian glutamine synthetases) (Shankar and Anderson, 1985). The very low K_m of the mitochondrial glutamine synthetase for (NH_3/NH_4^+), the co-localization of glutamine synthetase and CPSase III in the mitochondrial matrix, and the fact that under optimal conditions the units of glutamine synthetase are about five fold higher than the units of CPSase III (1.2 versus 0.23 μmoles/min/g tissue) (Shankar and Anderson, 1985) together probably account for the observed very rapid and stoichiometric conversion by isolated mitochondria of ammonia at concentrations as low as 70 μM into citrulline. An additional contributing factor is that the K_m of CPSase III for glutamine is 0.16 mM under optimal conditions. This value is an order of magnitude lower than the K_m values for (NH_3/NH_4^+) for

CPSase I. For these reasons the glutamine synthetase–CPSase III coupled reaction is probably a more efficient ammonia-assimilating system than the mammalian CPSase I system. It has been suggested that this may represent an adaptive mechanism for sequestering low concentrations of ammonia from blood for urea synthesis (ureoosmotic physiology), as opposed to loss of ammonia from the blood via the gills (ammonotelic physiology), thus explaining why an extra energy-requiring step is utilized for urea synthesis (Anderson and Casey, 1984; Anderson, 1991; Campbell and Anderson, 1991). Considerable quantities of ammonia are released from muscle into the circulatory system of *Squalus acanthias* before and during starvation, and the prebranchial plasma concentrations of ammonia are relatively high (Leech *et al.,* 1979); these authors suggested that this is the major source of ammonia for urea synthesis.

One possible consequence of the likely use of most of the glutamine formed intramitochondrially for urea synthesis could be a lack of glutamine in liver for other amidotransferase reactions, such as occurs in the purine and pyrimidine nucleotide biosynthetic pathways. The observed absence of aspartate transcarbamylase activity in liver suggests that the pyrimidine nucleotide biosynthetic pathway is, in fact, absent (Anderson, 1989). However, all enzymes of the pyrimidine pathway, including CPSase II, are present in extrahepatic tissues; the first three enzymes of the pathway and glutamine synthetase are localized in the cytosol of spleen (Anderson, 1989). Thus, *Squalus acanthias* has two isozymes of glutamine synthetase, which Campbell and coworkers have shown are coded for by a single gene (Smith *et al.,* 1987; Campbell and Anderson, 1991).

The significance of the localization of arginase in the mitochondrial matrix is not known. One consequence is that urea is formed inside the mitochondria. Even though urea is permeable to the mitochondrial membrane (Anderson, 1986; Ballantyne and Moon, 1986; Moyes *et al.,* 1986), its formation in the mitochondrial matrix could result in a higher matrix concentration than is present in the cytosol. It has been suggested that this could be of importance for osmoregulation, resulting in feedback inhibition of the CPSase III (Anderson, 1991; Campbell and Anderson, 1991). The mitochondrial CPSase III and, to a lesser extent, the glutamine synthetase from *Squalus acanthias* are subject to significant inhibition by physiological concentrations of urea (Anderson, 1981; Shankar and Anderson, 1985). The observation that mitochondrial citrulline synthesis, but not respiration (under the same conditions), is inhibited by urea support this view (Anderson, 1986). That this is uniquely related to osmoregulation is suggested by the fact that the CPSase III from largemouth bass is not as significantly inhibited by urea.

Another consequence of the localization of arginase in the mitochondrial matrix is that ornithine availability for citrulline synthesis is dependent on the transport of arginine into the mitochondria and its subsequent hydrolysis to ornithine catalyzed by arginase. Since arginase is subject to significant product inhibition by ornithine, a reduced rate of carbamoyl phosphate synthesis could result in ornithine accumulation and inhibition of arginase as a regulatory mechanism. Stoichiometric citrulline synthesis by isolated mitochondria proceeds at equal rates when using equivalent concentrations of either arginine or ornithine, even at very low concentrations (e.g., 0.08 m*M*), indicating that arginine and ornithine appear to be equally permeable to the mitochondrial membrane. The rate of hydrolysis of arginine by arginase in intact, respiring mitochondria is not altered by conditions that would reduce carbamoyl phosphate availability and, consequently, permit accumulation of ornithine, suggesting that ornithine can apparently rapidly exit the mitochondria if not utilized for citrulline synthesis and arginase activity is sufficiently high so that formation of ornithine from arginine is not rate-limiting for citrulline synthesis. In contrast to rat liver mitochondria (Gamble and Lehninger, 1973), neither arginine nor ornithine permeability is dependent on active respiration. Thus, the mitochondrial localization of arginase does not appear to reflect a mechanism for regulating ornithine availability (Casey and Anderson, 1985).

With respect to nitrogen excretory pathways initiated in liver, elasmobranchs appear to reflect characteristics of both uricotelism (mitochondrial arginase and glutamine synthetase) and ureotelism (mitochondrial CPSase and ornithine transcarbamylase). The unique co-functioning of glutamine synthetase and CPSase III in ammonia assimilation in the mitochondrial matrix probably reflects the adaptation of urea synthesis for the dual role of ureoosmotic and ureotelic functions.

2. Teleosts

In contrast to *Squalus acanthias,* glutamine synthetase in liver of *Micropterus salmoides* is localized in the cytosol along with CPSase II, aspartate transcarbamylase, and dihydro-orotase. CPSase III, ornithine transcarbamylase, and arginase, as in the dogfish shark, are localized in the mitochondria (Cao *et al.,* 1991). Isolated, actively respiring mitochondria from bass liver are not able to synthesize citrulline from glutamate plus ammonia as nitrogen-donating substrate (as expected, since the glutamine synthetase is localized in the cytosol) or from glutamine (unexpected, since CPSase III is present in the mitochondria). In contrast to shark mitochondria, glutamine does not serve as a substrate for respiration (Cao *et al.,* 1991). These observations suggest that the low levels of CPSase

III activity in bass liver mitochondria may be physiologically unimportant in the adult fish.

Little information is available concerning citrulline synthesis and the physiological function of CPSase III in most other species where CPSase III is present. Mommsen and Walsh (1989) reported that the submitochondrial localization of glutamine synthetase is quite variable (mitochondrial or cytosolic) between different teleost species that have CPSase III activity, but that arginase is localized in the mitochondria of all species that have CPSase III activity. Recent studies, however, indicate that arginase and glutamine synthetase are present in both the mitochondrial and cytosolic compartments in the toadfish *Opsanus beta* and in the related species *Porichthys notatus* (midshipman); like bass, the enzymes of the pyrimidine nucleotide biosynthetic pathway are present in the cytosol in these species (Anderson and Walsh, 1995).

III. UREA SYNTHESIS IN FISH

The function and pathway of urea synthesis in ureoosmotic elasmobranchs is well established. Although some teleost fishes are able to synthesize urea via the urea cycle and possess CPSase III activity, it appears that this is significant in only a few species (at least for adult fish) and is correlated with unusual environment, adaptive, or life cycle circumstances considered in this section. However, even when exposed to apparently similar circumstances, the physiological responses or adaptations appear to vary and the factors controlling these responses have not been investigated. Molecular studies with these groups of fishes should help clarify the physiological and biochemical basis for expression of CPSase III and ureotelism in teleosts.

A. Amphibious Fish

Amphibious fish exhibit a broad diversity of habitat and structural and functional adaptations to life out of water (Gordon, 1970; Davenport and Sayer, 1986; Ramaswamy and Reddy, 1983; Saha and Ratha, 1989). One expected consequence of spending extensive periods of time out of water might be changes in the pattern of end-product nitrogen metabolism or excretion.

Gordon and coworkers (Gordon, 1970; Gordon *et al.,* 1969, 1970, 1978) investigated the changes in nitrogen excretion in an East African mudskipper (*Periophthalmus cantonensis*) and the Chilean clingfish (*Sicy-*

ases sanguineus). Their findings indicated a shift toward ureotelism while out of water. Subsequent studies by Morii and coworkers (Morii, 1979; Morii *et al.*, 1978, 1979) with two mudskipper species (*Periophthalmus cantonensis* and *Boleophthalmus pectinirostris*) concluded that a shift from ammonotelism to ureotelism during the period out of water does not occur. The explanation for these differences has not been resolved. In both series of studies, however, urea was formed and excreted, but the source of urea was not investigated. Chew and Ip (1987) reported that glutamine synthetase activity could not be detected in two mudskipper species (*Periophthalmodon schlosseri* and *Boleophthalmus boddaerti*), which would seem to rule out uricolysis or glutamine-dependent CPSase III and the urea cycle as sources of urea. Gregory (1977) found that two members of the mudskipper family (*Periophthalmus expeditonium* and *Periophthalmus gracilis*) and one member of the amphibious Scartelaos family (*Scartelaos histophorus*) excreted both urea (up to 33% of total nitrogen excreted) and ammonia, and that liver extracts of the two mudskippers had sufficient uricolytic enzyme activity to account for the urea formed, but that of the five required urea cycle enzymes only arginase and ornithine transcarbamylase could be detected (glutamine synthetase activity was not measured). However, CPSase is not active under the assay conditions described (10 mM ATP, 6 mM Mg^{2+}) (Anderson, 1981; Casey and Anderson, 1983). Another amphibious marine teleost, *Blennius pholis* (L.) (blenny), was found to be predominantly ammonotelic in seawater and during periods of aerial exposure (Davenport and Sayer, 1986). The major route of nitrogen elimination during aerial exposure was via ammonia in mucous secretions. Also, in contrast to reports from studies of other amphibious fish, the blenny (1) apparently does not store nitrogen during aerial exposure and release it as a burst of ammonia and urea after reimmersion and (2) continues to excrete nitrogen waste when exposed to air.

A clearer picture of adaptation by amphibious fishes has emerged from studies with several species of freshwater air-breathing teleosts that live on the Indian subcontinent. Ramaswamy and Reddy (1983) found a marked shift toward ureotelism in two obligate air-breathing teleosts (*Anabas scandens* and *Channa gachua*) but not in one facultative air-breathing teleost (*Mystus vittatus*) when exposed to air for 5 or 10 h. Saha and Ratha (1987, 1989) found that four out of five species of air-breathing teleosts investigated had high levels of urea cycle enzyme activities (*Heteropneustes fossilis, Anabas testudineus, Clarias batrachus,* and *Amphinous cuchia* had all five enzymes, only argininosuccinate synthase was absent in *Channa punctatus*). These species also had all enzymes of the uricolytic pathway in liver extracts. Ammonia was found to be the major excretory product in all five species while in water, but the rate of urea excretion was higher than found in exclusively freshwater teleosts. Like the ureoosmotic elasmobranchs, the levels of glutamine synthetase in liver were high and the enzyme was localized in the mitochondria in liver and in the cytosol in brain (Chakravorty *et al.*, 1989). The subcellular distribution of all five urea cycle enzymes was found to be analogous to that of elasmobranchs (Dkher *et al.*, 1991); these authors also reported as unpublished observations that the CPSase activity was a CPSase III. *H. fossilis* was found to be remarkably tolerant to high concentrations of ammonia (e.g., no signs of stress after exposure to 75 m*M* NH_4Cl for 28 days); the rate of urea excretion was reported to increase twofold after 10–12 days of exposure (Saha and Ratha, 1990) and this was accompanied by a corresponding increase in the levels of several urea cycle enzymes (Saha and Ratha, 1986). Thus, this significant series of studies appears to have established that at least one of these air-breathing freshwater teleosts probably has a functional glutamine-dependent urea cycle as found in elasmobranchs. It will be of considerable interest to establish the nature of the regulatory mechanisms that apparently turn urea synthesis on when these fish are exposed to air.

B. Lungfish and Coelacanths

The African lungfish (*Protopterus aethiopicus*) is entirely dependent on aerial respiration. During drought periods the fish can survive for long periods of time by estivation in the mud surrounded by a hardened mucous cocoon, which is connected by a tube to the surface for breathing (Smith, 1930). While in water the fish excrete approximately equal amounts of ammonia and urea as end products of nitrogen metabolism. However, during estivation, to conserve water and preclude ammonia accumulation, ammonia formation ceases and only urea is formed, which is stored in body tissues (accumulating to levels as high as 3% of the body weight during long periods of estivation) and released when the fish returns to an aqueous environment (Smith, 1930). Although all enzymes of both the uricolytic and urea cycle pathways are present, virtually all urea is formed by the urea cycle; the rate of urea formation does not change significantly during the switch from an aqueous environment to estivation (Janssens, 1964; Brown *et al.*, 1966; Forster and Goldstein, 1966; Janssens and Cohen, 1966, 1968b). Mommsen and Walsh (1989) reported that the CPSase activity is a CPSase I, not a CPSase III, in this species of lungfish and that the glutamine synthetase and arginase activities are localized in the cytosol. Janssens and Cohen (1968a) were not able to detect glutamine synthetase activity in liver. These observations seem to clearly indicate that the ammonia-dependent urea cycle characteristic of higher vertebrates operates in lungfish.

The Australian lungfish (*Neoceratodus forsteri*) uses its lung only as an accessory breathing organ and cannot survive deprivation of water by estivation. Accordingly, the level of the urea cycle enzymes and the rate of urea synthesis are dramatically lower than in the African lungfish (Goldstein *et al.*, 1967). Results similar to those described for the African lungfish have been reported for the South American lungfish (*Lepidosiren paradoxa*), which estivates in a moist cocoon, except that the rates of urea synthesis and levels of urea cycle enzymes were lower, which is consistent with the intermediate environmental position between the African lungfish (estivates in a dry cocoon) and the Australian lungfish (does not estivate) (Carlisky and Barrio, 1972; Funkhouser *et al.*, 1972). The latter authors suggested that accumulation of urea during estivation may serve a second useful function of elevating the vapor pressure and thereby minimizing water loss by evaporation.

The coelacanth *Latimeria chalumnae* is the only living representative of the crossopterygians (Cloutier, 1991; Musick *et al.*, 1991). The coelacanth has high levels of urea and trimethylamine oxide in its tissues (Brown and Brown, 1967; Pickford and Grant, 1967; Lutz and Robertson, 1971), as well as high levels of all five urea cycle enzymes in liver (Brown and Brown, 1967; Goldstein *et al.*, 1973), comparable to levels in marine elasmobranchs. Mommsen and Walsh (1989) reported that the CPSase activity is due to the presence of a CPSase III and that the liver arginase was localized in the mitochondria. Therefore, the coelacanth appears to be closely related to elasmobranchs with respect to the retention of urea as an osmolyte and with respect to the existence of an apparent glutamine-dependent urea cycle. This might suggest that the coelacanth is closely related to chondrichthyans, but sequence analysis of 28S ribosomal RNA from the coelacanth clearly supports its close relationship to the sarcopterygians and not the Chondrichthyes (Hillis *et al.*, 1991).

C. Toadfish

The earlier report by Read (1971) of high levels of all five urea cycle enzymes in oyster toadfish (*Opsanus tau*) has been confirmed by Mommsen and Walsh (1989), who also demonstrated that in both oyster toadfish and the related gulf toadfish (*Opsanus beta*) the CPSase III, glutamine synthetase, and arginase activities are localized in the mitochondria and isolated hepatocytes have a high capacity for [^{14}C]urea synthesis when incubated with [^{14}C]bicarbonate and other appropriate substrates. Thus, these marine teleost species appear to have a high capacity for glutamine-dependent urea synthesis analogous to that of elasmobranchs (Fig. 3).

The physiological role of the high capacity for urea synthesis is not clearly understood. The concentration of urea in plasma is relatively low (1–10 m*M*), indicating that urea does not play a significant osmoregulatory role (Mommsen and Walsh, 1989; Walsh *et al.*, 1990). Studies by Walsh and coworkers (1989; Barber and Walsh, 1993) indicate that urea synthesis in toadfish does not appear to be related to regulation of acid–base balance. Urea excretion rates were found to be quite variable and increase significantly in response to high levels of external ammonia or exposure to air for extended periods of time (Walsh *et al.*, 1990). Physical confinement initiates a switch to ureagenesis within a day or two, which is accompanied by a significant increase in glutamine synthetase activity (Walsh *et al.*, 1994). These effects are not due to increases in ammonia concentration or to air exposure subsequent to confinement, suggesting that this is a behavioral response to stress. Regardless of the regulatory mechanisms, these studies indicate that urea synthesis in toadfish is related to variations in environmental habitat or stress.

D. Adaptation to Alkaline Waters

An alkaline environment (pH 8.5–10) causes severe physiological disturbances for most fish, initially inhibiting diffusion of ammonia across the gills, resulting in an increase in plasma ammonia concentration, among other effects (Wright *et al.*, 1990; Wood, 1993). Randall *et al.* (1989) reported that the Lake Magadi tilapia, *Oreochromis alcalicus grahami,* however, had adapted to life in an alkaline soda lake (pH 9.6–10) at 30–40°C by excreting virtually all its nitrogen waste as urea. Significant levels of all urea cycle enzymes measured, including CPSase III and glutamine synthetase, were present in liver extracts. These fish have been found to be considerably more tolerant to elevated external ammonia concentrations than most other teleosts, one response being an immediate increase in urea excretion, presumably due to high urea cycle capability (Randall *et al.*, 1989; Walsh *et al.*, 1993).

Other fish adapted to alkaline waters, including Lahontan cutthroat trout (*Oncorhynchus clarki henshawi*) adapted to the highly alkaline (pH 9.4) Pyramid Lake in Nevada (Wright *et al.*, 1993; Wilkie *et al.*, 1993) and the cyprinid *Chalcalburnus tarichi* endemic to the extremely alkaline (pH 9.8) Lake Van in Turkey (Danulat and Kempe, 1992), do not excrete predominantly urea and do not have high levels of urea enzymes or glutamine synthetase.

E. Embryogenesis

Several reports suggest that urea synthesis and expression of urea cycle enzymes may occur during embryogenesis. Dépêche *et al.* (1979) noted increases in urea concentrations during specific stages of embryonic development in *Poecilia reticulata* (guppy) and *Salmo gairdneri* (rainbow trout), which was associated with an increase in urea cycle activity as measured by formation of [^{14}C]urea from [^{14}C]bicarbonate and was stimulated by exposure to hyperosmotic media. Urea levels declined toward values found in adults at the end of embryogenesis and urea cycle activity could not be detected in adults. Similar observations were made by Rice and Stokes (1974) in rainbow trout, who also reported an increase in ornithine transcarbamylase activity through development until the yolk was nearly absorbed, after which the activity decreased to very low levels. Arginase activity increased rapidly after the yolk was absorbed. CPSase and argininosuccinate synthetase activities could not be detected. Very low levels of CPSase III have recently been shown to coincide with the expression of ornithine transcarbamylase (Wright *et al.*, 1994). DeVlaming *et al.* (1983) reported that the urea concentration in pregnant female ovarian fluid of viviparous embiotocid fishes is very high compared to that in the maternal serum; citrulline levels were also elevated, suggesting that the source of urea was via the urea cycle.

Griffith (1991) has proposed a scenario for the evolution of urea retention in fishes that involves the existence of a functional urea cycle in early gnathostomes that was expressed only during early embryogenesis as a means of detoxifying ammonia. The pedogenic retention of the urea cycle pathway by adults that invaded the marine habitat is considered to explain the origin of urea retention in the ancestor(s) of the extant ureoosmotic fishes. This model suggests that extensions of the study of Dépêche *et al.* (1979) to a variety of other species would demonstrate that the role of urea synthesis in ammonia detoxification is a general phenomenon in the early development of fish (Griffith, 1991). This could prove to be a significant area of investigation, providing a basis for the probable presence of urea cycle enzymes in the genomes of all fish.

ACKNOWLEDGMENTS

Thanks are extended to C. Helbing and D. E. Atkinson and to P. J. Walsh, B.C. Tucker, and T. E. Hopkins for supplying copies of manuscripts in press, to P. A. Wright, W. L. Salo, and P. J. Walsh for reviewing the manuscript or parts of the manuscript, and to the National Science Foundation for research grant support (DCB-91057997).

REFERENCES

Anderson, P. M. (1976). A glutamine- and *N*-acetyl-L-glutamate-dependent carbamyl phosphate synthetase activity in the teleost *Micropterus salmoides. Comp. Biochem. Physiol. B* **54,** 261–263.

Anderson, P. M. (1980). Glutamine- and *N*-acetylglutamate-dependent carbamoyl phosphate synthetase in elasmobranchs. *Science* **208,** 291–293.

Anderson, P. M. (1981). Purification and properties of the glutamine- and *N*-acetyl-L-glutamate-dependent carbamoyl phosphate synthetase from liver of *Squalus acanthias. J. Biol. Chem.* **256,** 12228–12238.

Anderson, P. M. (1986). Effects of urea, trimethylamine oxide, and osmolality on respiration and citrulline synthesis by isolated hepatic mitochondria from *Squalus acanthias. Comp. Biochem. Physiol. B* **85,** 783–788.

Anderson, P. M. (1989). Glutamine-dependent carbamoyl-phosphate synthetase and other enzyme activities related to the pyrimidine pathway in spleen of *Squalus acanthias* (spiny dogfish). *Biochem. J.* **261,** 523–529.

Anderson, P. M. (1991). Glutamine-dependent urea synthesis in elasmobranch fishes. *Biochem. Cell Biol.* **69,** 317–319.

Anderson, P. M., and Casey, C. A. (1984). Glutamine-dependent synthesis of citrulline by isolated hepatic mitochondria from *Squalus acanthias. J. Biol. Chem.* **259,** 456–462.

Anderson, P. M., and Walsh, P. J. (1995) Subcellular localization and biochemical properties of the enzymes of carbamoyl phosphate and urea synthesis in the batrachoidid fishes *Opsanus beta, Opsanus tau,* and *Porichthys notatus. J. Exp. Biol.* **198,** (in press).

Atkinson, D. E. (1992). Functional roles of urea synthesis in vertebrates. *Physiol. Zool.* **65,** 243–267.

Ballantyne, J. S., and Moon, T. W. (1986). Solute effects on mitochondria from an elasmobranch (*Raja erinacea*) and a teleost (*Pseudopleuronectes americanus*). *J. Exp. Zool.* **239,** 319–328.

Barber, M. L., and Walsh, P. J. (1993). Interactions of acid–base status and nitrogen excretion and metabolism in the ureogenic teleost *Opsanus tau. J. Exp. Biol.* **185,** 87–105.

Bein, K. Simmer, J. P., and Evans, D. R. (1991). Molecular cloning of a cDNA encoding the amino end of the mammalian multifunctional protein CAD and analysis of the 5′ flanking region of the CAD. *J. Biol. Chem.* **266,** 3791–3799.

Brown, G. W., Jr. (1964). Urea synthesis in elasmobranchs. *In* "Taxonomic Biochemistry and Serology" (C. A. Leone, ed.), pp. 407–416. Ronald Press, New York.

Brown, G. W., Jr., and Brown, S. G. (1967). Urea and its formation in coelacanth liver. *Science* **155,** 570–573.

Brown , G. W., Jr., and Cohen, P. P. (1960). Comparative biochemistry of urea synthesis. III. Activities of urea-cycle enzymes in various higher and lower vertebrates. *Biochem. J.* **75,** 82–91.

Brown, G. W., Jr., James, J., Henderson, R. J., Thomas, W. N., Robinson, R. O., Thompson, A. L., Brown, E., and Brown, S. G. (1966). Uricolytic enzymes in liver of the dipnoan *Protopterus aethiopicus. Science* **153,** 1653–1654.

Campbell, J. W. (1991). Excretory nitrogen metabolism. *In* "Environmental and Metabolic Animal Physiology" (C. L. Prosser, ed.), Chapter 7, pp. 277–324. Wiley-Liss, New York.

Campbell, J. W., and Anderson, P. M. (1991). Evolution of mitochondrial enzyme systems in fish: The mitochondrial synthesis of glutamine and citrulline. *In*"Biochemistry and

Molecular Biology of Fishes" (P. W. Hochachka and T. P. Mommsen, eds.), Vol. 1, pp. 43–76. Elsevier, Amsterdam.

Cao, X., Kemp, J. R., and Anderson, P. M. (1991). Subcellular localization of two glutamine-dependent carbamoyl-phosphate synthetases and related enzymes in liver of *Micropterus salmoides* (largemouth bass) and properties of isolated liver mitochondria: Comparative relationships with elasmobranchs. *J. Exp. Zool.* **258,** 24–33.

Carlisky, N. J., and Barrio, A. (1972). Nitrogen metabolism of the South American lungfish *Lepidosiren paradoxa. Comp. Biochem. Physiol., B* **41,** 857–873.

Casey, C. A., and Anderson, P. M. (1982). Subcellular location of glutamine synthetase and urea cycle enzymes in liver of spiny dogfish (*Squalus acanthias*). *J. Biol. Chem.* **257,** 8449–8453.

Casey, C. A., and Anderson, P. M. (1983). Glutamine- and *N*-acetyl-L-glutamate-dependent carbamoyl phosphate synthetase from *Micropterus salmoides*. Purification, properties, and inhibition by glutamine analogs. *J.Biol. Chem.* **258,** 8723–8732.

Casey, C. A., and Anderson, P. M. (1985). Submitochondrial localization of arginase and other enzymes associated with urea synthesis and nitrogen metabolism, in liver of *Squalus acanthias. Comp. Biochem. Physiol., B* **82,** 307–315.

Chakravorty, J., Saha, N., and Ratha, B. K. (1989). A unique pattern of tissue distribution and sub-cellular localization of glutamine synthetase in a freshwater air-breathing teleost, *Heteropneustes fossilis* (Bloch). *Biochem. Int.* **19,** 519–527.

Cheung, C. -W., and Raijman, L. (1980). The regulation of carbamyl phosphate synthetase (ammonia) in rat liver mitochondria. Effects of acetylglutamate and ATP translocation. *J. Biol. Chem.* **255,** 5051–5057.

Cheung, C. -W., and Raijman, L. (1981). Arginine, mitochondrial arginase, and the control of carbamyl phosphate synthesis. *Arch. Biochem. Biophys.* **209,** 643–649.

Chew, S. F., and Ip, Y. K. (1987). Ammoniagenesis in mudskippers *Boleophthalmus boddaerti* and *Periophthalmodon schlosseri. Comp. Biochem. Physiol., B* **87,** 941–948.

Chiu, Y. N., Austic, R. E., and Rumsey, G. L. (1986). Urea cycle activity and arginine formation in rainbow trout (*Salmo gairdneri*). *J. Nutr.* **116,** 1640–1650.

Clarke, S. J. (1976). A major polypeptide component of rat liver mitochondria: Carbamyl phosphate synthetase. *J. Biol. Chem.* **251,** 950–961.

Cloutier, R. (1991). Patterns, trends, and rates of evolution within the Actinistia. *Environ. Biol. Fishes* **32,** 23–58.

Cohen, N. S., Kyan, F. S., Kyan, S. S., Cheung, C.-W., and Raijman, L. (1985). The apparent K_m of ammonia for carbamoyl phosphate synthetase (ammonia) *in situ. Biochem. J.* **229,** 205–211.

Cohen, P. P. (1976). Evolutionary and comparative aspects of urea biosynthesis. *In* "The Urea Cycle" (S. Grisolia, R. Báguena, and F. Mayor, eds.), pp. 21–38. Wiley, New York.

Cvancara, V. A. (1969). Comparative study of liver uricase activity in fresh-water teleosts. *Comp. Biochem. Physiol.* **28,** 725–732.

Cvancara, V. A. (1974). Liver carbamoyl phosphate synthetase in the primitive freshwater bony fishes (Chondrostei, Holostei). *Comp. Biochem. Physiol., B* **49,** 785–787.

Danulat, E., and Kempe, S. (1992). Nitrogenous waste and accumulation of urea and ammonia in *Chalcalburnus tarichi* (Cyprinidae), endemic to the extremely alkaline Lake Van (eastern Turkey). *Fish Physiol. Biochem.* **9,** 377–386.

Davenport, J., and Sayer, M. D. J. (1986). Ammonia and urea excretion in the amphibious teleost *Blennius pholis* (L.) in sea-water and in air. *Comp. Biochem. Physiol., A* **84,** 189–194.

Dépêche, J., Gilles, R., Daufresne, S., and Chiapello, H. (1979). Urea content and urea

production via the ornithine-urea cycle pathway during the ontogenic development of two teleost fishes. *Comp. Biochem. Physiol., A* **63,** 51–56.
deVlaming, V. L., Baltz, D., Anderson, S., Fitzgerald, R., Delahunty, G., and Barkley, M. (1983). Aspects of embryo nutrition and excretion among viviparous embiotocid teleosts: Potential endocrine involvements. *Comp. Biochem. Physiol., A* **76,** 189–198.
Dkher, J., Saha, N., and Ratha, B. K. (1991). Ureogenesis in a freshwater teleost: An unusual sub-cellular localization of ornithine-urea cycle enzymes in the freshwater air-breathing teleost *Heteropneustes fossilis. Biochem. Int.* **25,** 1061–1069.
Dragojević, M., and Dida, B. X. (1977). Effect of various media on carbamoyl-phosphosynthetase concentrations in carp liver. *Acta Biol. Med. Exp.* **2,** 111–113.
Evans, D. E. (1986). CAD, a chimeric protein that initiates *de novo* pyrimidine biosynthesis in higher eukaryotes. *In* "Multidomain Proteins: Structure and Function" (D. G. Hardie, ed.), Chapter 9, pp. 283–331. Elsevier, Amsterdam.
Evans, D. R., Bein, K., Guy, H. I., Liu, X., Molina, L. A., and Zimmerman, B. H. (1993). CAD gene sequence and the domain structure of the mammalian multifunctional protein CAD. *Biochem. Soc. Trans.* **21,** 186–191.
Forster, R. P., and Goldstein, L. (1966). Urea synthesis in the lungfish: Relative importance of purine and ornithine cycle pathways. *Science* **153,** 1650–1652.
Funkhouser, D., Goldstein, L., and Forster, R. P. (1972). Urea biosynthesis in the South American lungfish, *Lepidosiren paradoxa:* Relation to its ecology. *Comp. Biochem. Physiol., A* **41,** 439–443.
Gamble, J. G., and Lehninger, A. L. (1973). Transport of ornithine and citrulline across the mitochondrial membrane. *J. Biol. Chem.* **248,** 610–618.
Goldstein, L. (1967). Urea biosynthesis in elasmobranchs. *In* "Sharks, Skates, and Rays" (P. W. Gilbert, R. F. Mathewson, and D. P. Ralls, eds.), pp. 207–214. Johns Hopkins Univ. Press, Baltimore.
Goldstein, L. (1970). Urea metabolism and osmoregulation in euryhaline elasmobranchs and amphibia. *In* "Urea and the Kidney" (B. Schmidt-Nielsen and D. W. S. Kerr, eds.), pp. 243–251. Excerpta Med. Found., Amsterdam.
Goldstein, L. (1972). Adaptation of urea metabolism in aquatic vertebrates. *In* "Nitrogen Metabolism and the Environment" (J. W. Campbell and L. Goldstein, eds.), pp. 55–77. Academic Press, New York.
Goldstein, L, and Forster, R. P. (1965). The role of uricolysis in the production of urea in fishes and other aquatic vertebrates. *Comp. Biochem. Physiol.* **14,** 567–576.
Goldstein, L., and Forster, R. P. (1970). Nitrogen metabolism in fishes. *In* "Comparative Biochemistry of Nitrogen Metabolism" (J. W. Campbell, ed.), Vol. 2, pp. 495–518, Academic Press, New York.
Goldstein, L., and Forster, R. P. (1971a). Osmoregulation and urea metabolism in the little skate *Raja erinacea. Am. J. Physiol.* **220,** 742–746.
Goldstein, L., and Forster, R. P. (1971b). Urea biosynthesis and excretion in freshwater and marine elasmobranchs. *Comp. Biochem. Physiol., B* **39,** 415–421.
Goldstein, L., Janssens, P. A., and Forster, R. P. (1967). Lungfish *Neoceratodus forsteri:* Activities of ornithine-urea cycle and enzymes. *Science* **157,** 316–317.
Goldstein, L., Harley-DeWitt, S., and Forster, R. P. (1973). Activities of ornithine-urea cycle enzymes and of trimethylamine oxidase in the coelacanth, *Latimeria chalumnae. Comp. Biochem. Physiol., B* **44,** 357–362.
Gordon, M. S. (1970). Patterns of nitrogen excretion in amphibious fishes. *In* "Urea and the Kidney" (B. Schmidt-Nielsen and D. W. S. Kerr, eds.), pp. 238–242. Excepta Med. Found., Amsterdam.
Gordon, M. S., Boëtius, I., Evans, D. H., McCarthy, R., and Oglesby, L. C. (1969).

Aspects of the physiology of terrestrial life in amphibious fishes. I. The mudskipper, *Periophthalmus sobrinus*. *J. Exp. Biol.* **50,** 141–149.

Gordon, M. S., Fisher, S., and Tarifeño S. E. (1970). Aspects of the physiology of terrestrial life in amphibious fishes. II. The Chilean clingfish, *Sicyases sanguineus*. *J. Exp. Biol.* **53,** 559–572.

Gordon, M. S., Ng, W. W. -s., and Yip, A. Y. -w. (1978). Aspects of the physiology of terrestrial life in amphibious fishes. III. The Chinese mudskipper *Periophthalmus cantonensis*. *J. Exp. Biol.* **72,** 57–75.

Gregory, R. B. (1977). Synthesis and total excretion of waste nitrogen by fish of the *Periophthalmus* (mudskipper) and *Scartelaos* families. *Comp. Biochem. Physiol., A* **57,** 33–36.

Griffith, R. W. (1981). Composition of the blood serum of deep-sea fishes. *Biol. Bull.* (*Woods Hole, Mass.*) **160,** 250–264.

Griffith, R. W. (1991). Guppies, toadfish, lungfish, coelacanths and frogs: A scenario for the evolution of urea retention in fishes. *Environ. Biol. Fishes* **32,** 199–218.

Haraguchi, Y., Uchino, T., Takiguchi, M., Endo, F., Mori, M., and Matsuda, I. (1991). Cloning and sequence of a cDNA encoding human carbamoyl phosphate synthetase I: Molecular analysis of hyperammonemia. *Gene* **107,** 335–340.

Hayashi, S., Fujiwara, S., and Noguchi, T. (1989). Degradation of uric acid in fish liver peroxisomes. Intraperoxisomal localization of hepatic allantoicase and purification of its peroxisomal membrane-bound form. *J. Biol. Chem.* **264,** 3211–3215.

Helbing, C., and Atkinson, B. G. (1994). 3,5,3′-Triiodothyronine-induced carbamyl-phosphate synthetase gene expression is stabilized in the liver of *Rana catesbeiana* tadpoles during heat shock. *J. Biol. Chem.* **269,** 11743–11750.

Hillis, D. M., Dixon, M. T., and Ammerman, L. K. (1991). The relationships of the coelacanth *Latimeria chalumnae:* Evidence from sequences of vertebrate 28S ribosomal RNA genes. *Environ. Biol. Fishes* **32,** 119–130.

Hong, J., Salo, W. L., Lusty, C. A., and Anderson, P. M. (1994). Carbamyl phosphate synthetase III, an evolutionary intermediate in the transition between glutamine-dependent and ammonia-dependent carbamyl phosphate synthetases. *J. Mol. Biol.* **243,** 131–140.

Huggins, A. K., Skutsch, G., and Baldwin, E. (1969). Ornithine-urea cycle enzymes in teleostean fish. *Comp. Biochem. Physiol.* **28,** 587–602.

Janssens, P. A. (1964). The metabolism of the aestivating African lungfish. *Comp. Biochem. Physiol.* **11,** 105–117.

Janssens, P. A., and Cohen, P. P. (1966). Ornithine-urea cycle enzymes in the African lungfish, *Protopterus aethiopicus*. *Science* **152,** 358–359.

Janssens, P. A., and Cohen, P. P. (1968a). Nitrogen metabolism in the African lungfish. *Comp. Biochem. Physiol.* **24,** 879–886.

Janssens, P. A., and Cohen, P. P. (1968b). Biosynthesis of urea in the estivating African lungfish and in *Xenopus laevis* under conditions of water-shortage. *Comp. Biochem. Physiol.* **24,** 887–898.

Jones, M. E. (1980). Pyrimidine nucleotide biosynthesis in animals: Genes, enzymes, and regulation of UMP biosynthesis. *Annu. Rev. Biochem.* **49,** 253–280.

Leech, A. R., Goldstein, L., Cha, C., and Goldstein, J. M. (1979). Alanine biosynthesis during starvation in skeletal muscle of the spiny dogfish, *Squalus acanthias*. *J. Exp. Zool.* **207,** 73–80.

Lutz, P. L., and Robertson, J. D. (1971). Osmotic constituents of the coelacanth *Latimeria chalumnae* Smith. *Biol. Bull.* (*Woods Hole, Mass.*) **141,** 553–560.

Marshall, M. (1976). Carbamyl phosphate synthetase I from frog liver. *In* "The Urea Cycle" (S. Grisolea, R. Báguena, and F. Mayor, eds.), pp. 133–142. Wiley, New York.

Meijer, A. J., Lamers, W. H., and Chamuleau, R. A. F. M. (1990). Nitrogen metabolism and ornithine cycle function. *Physiol. Rev.* **70,** 701–748.

Meister, A. (1989). Mechanism and regulation of the glutamine-dependent carbamyl phosphate synthetase of *Escherichia coli. Adv. Enzymol. Relat. Areas Mol. Biol.* **62,** 315–374.

Mommsen, T. P., and Walsh, P. J. (1989). Evolution of urea synthesis in vertebrates: The piscine connection. *Science* **243,** 72–75.

Mommsen, T. P., and Walsh, P. J. (1991). Urea synthesis in fishes: Evolutionary and biochemical perspectives. *In* "Biochemistry and Molecular Biology of Fishes" (P. W. Hochachka and T. P. Mommsen, eds.), Vol. 1, pp. 137–163. Elsevier, Amsterdam.

Mommsen, T. P., and Walsh, P. J. (1992). Biochemical and environmental perspectives on nitrogen metabolism in fishes. *Experientia* **48,** 583–593.

Morii, H. (1979). Changes with time of ammonia and urea concentrations in the blood and tissue of mudskipper fish, *Periophthalmus cantonensis* and *Boleophthalmus pectinirostris* kept in water and on land. *Comp. Biochem. Physiol., A* **64,** 235–243.

Morii, H., Nishikata, K., and Tamura, O. (1978). Nitrogen excretion of mudskipper fish *Periophthalmus cantonensis* and *Boleophthalmus pectinirostris* in water and on land. *Comp. Biochem. Physiol., A* **60,** 189–193.

Morii, H., Nishikata, K., and Tamura, O. (1979). Ammonia and urea excretion from mudskipper fishes *Periophthalmus cantonensis* and *Boleophthalmus pectinirostris* transferred from land to water. *Comp. Biochem. Physiol., A* **63,** 23–28.

Morita, T., Mori, M., and Tatibana, M. (1982). Regulation of *N*-acetyl-L-glutamate degradation in mammalian liver. *J. Biochem. (Tokyo)* **91,** 563–569.

Morris, S. M., Jr. (1992). Regulation of enzymes of urea and arginine synthesis. *Annu. Rev. Nutr.* **12,** 81–101.

Moyes, C. D., Moon, T. W., and Ballantyne, J. S. (1986). Osmotic effects on amino acid oxidation in skate liver mitochondria. *J. Exp. Biol.* **125,** 181–195.

Musick, J. A., Bruton, M. N., and Balon, E. K., eds. (1991). "The Biology of *Latimeria chalumnae* and Evolution of Coelacanths." Kluwer Academic Publishers, Boston [reprinted from *Environ. Biol. Fishes* **32**(1–4) (1991)].

Nyunoya, H., and Lusty, C. J. (1983). The *carB* gene of *Escherichia coli:* A duplicated gene coding for the large subunit of carbamoyl-phosphate synthetase. *Proc. Natl. Acad. Sci. U.S.A.* **80,** 4629–4633.

Olson, K. R., and Fromm, P. O. (1971). Excretion of urea by two teleosts exposed to different concentrations of ambient ammonia. *Comp. Biochem. Physiol., A* **40,** 999–1007.

Perlman, D. F., and Goldstein, L. (1988). Nitrogen metabolism. *In* "Physiology of Elasmobranch Fishes" (T. J. Shuttleworth, ed.), pp. 253–276. Springer-Verlag, Berlin.

Pickford, G. E., and Grant, F. B. (1967). Serum osmolality in the coelacanth, *Latimeria chalumnae:* Urea retention and ion regulation. *Science* **155,** 568–570.

Quinn, C. L., Stephenson, B. T., and Switzer, R. L. (1991). Functional organization and nucleotide sequence of the *Bacillus subtilus* pyrimidine biosynthetic operon. *J. Biol. Chem.* **266,** 9113–9127.

Ramaswamy, M., and Reddy, T. G. (1983). Ammonia and urea excretion in three species of air-breathing fish subjected to aerial exposure. *Proc.—Indian Acad. Sci. [Ser.]: Anim. Sci.* **92,** 293–297.

Randall, D. J., and Wright, P. A. (1987). Ammonia distribution and excretion in fish. *Fish Physiol. Biochem.* **3,** 107–120.

Randall, D. J., Wood, C. M., Perry, S. F., Bergman, H., Maloiy, G. M. O., Mommsen, T. P., and Wright, P. A. (1989). Urea excretion as a strategy for survival in a fish living in a very alkaline environment. *Nature (London)* **337,** 165–166.

Read, L. J. (1967). Enzymes of the ornithine-urea cycle in the chimaera *Hydrolagus colliei*. *Nature (London)* **215**, 1412–1413.

Read, L. J. (1970). Nitrogen metabolism in chondrichthyean and agnathan fishes. *In* "Urea and the Kidney" (B. Schmidt-Nielsen and D. W. S. Kerr, eds.), pp. 23–34. Excerpta Med. Found., Amsterdam.

Read, L. J. (1971). The presence of high ornithine-urea cycle enzyme activity in the teleost *Opsanus tau*. *Comp. Biochem. Physiol., B* **39**, 409–413.

Rice, S. D., and Stokes, R. M. (1974). Metabolism of nitrogenous wastes in the eggs and alevins of rainbow trout, *Salmo gairdneri* Richardson. *In* "The Early Life History of Fish" (J. H. S. Baxter, ed.), pp. 325–337. Springer-Verlag, New York.

Saha, N., and Ratha, B. K. (1986). Effect of ammonia stress on ureogenesis in a freshwater air-breathing teleost. *Heteropneustes fossilis*. *In* "Contemporary Themes in Biochemistry" (O. L. Kon, M. C.-M. Chung, P. L. H. Hwang, S.-F. Leong, K. H. Loke, P. Thiyagarajah, and P. T.-H. Wong, eds.) Volume 6, pp. 342–343. Cambridge University Press, London.

Saha, N., and Ratha, B. K. (1987). Active ureogenesis in a freshwater air-breathing teleost, *Heteropneustes fossilis*. *J. Exp. Zool.* **241**, 137–141.

Saha, N., and Ratha, B. K. (1989). Comparative study of ureogenesis in freshwater, air-breathing teleosts. *J. Exp. Zool.* **252**, 1–8.

Saha, N., and Ratha, B. K. (1990). Alterations in excretion pattern of ammonia and urea in a freshwater air-breathing teleost, *Heteroptneustes fossilis* (Bloch) during hyper-ammonia stress. *Ind. J. Exp. Biol.* **28**, 597–599.

Schooler, J. M., Goldstein, L., Hartman, S. C., and Forster, R. P. (1966). Pathways of urea synthesis in the elasmobranch, *Squalus acanthias*. *Comp. Biochem. Physiol.* **18**, 271–281.

Shankar, R. A., and Anderson, P. M. (1985). Purification and properties of glutamine synthetase from liver of *Squalus acanthias*. *Arch. Biochem. Biophys.* **239**, 248–259.

Simmer, J. P., Kelly, R. E., Rinker, A. G., Scully, J. L., and Evans, D. R. (1990). Mammalian carbamyl phosphate synthetase (CPS). cDNA seqeunces and evolution of the CPS domain of the Syrian hamster multifunctional protein CAD. *J. Biol. Chem.* **265**, 10395–10402.

Smith, D. D., Jr., Ritter, N. M., and Campbell, J. W. (1987). Glutamine synthetase isozymes in elasmobranch brain and liver tissues. *J. Biol. Chem.* **262**, 198–202.

Smith, H. (1930). Metabolism of the lung-fish, *Protopterus aethiopicus*. *J. Biol. Chem.* **88**, 97–130.

Thomson, K. S. (1971). The adaptation and evolution of early fishes. *Q. Rev. Biol.* **46**, 139–166.

Trammel, P. R., and Campbell, J. W. (1970). Carbamyl phosphate synthesis in a land snail, *Strophocheilus oblongus*. *J. Biol. Chem.* **245**, 6634–6641.

Trammel, P. R., and Campbell, J. W. (1971). Carbamyl phosphate synthesis in invertebrates. *Comp. Biochem. Physiol., B* **40**, 395–406.

Vellas, F., and Serfaty, A. (1974). L'ammoniaque et l'urée chez un téléostéen d'eau douce: la carpe (*Cyprinus carpio* L.). *J. Physiol.* (Paris) **68**, 591–614.

Vorhaben, J. E., and Campbell, J. W. (1972). Glutamine synthetase. A mitochondrial enzyme in uricotelic species. *J. Biol. Chem.* **247**, 2763–2767.

Walsh, P. J., Parent, J. J., and Henry, R. P. (1989). Carbonic anhydrase supplies bicarbonate for urea synthesis in toadfish (*Opsanus beta*) hepatocytes. *Physiol. Zool.* **62**, 1257–1272.

Walsh, P. J., Danulat, E., and Mommsen, T. P. (1990). Variation in urea excretion in the gulf toadfish (*Opsanus beta*). *Mar. Biol. (Berlin)* **106**, 323–328.

Walsh, P. J., Bergman, H. L., Narahara, A., Wood, C. M., Wright, P. A., Randall, D. J., Maira, J. N., and Laurent, P. (1993). Effects of ammonia on survival, swimming, and activities of enzymes of nitrogen metabolism in the Lake Magadi tilapia *Oreochromis alcalicus grahamii*. *J. Exp. Biol.* **180,** 323–387.

Walsh, P. J., Tucker, B. C., and Hopkins, T. E. (1994). Effects of confinement and crowding on ureogenesis in the gulf toadfish *Opsanus beta*. *J. Exp. Biol.* **191,** 195–206.

Watford, M. (1991). The urea cycle: A two-component system. *Essays Biochem.* **26,** 49–58.

Watts, D. C., and Watts, R. L. (1966). Carbamoyl phosphate synthetase in the elasmobranchii: Osmoregulatory function and evolutionary implications. *Comp. Biochem. Physiol.* **17,** 785–798.

Watts, R. L., and Watts, D. C. (1973). Nitrogen metabolism in fishes. *Chem. Zool.* **8,** 369–449.

Webb, J. T., and Brown, G. W., Jr. (1980). Glutamine synthetase: An assimilatory role in liver as related to urea retention in marine chondrichthyes. *Science* **208,** 293–295.

Wilkie, M. P., Wright, P. A., Iwama, G. K., and Wood, C. M. (1993). The physiological responses of the Lahontan cutthroat trout (*Oncorhynchus clarki henshawi*), a resident of highly alkaline Pyramid Lake (pH 9.4), to challenge at pH 10. *J. Exp.Biol.* **175,** 173–194.

Wilson, R. P. (1973). Nitrogen metabolism in channel catfish *Ictalurus punctatus*. II. Evidence for an apparent incomplete ornithine-urea cycle. *Comp. Biochem. Physiol., B* **46,** 625–634.

Wood, C. M. (1993). Ammonia and urea metabolism and excretion. *In* "The Physiology of Fishes" (D. H. Evans, ed.), pp. 379–425. CRC Press, Boca Raton, FL.

Wright, P. A. (1993). Nitrogen excretion and enzyme pathways for ureagenesis in freshwater tilapia (*Oreochromis niloticus*). *Physiol. Zool.* **66,** 881–901.

Wright, P. A., Perry, S. F., Randall, D. J., Wood, C. M., and Bergman, H. (1990). The effects of reducing water pH and total CO_2 on a teleost fish adapted to an extremely alkaline environment. *J. Exp. Biol.* **151,** 361–369.

Wright, P. A., Iwama, G. K., and Wood, C. M. (1993). Ammonia and urea excretion in Lahontan cutthroat trout (*Oncorhynchus clarki henshawi*) adapted to highly alkaline Pyramid Lake (pH 9.4). *J. Exp. Biol.* **175,** 153–172.

Wright, P. A., Felskie, A., and Anderson, P. M. (1995). Induction of ornithine-urea cycle enzymes and nitrogen metabolism and excretion in rainbow trout (*Oncorhynchus mykiss*) during early life stages. *J. Exp. Biol.* **198,** 127–135.

Zalkin, H. (1993). The amidotransferases. *Adv. Enzymol. Relat. Areas Mol. Biol.* **66,** 203–309.

4

SECRETORY ELECTROLYTE TRANSPORT IN RENAL PROXIMAL TUBULES OF FISH

KLAUS W. BEYENBACH

I. INTRODUCTION

Nearly half of the 45,000 vertebrate species existing today are fish (Beyenbach and Baustian, 1989). In more than 300 million years of evolution, the adaptive radiation of fish has led to the present spectacular diversity of structure and function. In terms of structure, fish can be as large as the whale shark (12.1 m) or as small as gobies (0.008 m). They may be stream-lined cruisers built for efficient movement through water, or they may be denizens of the ocean floors, nearly sessile and shaped for deception and camouflage. The extent of their radiation in virtually every aquatic habitat documents a particularly successful interplay between physiology and environment. Actively swimming fish have effective organs for locomotion, circulation, and respiration. Parasitic fish funnel metabolic energy primarily into reproduction. Migrating fish command powerful mechanisms of salt and water balance to support life in diverse aquatic environments, and no other class of vertebrates commands as many and seemingly bizarre strategies of reproduction as fish.

Evidence of structural and functional diversity can also be found in organs of extracellular fluid homeostasis. For example, fish are the only vertebrates with kidneys able to produce urine by glomerular and aglomerular mechanisms. Thus, renal function in fish spans the spectrum from glomerular filtration to tubular secretion. The renal tubules of fish may consist of only one or two tubular segments, or they may include the full complement of the vertebrate nephron (Hentschel and Elger, 1989).

Despite their evolutionary success, growing economic importance, and impressive biological diversity, fish do not occupy a deserved position in the minds of men. Plato is said to have considered them as "senseless beings . . . which have received the most remote habitations as a punishment for their extreme ignorance" (Hickman *et al.,* 1984). Einstein is said to have attributed "the soul of a herring" to one of his scientific contemporaries (Guterl, 1994). Assertions creating doubt and suspicion are considered "fishy" in the English language.

It is not uncommon for popular bias to extend to scientific bias. Accordingly, the experimental models offered by fish have been underutilized. The subject of this review, secretion in kidneys, illustrates the point. To this date, the mechanisms by which the vertebrate kidney contributes to extracellular fluid homeostasis without glomerular filtration are unknown. Yet the mechanisms of aglomerular urine formation are clearly expressed in some 30 species of aglomerular fish. In the present review I will focus on possible epithelial transport mechanisms of aglomerular urine formation as we have observed them in, surprisingly, glomerular kidneys.

II. FIRST OBSERVATION OF FLUID SECRETION IN THE KIDNEY OF THE FLOUNDER

The development of *in vitro* microperfusion of renal tubules by Burg *et al.* in the 1960s has allowed renal functions in fish to be examined at the level of isolated tubules (Burg *et al.,* 1966). However, before we could apply *in vitro* microperfusion methods to renal tubules of fish, we first had to learn how to recognize specific tubule segments in small ($5 \times 5 \times 5$ mm), teased pieces of freshly isolated kidney. Identification of tubule segments is easy in kidneys of mammals and sharks, where the renal tubules are positioned along clearly recognizable anatomical boundaries that guide the dissection. In contrast, convenient reference points are lacking in the kidneys of teleost fish, where the organization of renal tubules appears as random as pasta in a dish of spaghetti. The lack of structural markers forced us to use functional markers such as the presence of magnesium (Mg) or sulfur (S) in the tubular fluid. Renal proximal tubules of marine fish are known to secrete Mg and S (Hickman *et al.,* 1984; Natochin and Gusev, 1970; Renfro, 1989). Hence high Mg and S concentrations in the luminal fluid, and the continued secretion of these divalent ions *in vitro,* can serve as a functional indicator of proximal tubules. To collect luminal fluid from freshly dissected renal tubules of the winter flounder *Pleuronectes americanus* (formerly *Pseudopleuronectes americanus*), we first had to expel it from the lumen. We attempted to do this by perfusing the tubule lumen with light paraffin oil (Beyenbach, 1982). As soon as perfusion of the tubule lumen with oil was stopped, the oil column in the lumen broke up as epithelial cells secreted an aqueous fluid into the lumen (Fig. 1). With time, the volume of aqueous fluid grew in the tubule lumen displacing the oil after approximately 75 minutes. While perfusion with oil had expelled the tubular fluid present *in vivo* at the time

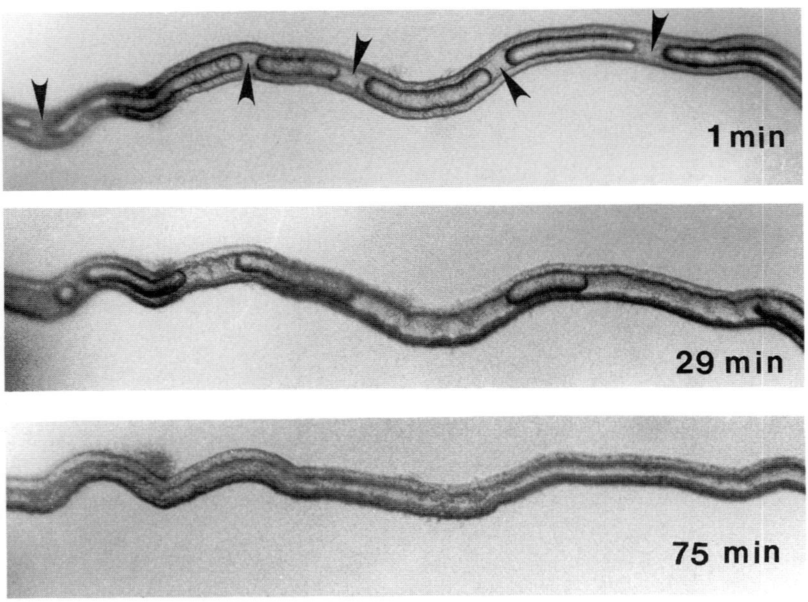

Fig. 1. First observation of fluid secretion in an isolated proximal tubule of the glomerular flounder *Pleuronectes americanus*. The tubule was perfused with light mineral oil from left to right at time zero. After 1 minute, transepithelial secretion of an aqueous fluid had split the luminal oil column at several points (arrows) along the perfused segment. After 75 minutes, epithelial secretion of fluid has caused oil to flow toward and out of the open end of the tubule on the right. [Reprinted with permission from *Nature* (Beyenbach, 1982). Copyright (1982) Macmillan Magazines Limited.]

of tubule isolation, it was epithelial secretion of new aqueous fluid that now expelled the oil from the lumen (Fig. 1).

The kidney of the winter flounder is glomerular. In glomerular kidneys the proximal tubule is expected to reabsorb fluid. Therefore we were surprised when renal tubules of the glomerular winter flounder showed signs of secretion *in vitro* (Fig. 1). Moreover, chemical analysis of secreted fluid revealed elevated Mg and S concentrations (Table I), identifying the perfused segments as proximal tubules (Beyenbach, 1982). Since our first observation of fluid secretion in renal proximal tubules of the winter flounder we have now observed fluid secretion in renal proximal tubules of two other glomerular fish, in the dogfish shark (Sawyer and Beyenbach, 1985) and in seawater- and freshwater-adapted killifish (Cliff and Beyenbach, 1992). Thus, fluid secretion is observed in renal proximal tubules of marine and euryhaline fish regardless of phylogenetic position and regardless of the osmotic and ionic loads of their external environment.

III. RATES OF TRANSEPITHELIAL FLUID SECRETION

Transepithelial fluid secretion can be studied in isolated proximal tubules by closing one end of the tubule and allowing secreted fluid to flow from the other end into a collecting pipette (Fig. 2). Timed measurements of secreted volume yield the fluid secretion rate normalized to tubule length (pl/min-mm). The rates of transepithelial volume flow, between 20 and 60 pl/min-mm (or $2\text{–}6 \times 10^{-8}$ ml/min-mm) are the lowest measured to date in any proximal tubule isolated from a vertebrate (Table I). Transepithelial volume flows are at least 5 times greater in proximal tubules of amphibians (Sackin and Boulpaep, 1981) and more than 10 times greater in proximal tubules of reptiles (Dantzler and Bentley, 1978) and mammals (Andreoli *et al.*, 1978). We are able to measure such low rates of fluid transport in fish proximal tubules, because we measure volume directly with the aid of calibrated picoliter (10^{-12} liter) pipettes (Beyenbach and Dantzler, 1990; Cliff and Beyenbach, 1992). If transepithelial volume flow were measured with the usual methods using volume markers such as inulin (Baustian and Beyenbach, 1993), fluid secretion would go undetected for lack of resolution. Lack of resolution of marker methods might be the reason why transepithelial secretion of fluid has not been observed in the study of isolated perfused renal proximal tubules of other vertebrates (Beyenbach *et al.*, 1986; Williams and Schafer, 1990).

Table I

Fluid Secretion across Isolated Renal Proximal Tubules of Glomerular and Aglomerular Fish

			Transepithelial fluid secretion (pl/min-mm)		Osmotic pressure ($mOsm/kg H_2O$)		Composition of secreted fluid (m*M*)				
Species	Kidney	Habitat	Rate	Range	Peri-tubular bath	Secreted fluid	Na	Cl	Mg	S	Reference[a]
Winter flounder	Glomerular	Seawater	36.6 ± 4.2 (53)	7.1–143	290	318 ± 5 (10)	152 ± 7 (22)	155 ± 4 (22)	26 ± 4 (22)	10 ± 1 (22)	1–3
Dogfish shark	Glomerular	Seawater	27.6 ± 3.9 (21)	9.1–86.9	873	905 ± 11 (6)	291 ± 14 (5)	272 ± 16 (5)	6 ± 0.5 (8)	1 ± 0.1 (3)	4,5
Killifish	Glomerular	Seawater	54 ± 6 (28)	14–164	290	n.m.[b]	127 ± 4 (28)	153 ± 2 (28)	28 ± 2 (28)	10 ± 1 (28)	6
	Glomerular	Fresh water	34 ± 5 (6)	7–117	290	n.m.	147 ± 7 (6)	148 ± 5 (6)	15 ± 4 (6)	11 ± 3 (6)	6
Toadfish	Aglomerular	Seawater	28 ± 5 (12)	8–74	300	n.m.	195 ± 4 (12)	171 ± 4 (12)	12 ± 3 (12)	9 ± 3 (12)	7

[a] (1) Beyenbach *et al.* (1986); (2) Cliff *et al.* (1986); (3) Beyenbach (1982); (4) Sawyer and Beyenbach (1985); (5) Beyenbach (1986); (6) Cliff and Beyenbach (1992); (7) Baustian and Beyenbach, unpublished observations, Cornell University.

[b] n.m., not measured.

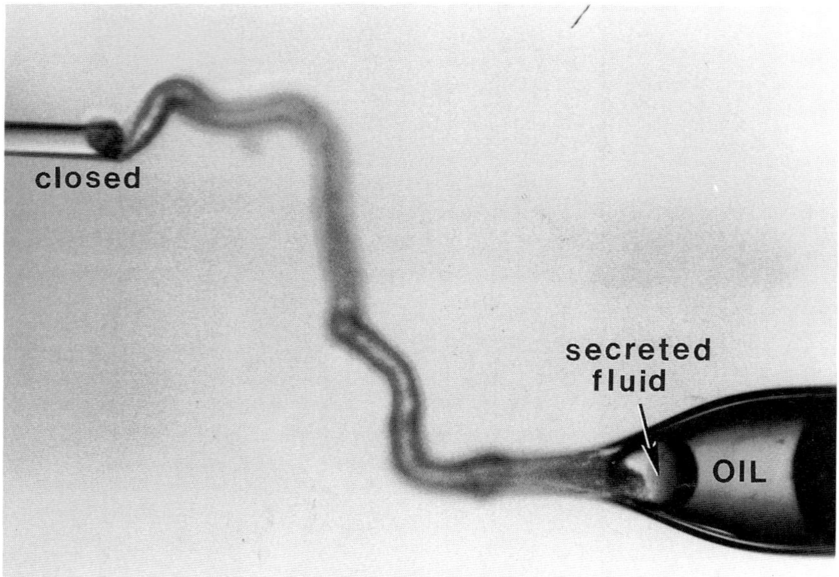

Fig. 2. Preparation of renal proximal tubules for the study of transepithelial fluid secretion. One end of the isolated tubule is crimped closed through a hairpin turn in a holding pipette (upper left). The other end opens into a collecting pipette, where secreted fluid accumulates under oil. Droplet volume is measured with calibrated picoliter pipettes. (From Beyenbach *et al.*, 1986, with permission from the American Physiological Society.)

IV. COMPOSITION OF SECRETED FLUID

In those tubules where transepithelial osmotic pressure differences were measured, secreted fluid is hyperosmotic to the peritubular medium (Table I). Secreted fluid is hyperosmotic to the peritubular bath in proximal tubules of the winter flounder and the dogfish shark by 28 and 32 mOsm, respectively (Table I). Lack of resolution of present-day osmotic pressure measurments leaves some uncertainty in the exact value of luminal hyperosmolarity. Even if methods with high resolution were available, the "effective" osmotic pressure existing across the tubule wall would still be uncertain. The "effective" osmotic pressure is the true osmotic pressure existing across a biological barrier, to be distinguished from the "absolute" osmotic pressure (Schafer and Andreoli, 1981). Measurements based on the colligative properties of solutions (boiling point, freezing point, vapor pressure) yield the "absolute" osmotic pressure, which is referenced to solute-free water (distilled water) across an ideally semipermeable barrier, permeable to water and impermeable to solutes (solute reflection coefficients, $r = 1$). However, biological membranes and epithelial tissues are not ideal semipermeable structures. They offer permeability to most solutes. Hence, solute reflection coefficients are usually less than unity ($r<1$), across biological barriers. Thus, the "effective" osmotic pressure is less than the "absolute" osmotic pressure:

$$\pi_{(\text{eff})} = r \cdot \pi_{(\text{abs})}. \tag{1}$$

Accordingly, measurements of the "absolute" osmotic pressures of secreted fluid and peritubular Ringer overestimate the "effective" osmotic pressures and therefore osmotic pressure differences existing across the tubule wall. In cases where membranes or epithelia separate solutions with solutes of different reflection coefficients, measures of "absolute" osmotic pressures may even give the wrong conclusion about the direction of osmotic water flow! Whatever the "effective" transepithelial osmotic pressure difference is in secretory proximal tubules, it must be great enough to account for transepithelial secretion of water and for the axial, downstream flow of tubular fluid (Figs. 1 and 2). At a minimum, the osmotic pressure in the tubule lumen must be sufficiently large to yield the hydrostatic pressure to drive flow downstream toward the open end of the tubule (Fig. 2). Assuming the conditions for laminar flow, a lumen radius of 10 μm (Beyenbach *et al.*, 1986), and a viscosity of secreted fluid of 1.002×10^{-2} dynes-sec/cm^2, an osmotic pressure of approximately 60 nOsm is required to drive flow of 36 pl/min through a 1-mm length of lumen (Poiseuille equation). Thus, the osmotic pressure needed to drive flow downstream in a tubule segment of 1-mm length is low, between 10^{-8} and 10^{-7} Osm, far below the limits of resolution of present osmotic pressure determinations. The osmotic pressure needed to drive flow down an entire nephron is, of course, much greater. For one reason, the resistance to flow increases with length of the nephron. More importantly, the lumen diameter of tubules may change depending on the tone of myoepithelial cells present in flounder proximal tubules (Fig. 3; Beyenbach *et al.*, 1986).

In view of the small volumes of fluid secreted by single segments of proximal tubule, very sensitive analytical methods must be employed for the analysis of its composition. We use two forms of electron microprobe analysis: energy-dispersive X-ray spectroscopy (EDS) and wavelength-dispersive X-ray spectroscopy (WDS; Williams and Beyenbach, 1983; Beyenbach and Dantzler, 1990). EDS identifies all elements in secreted

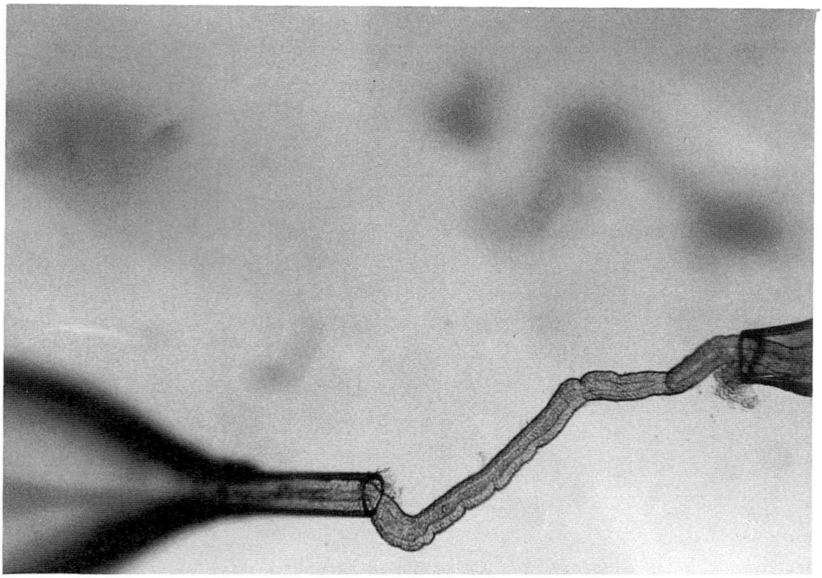

Fig. 3. Contractions of renal proximal tubules observed in the flounder *Pleuronectes americanus*. The contractions cause constrictions that reduce lumen diameter. (From Beyenbach *et al.*, 1986, with permission from the American Physiological Society.)

fluid with atomic numbers greater than 10 (neon). For example, EDS revealed Na, Cl, Mg, and S in fluid secreted by renal proximal tubules of killifish (*Fundulus heteroclitus*) adapted to either seawater or fresh water (Fig. 4). WDS offers direct quantitation of secreted Na, Cl, Mg, and S by counting the X rays emitted by each element in proportion to its concentration. Thus, WDS revealed Na and Cl as the dominant osmotic solutes in fluid secreted by proximal tubules of the winter flounder, dogfish shark, and killifish, all of which have glomerular kidneys (Table I). Finding Na and Cl in the tubule lumen at concentrations approaching or exceeding those in the peritubular Ringer bath was as surprising as the observation of fluid secretion itself. According to dogma in renal physiology, proximal tubules of glomerular nephrons are supposed to reabsorb Na and Cl, not secrete it. That Na and Cl are the principal electrolytes in fluid secreted by proximal tubules of three unrelated glomerular fish (flounder, shark, and killifish) indicates a widespread, perhaps fundamental transport phenomenon in proximal tubules of fish and, conceivably, other vertebrates.

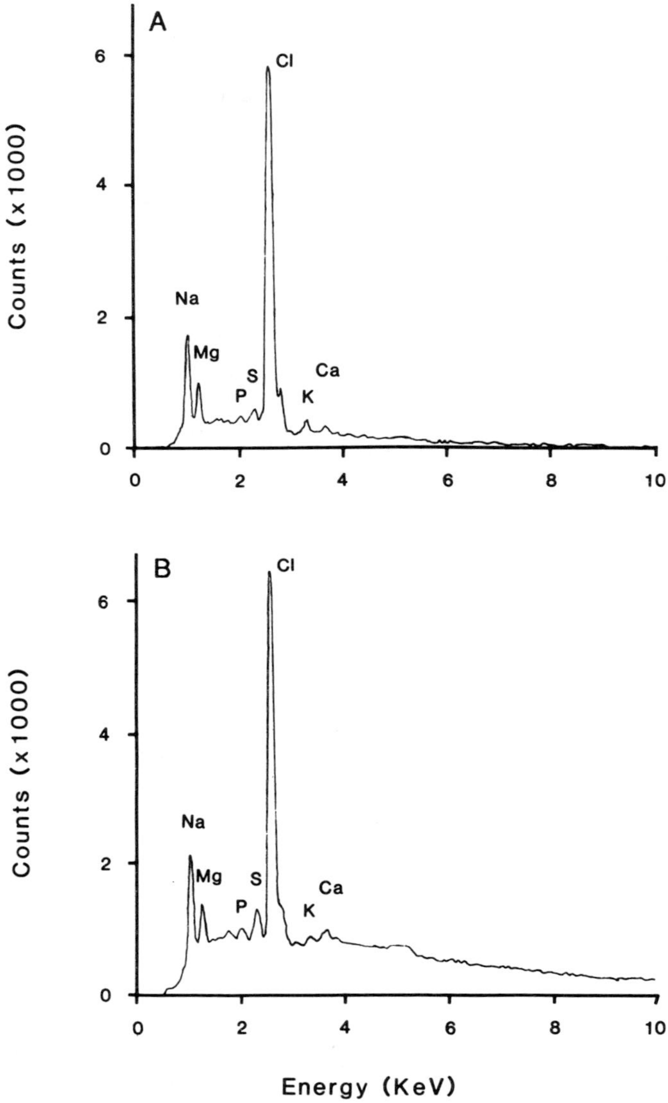

Fig. 4. Energy-dispersive spectra of fluid secreted by renal proximal tubules of killifish, *Fundulus heteroclitus,* adapted to seawater (A) or fresh water (B). (From Cliff and Beyenbach, 1992, with permission from the American Physiological Society.)

V. SECONDARY ACTIVE TRANSPORT OF CHLORIDE

It is axiomatic in biology that the movement of water across membranes and epithelia is secondary to the movement of osmotically active solutes. Accordingly, our investigations of the mechanisms of fluid secretion in fish proximal tubules focused on elucidating the mechanisms by which osmotic solutes are secreted into the tubule lumen. Secretion of Na and Cl received our first attention. Our studies of fluid secretion in proximal tubules of the dogfish shark *Squalus acanthias* most clearly illustrate secondary active transport of Cl as one mechanism for driving fluid secretion (Beyenbach and Frömter, 1985; Sawyer and Beyenbach, 1985). Inhibition of transepithelial fluid secretion by furosemide added to the peritubular bath suggests the presence of Na–K–2Cl cotransport in the basolateral membrane of proximal epithelial cells, bringing Cl into the cell (Fig. 5). Entry of Cl into the cell is called "secondary active transport" because

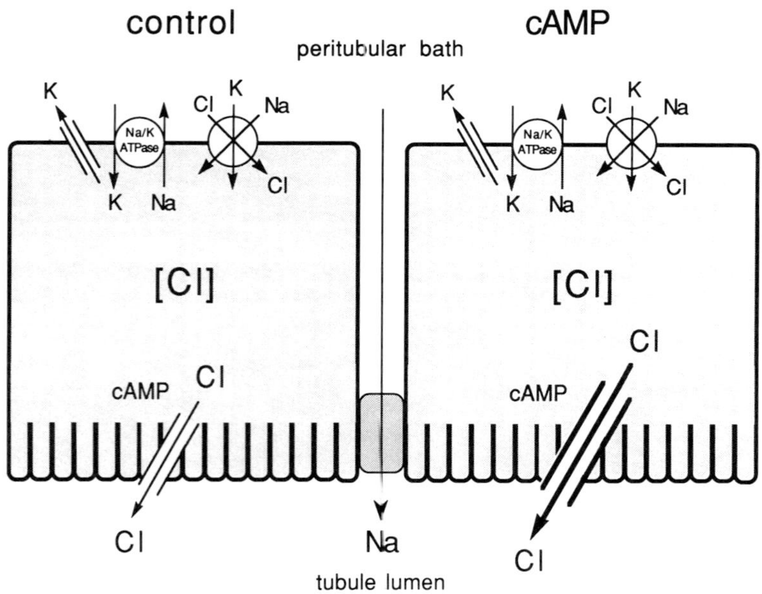

Fig. 5. Mechanism of secondary active transport of Cl across renal proximal tubules of the dogfish shark *Squalus acanthias*. Entry of Cl into the epithelial cell of the proximal tubule is driven by the transmembrane gradient for Na. Exit of Cl into the tubule lumen is mediated via cAMP-regulated Cl channels present in the apical membrane. (From Sawyer and Beyenbach, 1985, with permission from the American Physiological Society.)

it is dependent on the transmembrane Na concentration difference generated by primary active Na transport via the Na/K-ATPase (Fig. 5). The energy present in the Na gradient is thought to lift intracellular Cl to an electrochemical potential above that in the tubule lumen. Since the apical membrane of shark proximal epithelial cells is permeable to Cl, Cl electrodiffuses out of the cell into the tubule lumen (Fig. 5). Cyclic AMP, a stimulator of transepithelial fluid secretion in shark proximal tubules (Sawyer and Beyenbach, 1985), increases the Cl conductance of the apical membrane (Beyenbach and Frömter, 1985) as one mechanism of regulating transepithelial Cl secretion (Fig. 5). Hence, Cl channels at the apical membrane and Na–K–2Cl cotransport at the basolateral membrane provide the mechanisms for transcellular secretion of Cl into the tubule lumen. Electrical coupling of the transcellular pathway for Cl to the paracellular pathway for Na is responsible for the secretion of Na with each secreted Cl. Accordingly, the solutions on both sides of the epithelium remain electrically neutral in spite of net transepithelial transport of cations and anions. Nearly equimolar concentrations of Na and Cl in secreted fluid reflect the close stoichiometric relationship between transepithelial Na and Cl secretion, especially when rates of transepithelial Mg and S secretion are low (Table I).

Though cAMP-regulated Cl secretion provides one mechanism for driving transepithelial secretion of NaCl and water, this transport mechanism is not present in all proximal tubules of fish kidneys. Fluid secretion does not respond to cAMP in flounder proximal tubules, but it does respond in proximal tubules of the killifish (Cliff and Beyenbach, 1988). Although the significance of this species difference is unknown, it does indicate additional mechanisms of transepithelial fluid secretion.

VI. ACTIVE SECRETION OF OSMOLYTES

Evaluation of transepithelial electrochemical potentials of divalent ions makes it clear that renal proximal tubules of fish possess active transport mechanisms other than secondary active transport of Cl (Table II). Mg is secreted into the tubule lumen against electrochemical potentials in every proximal tubule observed to secrete fluid *in vitro* (Table II). The mechanism of epithelial Mg secretion is largely unknown (Beyenbach *et al.*, 1993). Since luminal Mg concentrations are high compared to peritubular concentrations, the epithelium must have a low passive permeability to Mg, that is, the Mg reflection coefficient must be close to 1. Thus, Mg secreted into the tubule lumen by active transport mechanism(s) is

Table II
Transepithelial Electrochemical Potentials across Isolated Renal Proximal Tubules of Glomerular Fish

	Concentration in peritubular Ringer solution (m*M*)				Concentration in secreted fluid (m*M*)				Transepithelial voltage resistance		Transepithelial electrochemical potential (mV)[a]				
Species	Na	Cl	Mg	S	Na	Cl	Mg	S	(mV)	(Ω cm^2)	Na	Cl	Mg	S	Reference[b]
Pseudopleuronectes americanus (winter flounder)	168.6	152	1	1	165.2	162.9	25.8	9.9	−1.9	25.6	−2.4	+3.7	+40.4	+31.8	1
Squalus acanthias (dogfish shark)	280	285	3	0.5	291	272	6.2	1.1	1.2	36.3	+2.2	−2.4	+17.6	+9.1	2,3
Fundulus heteroclitus (killifish)	152.1	152	1	1	172	153	28	10	−2.0	24.2	+1.2	+2.2	+41.4	+32.0	4

[a] Positive values indicate transport into the lumen against the electrochemical potential; negative values indicate transport down the electrochemical potential.

[b] (1) Beyenbach *et al.* (1986); (2) Beyenbach and Frömter (1985); (3) Sawyer and Beyenbach (1985); and (4) Cliff and Beyenbach (1992).

"trapped" in the lumen where it exerts an osmotic pressure drawing water from the peritubular medium. Similarly, active secretion of any other osmolyte, such as Cl (see foregoing), SO_4 (Renfro, 1989), taurine (King *et al.*, 1982), creatine (Brull and Nizet, 1953), organic cations and anions (Pritchard and Miller, 1993), or foreign solutes such as phenol red (Kinter, 1966) or *p*-aminohippuric acid (Eveloff *et al.*, 1979), generates osmotic pressure in the tubule lumen as a driving force for transepithelial secretion of water.

VII. PASSIVE SECRETION OF OSMOLYTES

Renal proximal tubules of fish share with other renal proximal tubules a high transepithelial water permeability (Schafer and Andreoli, 1981). Peritubular and luminal fluids have an absolute osmotic pressure of approximately 320 mOsm. Since renal proximal tubules of marine fish are thought to function primarily as secretors of divalent ions (Hickman and Trump, 1969), a 320-mOsm tubular fluid would be expected to contain at least 160 m*M* $MgSO_4$. However, in experiments we measure on average concentrations less than 30 m*M* for Mg and S (Table I). Moreover, even in Mg- and S-secreting proximal tubules, Na and Cl are the dominant osmotic and ionic solutes in secreted fluid. There are at least two reasons for the presence of NaCl in secreted fluid. First, the active transport mechanisms for transepithelial Cl, Mg, and S secretion may all be present in the same segment of proximal tubule (Beyenbach, 1986). Second, and perhaps more importantly, the transepithelial shunt, presumably the paracellular pathway, is highly permeable to Na and Cl in proximal tubules of fish (Beyenbach *et al.*, 1986; Cliff and Beyenbach, 1992). It allows the transepithelial passage of Na and Cl between epithelial cells. Thus, should active transepithelial transport of $MgSO_4$ deliver a $MgSO_4$-rich solution into the tubule lumen, Na and Cl will diffuse into this fluid from the peritubular medium (or renal interstitium). Relative rates of transepithelial active Mg and SO_4 transport and passive Na and Cl transport will determine the concentrations of Na, Mg, Cl, and S in secreted fluid. An active transport rate of $MgSO_4$ higher than the rate of NaCl diffusion will generate high $MgSO_4$ and low NaCl concentrations in secreted fluid. An active transport rate of $MgSO_4$ lower than the rate of NaCl diffusion will generate low $MgSO_4$ and high NaCl concentrations in secreted fluid. Thus, relative rates of active and passive transport determine the ionic composition of a fluid limited to an osmotic ceiling of approximately 320 mOsm.

VIII. ROLE OF DONNAN EFFECT IN TRANSEPITHELIAL FLUID SECRETION

The secretion of Mg into the lumen of proximal tubules with high Na and Cl shunt permeabilities has caused me to consider Donnan effects to generate a lumen oncotic pressure as the driving force of transepithelial fluid secretion. To illustrate the point, I will first review an example of perfect Donnan equilibrium. I will then extend the example to renal tubules that "try" to reach Donnan equilibrium (Fig. 6).

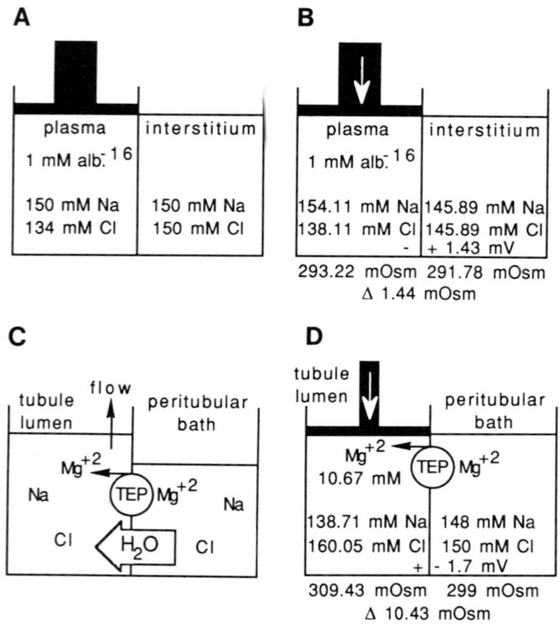

Fig. 6. Systems at Donnan equilibrium (B, D) or below Donnan equilibrium (A, C). (A) initial condition for the development of Donnan equilibrium across the capillary endothelium permeable to Na, Cl, and water, but impermeable to albumin (alb.). (B) Donnan equilibrium. The voltage existing across the capillary endothelium (1.43 mV) is the Nernst equilibrium voltage for both Na and Cl. (C) Mg as oncotic agent in renal proximal tubules under free flow conditions. Mg is secreted into the tubule lumen by active transepithelial (TEP) transport mechanism(s). Since Mg carries charge and cannot diffuse back into the peritubular bath, Mg is an oncotic agent like albumin in example (A). However, Donnan equilibrium is not reached because the oncotic pressure in secreted fluid is unopposed. As a result, secretory flow into the lumen and the downstream flow of fluid occurs. (D) Estimate of luminal oncotic pressure (10.43 mOsm) if the Mg-secreting renal tubule were allowed to go to Donnan equilibrium. At Donnan equilibrium, the expected transepithelial voltage, the Nernst potential for both Na and Cl, is 1.7 mV lumen-positive.

The example illustrated in Fig. 6A assumes ideal conditions for Donnan equilibrium: two fluid compartments, plasma and interstitium, are separated by a capillary endothelium that is impermeable to albumin ($r = 1$) and permeable to Na,Cl, and water. Osmotic water flow between plasma and interstitial fluid is prevented by allowing no volume change in the plasma (piston in Figs. 6A and 6B). The presence of approximately 1 m*M* albumin in the plasma, carrying a total negative charge of -16, causes Na and Cl to distribute themselves across the capillary endothelium in accordance with their lowest level of energy (electrochemical equilibrium). This redistribution of Na and Cl proceeds across the capillary endothelium until the electrochemical potential difference for Cl across the capillary endothelium is equal to the electrochemical potential difference for Na (Fig. 6B). The equivalence of the two electrochemical potentials defines Donnan equilibrium, where the voltage across the capillary endothelium (1.43 mV) is the Nernst equilibrium potential for both Na and Cl (Fig. 6B). As a consequence of Donnan equilibrium, the plasma is now hyperosmotic to the interstitial fluid by 1.44 mOsm, the so-called plasma oncotic pressure (Fig. 6B). An oncotic pressure of 1.44 mOsm is equivalent to a pressure of 27 mmHg which draws fluid from the interstitium into the plasma. However, a fluid transfer does not take place, because the plasma is under hydrostatic pressure. This hydrostatic pressure is generated by the heart to drive the downstream flow of blood. It also causes filtration of fluid from the plasma to the interstitium. Since on average the plasma hydrostatic pressure (27 mmHg) equals the plasma oncotic pressure, there is no net driving force and hence no net transfer of fluid between capillary and interstitium, and Donnan equilibrium prevails.

Mg in the tubule lumen is analogous to albumin in the plasma. Both molecules are charged, and neither is free to leave the fluid compartment into which it was placed: albumin is placed into the plasma by the liver, and Mg is placed into the tubule lumen by an active transport mechanism of epithelial cells (Fig. 6). The presence of charged Mg in the tubule lumen is expected to cause the transepithelial redistribution of Na and Cl in accordance with Donnan equilibrium (Figs. 6C and 6D). As a result, the luminal fluid acquires an oncotic pressure that draws water from the peritubular bath (or renal interstitium), i.e., water is secreted into the tubule lumen (Fig. 6C). Since the tubule lumen is open to the atmosphere—in contrast to the capillary lumen—the oncotic pressure that develops in the lumen is unopposed and causes flow towards the open end of the tubule. Thus, a critical condition for Donnan equilibrium, no radial (lateral) volume flow, is not satisfied in renal tubules, and the tubule will not reach Donnan equilibrium (Fig. 6C). However, it is the attempt of the tubule to go to Donnan equilibrium, which accounts for a lumen oncotic pressure, which in turn drives secretory (radial) and downstream (axial) flow. Obviously, this oncotic pressure is less under free flow conditions than under conditions of no flow (Donnan equilibrium, Figs. 6C and 6D). How close lumen oncotic pressure comes to the maximum oncotic pressure expected at Donnan equilibrium depends on the rate of axial flow out of the lumen. Maximum oncotic pressure would be reached when axial flow down the tubule lumen is stopped, yielding the conditions for Donnan equilibrium. This maximum oncotic pressure can be calculated for physiological Na, Cl, and Mg concentrations (Fig. 6D), yielding 10.4 mOsm which is equivalent to an oncotic pressure of 198 mmHg. Thus, under the no flow conditions of Donnan equilibrium, lumen oncotic pressure is approximately 0.25 atmospheres. However, under free flow conditions, when axial flow down the tubule lumen occurs, the oncotic pressure is obviously less, but probably still several orders of magnitude greater than the small hydrostatic pressure (10^{-7} Osm, equivalent to 1.9×10^{-3} mmHg) needed to drive flow of luminal fluid downstream (vide supra).

These considerations reveal the oncotic pressure of secreted fluid as an important driving force of transepithelial fluid secretion and downstream tubular flow. Furthermore, considerations of oncotic pressure due to nondiffusible, actively secreted Mg lead to the prediction that the higher the concentration of Mg in the tubule lumen, the greater the contribution of oncotic pressure to the transepithelial secretion of water. Indeed, high rates of transepithelial fluid secretion are observed in tubules with high concentrations of Mg in secreted fluid (Fig. 7). Finally, considerations of Donnan phenomena in fish renal proximal tubules may account for the not infrequently observed hyperosmotic urine in teleost fish (Table I; Stanley and Fleming, 1964; Elger *et al.*, 1987).

IX. FLUID SECRETION IN AGLOMERULAR PROXIMAL TUBULES

The mechanisms of aglomerular urine formation in fish that lack glomeruli have attracted my curiosity since I first studied renal Mg excretion in rainbow trout in the laboratory of Professor L. B. Kirschner (Beyenbach and Kirschner, 1975). Nearly 20 years later we now have the opportunity to study this intriguing problem in the toadfish *Opsanus tau*. The kidney of the aglomerular fish is thought to consist primarily of proximal tubules, in particular proximal segment II, which is the Mg- and S-secreting segment (Hickman and Trump, 1969; Hentschel and Elger, 1989). Our studies in aglomerular toadfish suggest that their renal tubules are not as homoge-

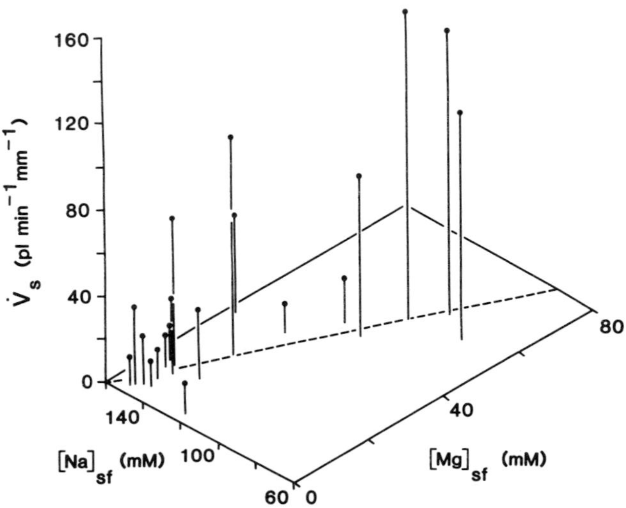

Fig. 7. Relationships between Na and Mg concentrations in secreted fluid and rates of transepithelial fluid secretion. As Mg concentration in secreted fluid increases, Na concentration decreases and rate of transepithelial fluid secretion increases. V_s is the rate of transepithelial fluid secretion. (From Cliff *et al.*, 1986, with permission from the American Physiological Society.)

neous as they are thought to be. Not all proximal tubules that we dissect for study oblige us with secretion of fluid *in vitro*. Only a subpopulation of renal tubules secretes fluid spontaneously when they are prepared for study as shown in Fig. 2. Fluid secretion in some but not all proximal tubules is reminiscent of filtration in some but not all glomeruli of fish kidneys (Brown *et al.*, 1993), suggesting tubular intermittency analogous to glomerular intermittency. Moreover, proximal tubules of toadfish that do secrete fluid do not respond to cAMP, like proximal tubules of the flounder and unlike proximal tubules of dogfish shark (Sawyer and Beyenbach, 1985) and killifish (Cliff and Beyenbach, 1988).

Those proximal tubules of aglomerular toadfish that did secrete fluid *in vitro* did so at rates similar to those of their glomerular counterparts (Table I). The elemental analysis of secreted fluid by EDS indicated Na, Cl, Mg, and S as the principal osmotic solutes (Table I). The quantitative analysis by WDS spectroscopy revealed the familiar composition: Na and Cl concentrations much higher than those of Mg and S (Table I). Thus, aglomerular proximal tubules secrete a fluid strikingly similar to that secreted by glomerular proximal tubules. The similarities would be expected if oncotic pressure, due to Mg or other oncotic agents that both glomerular and aglomerular tubules secrete, is an important driving force for transepithelial fluid secretion (Fig. 6).

X. REABSORPTIVE AND SECRETORY VOLUME FLOWS IN FISH PROXIMAL TUBULES

Transepiethelial volume flow is the difference between absorptive and secretory flow, even in tubules prepared for study of secretion (Fig. 2). When secreted fluid exits the tubule lumen, it has been operated upon by both secretory and reabsorptive transport mechanisms. What is measured are net effects. In a filtering nephron, large quantities of filtered Na, Cl, HCO_3, glucose, amino acids, small vitamins, and other solutes, which the animal can ill afford to lose, are presented to the apical membranes of proximal epithelial cells for absorption. Faced with the task of reabsorbing large quantities of solute, it is obvious why proximal tubules perfused either *in vivo* with ultrafiltrate or *in vitro* with Ringer solution transport fluid in a net reabsorptive direction. Under perfusion of the lumen the tubular transport load presented from the luminal side is much greater than that presented from the peritubular side. Hence, net reabsorption is the expected and also the observed result. Take away glomerular filtration, either through glomerular intermittency or through the evolutionary dismissal of glomeruli, and reabsorptive loads go to zero. Under these conditions, secretory loads and secretory transport mechanisms become unmasked. It is no surprise, therefore, that nonperfused proximal tubules of glomerular kidneys can be observed to secrete fluid when the opportunity for reabsorptive transport is missing (Figs. 1 and 2).

The secretory rates measured in isolated proximal tubules so far appear to be rather small (Tables I and II). They are small because the tubules are bathed in a Ringer solution, the composition of which aims at providing a life-sustaining medium. The tubules are not bathed in a fluid containing substances that need to be eliminated from the body. Accordingly, the small rates of transepithelial secretion that we have measured to date represent secretion rates in proximal tubules unchallenged to secrete. However, renal proximal tubules of fish possess active transport mechanisms for the secretion of organic acids, organic alkali, creatinine, creatine, taurine, and other solutes (vide supra). Without challenging these secretory transport mechanisms, the upper limit of transepithelial fluid secretion rates in glomerular and aglomerular renal proximal tubules remains unknown.

Regarding transport mechanisms driving secretory volume flow in renal proximal tubules of fish, we first discovered secondary active transport of Cl in shark and killifish proximal tubules as one mechanism for driving transepithelial secretory osmosis (Sawyer and Beyenbach, 1985; Cliff and Beyenbach, 1988). Second, active transepithelial transport of Mg, S, taurine, and other inorganic and organic solutes increases the osmotic pressure of secreted fluid in proportion to their transport rates, thereby increasing the osmotic driving force of transepithelial fluid secretion. Third, high Na and Cl permeabilities of epithelial shunts add additional osmotic solutes (Na and Cl) to secreted fluid, increasing the osmotic pressure in support of transepithelial fluid secretion. Finally, active secretion of oncotic agents (Mg, S, or other charged solutes with high reflection coefficients) into the tubule lumen, from which they cannot escape, can be expected to generate substantial oncotic pressures driving transepithelial osmotic water secretion. In view of these considerations, net transepithelial fluid secretion as we have observed it in glomerular proximal tubules of the flounder, shark, and killifish should not be surprising at all. On the contrary, net transepithelial fluid secretion should be expected under the special conditions of our *in vitro* study, where, in the absence of either glomerular filtration or luminal perfusion, the tubule lumen does not receive solutes that can be reabsorbed. Furthermore, the similarities of fluid secretion observed in glomerular and aglomerular proximal tubules suggest that the functional elements of aglomerular urine formation are not uniquely different from those of glomerular urine formation at the level of the proximal tubule.

ACKNOWLEDGMENTS

Over the years, work presented in this chapter has been supported by grants from the National Institutes of Health, the National Science Foundation, and the Humboldt Foundation of Germany. The experimental work goes to the credit of Dr. David Petzel, flounder proximal tubules; Dr. Eberhard Frömter and Dr. Douglas Sawyer, shark proximal tubules; Dr. William H. Cliff, killifish proximal tubules; and Mark Baustian, toadfish proximal tubules. Special thanks go to Mark Baustian for insightful discussions and to Carolina Freire and Steven Wang for the editorial review of this writing.

REFERENCES

Andreoli, T. E., Schafer, J. A., and Troutman, S. L. (1978). Perfusion rate-dependence of transepithelial osmosis in isolated proximal convoluted tubules: Estimation of the hydraulic conductance. *Kidney Int.* **14,** 263–269.

Baustian, M. D., and Beyenbach, K. W. (1993). Isolated renal proximal tubules: Studies of transepithelial secretory mechanisms in teleost and elasmobranch fish. *In* "New Insights in Vertebrate Kidney Function" (J. A. Brown, R. J. Balment, and J. C. Rankin, eds.), pp. 45–63. Cambridge Univ. Press, Cambridge, UK.

Beyenbach, K. W. (1982). Direct demonstration of fluid secretion by glomerular renal tubules in a marine teleost. *Nature (London)* **299,** 54–56.

Beyenbach, K. W. (1986). Secretory NaCl and volume flow in renal tubules. *Am. J. Physiol.* **250,** R753–R763.

Beyenbach, K. W., and Baustian, M. D. (1989). Comparative physiology of the proximal tubule. *In* "Comparative Physiology, Structure and Function of the Kidney" (R. K. H. Kinne, E. Kinne-Saffran, and K. W. Beyenbach, eds.), Vol. 1, pp. 104–142. Karger, Basel.

Beyenbach, K. W., and Dantzler, W. H. (1990). Comparative kidney tubule sources, isolation, perfusion, function. *In* "Methods in Enzymology" (S. Fleischer and B. Fleischer eds.), Vol. 191, pp. 167–226. Academic Press, San Diego, CA.

Beyenbach, K. W., and Frömter, E. (1985). Electrophysiological evidence for Cl secretion in shark renal proximal tubules. *Am. J. Physiol.* **248,** F282–F295.

Beyenbach, K. W., and Kirschner, L. B. (1975). Kidney and urinary bladder functions of the rainbow trout in Mg and Na excretion. *Am. J. Physiol.* **229,** 389–393.

Beyenbach, K. W., Petzel, D. H., and Cliff, W. H. (1986). Renal proximal tubule of flounder. I. Physiological properties. *Am. J. Physiol.* **250,** R608–R615.

Beyenbach, K. W., Freire, C. A., Kinne, R. H. K., and Kinne-Saffran, E. (1993). Epithelial transport of magnesium in the kidney of fish. *Miner. Electrolyte Metab.* **19,** 241–249.

Brown, J. A., Rankin, J. C., and Yokota, S. D. (1993). Glomerular hemodynamics and filtration in single nephrons of non-mammalian vertebrates. *In* "New Insights in Vertebrate Kidney Function" (J. A. Brown, R. J. Balment, and J. C. Rankin, eds.) pp. 1–44. Cambridge Univ. Press, Cambridge, UK.

Brull, N., and Nizet, E. (1953). Blood and urine constituents of *Lophius piscatorius*. *J. Mar. Biol. Assoc. U.K.* **32,** 321–328.

Burg, M. J., Grantham, J., Abramow, M., and Orloff, J. (1966). Preparation and study of fragments of single rabbit nephrons. *Am. J. Physiol.* **217,** 1053–1056.

Cliff, W. H., and Beyenbach, K. W. (1988). Secretory renal proximal tubules in seawater- and freshwater-adapted killifish. *Am. J. Physiol.* **262,** F108–F116.

Cliff, W. H., and Beyenbach, K. W. (1992). Fluid secretion in glomerular renal proximal tubules of freshwater-adapted fish. *Am. J. Physiol.* **254,** R154–R158.

Cliff, W. H., Sawyer, D. B., and Beyenbach, K. W. (1986). Renal proximal tubule of flounder. II. Transepithelial Mg secretion. *Am. J. Physiol.* **250,** R616–R624.

Dantzler, W. H., and Bentley, S. K. (1978). Fluid absorption with and without sodium in isolated perfused snake proximal tubules. *Am. J. Physiol.* **234,** F68–F78.

Elger, E., Elger, B., Hentschel, H., Stolte, H. (1987). Adaptation of renal function to hypotonic medium in the winter flounder (*Pseudopleuronectes americanus*). *J. Comp. Physiol.* **157,** 21–30.

Eveloff, J., Kinne, R., and Kinter, W. B. (1979). *p*-Aminohippuric acid transport into brush border vesicles isolated from flounder kidney. *Am. J. Physiol.* **237,** F291–F298.

Guterl, F. (1994). Profile: Albert Einstein, keyhole of a genius. *Sci. Am.* **270,** 26–28.

Hentschel, E., and Elger, M. (1989). Morphology of glomerular and aglomerular kidneys. *In* "Comparative Physiology, Structure and Function of the Kidney" (R. K. H. Kinne, E. Kinne-Saffran, and K. W. Beyenbach, eds.), Vol. 1, pp. 1–72, Karger, Basel.

Hickman C. P., Jr., and Trump, B. F. (1969). The kidney. *In* "Fish Physiology" (W. S. Hoar and D. J. Randall, eds.), Vol. 1, pp. 91–239. Academic Press, New York.

Hickman, C. P., Jr., Roberts, L. S., and Hickman, F. M. (1984). "Integrated Principles of Zoology," The Fishes, 7th edition, Chapter 23, p. 518. Times Mirror/Mosby College Publishing, St. Louis, MO.

King, P. A., Goldstein, L., and Beyenbach, K. W. (1982). Taurine transport by isolated flounder renal tubules. *J. Exp. Zool.* **223,** 103–114.

Kinter, W. B. (1966). Chlorphenol red influx and efflux: Microspectrophotometry of flounder kidney tubules. *Am. J. Physiol.* **211,** 1152–1164.

Natochin, Y. V., and Gusev, G. P. (1970). The coupling of magnesium secretion to sodium reabsorption in the kidney of teleost. *Comp. Biochem. Physiol.* **37,** 107–111.

Pritchard, J. B., and Miller, D. S. (1993). Mechanisms mediating renal secretion of organic anions and cations. *Physiol. Rev.* **73,** 765–796.

Renfro, J. L. (1989). Adaptability of marine teleost renal inorganic sulfate excretion: Evidence for glucocorticoid involvement. *Am. J. Physiol.* **257,** R511–R516.

Sackin, H., and Boulpaep, E. L. (1981). Isolated perfused salamander proximal tubule: Methods, electrophysiology, and transport. *Am. J. Physiol.* **241,** F39–F52.

Sawyer, D. B., and Beyenbach, K. W. (1985). Mechanism of fluid secretion in isolated shark renal proximal tubules. *Am. J. Physiol.* **249,** F884–F890.

Schafer, J. A., and Andreoli, T. E. (1981). Driving forces for volume absorption along the proximal nephron. *Jpn. J. Nephrol.* **23,** 965–972.

Stanley, J. G. and Fleming, W. R. (1964). Excretion of hypertonic urine by a teleost. *Science* **144,** 63–64.

Williams, J. C., Jr., and Beyenbach, K. W. (1983). Differential effects of secretagogues on Na and K secretion in Malpighian tubules of *Aedes aegypti* (L.). *J. Comp.Physiol.* **149,** 511–517.

Williams, J. C., Jr., and Schafer, J. A. (1990). Measurement of transmural water flow in isolated perfused tubule segments. *In* "Methods in Enzymology" (S. Fleischer and B. Fleischer eds.), Vol. 191, pp. 232–252. Academic Press, San Diego, CA.

5

FUNCTIONAL MORPHOLOGY OF THE ELASMOBRANCH NEPHRON AND RETENTION OF UREA

ERIC R. LACY AND ENRICO REALE

I. INTRODUCTION

At least three organs are used to maintain osmotic homeostasis in elasmobranch fish: rectal gland, gill, and kidney (Henderson *et al.*, 1981; Shuttleworth, 1988). Although most information has accrued on the function of the rectal gland, the paucity of *in vivo* physiological data leaves open the question of the exact roles for each of these organs. It currently appears that the kidney, which is the subject of the current chapter, is largely responsible for eliminating excess water, aiding acid–base balance,

CELLULAR AND MOLECULAR APPROACHES
TO FISH IONIC REGULATION

regulating some ions, and retaining urea (Henderson *et al.*, 1988; Shuttleworth, 1988).

It has long been known that *marine* elasmobranchs excrete less than 15% of the urea that is filtered across the renal glomerulus and thus it is this efficiency in urea reabsorption that is the primary mechanism for maintaining high tissue levels of this solute (Marshall, 1930; Smith, 1931; Kempton, 1953; Boylan, 1967; Forster, 1967). However, because of the anatomical complexity of the renal tubule, neither the site nor the cellular mechanisms involved in urea reabsorption could be identified despite numerous microscopic studies of the renal tubule (for review, see Lacy and Reale, 1985a). Although early studies suggested that the nephron contained a "special segment" for urea reabsorption (Marshall, 1934), later studies showed that a putative countercurrent system within the kidney might be a likely candidate for such efficient urea reabsorption (Boylan, 1972). However, the extreme complexity of the renal tubule prevented the use of routine anatomical techniques to elucidate its configuration and determine if a countercurrent system did indeed exist in these kidneys. It was not until the advent of large mainframe computers, coupled with the appropriate software, capable of handling the massive amounts of data from digitized photomicrographs that the configuration of the elasmobranch nephron and particularly the countercurrent system was unequivocally proven from three-dimensional reconstructed serial sections of the little skate (*R. erinacea*) kidney (Lacy *et al.*, 1984, 1985). Despite these advances, less is known about the solute and water transport properties of the elasmobranch renal tubule than in any other phylogenetic group. Nevertheless, great strides have been made in the last decade as it is now clear that these fish also have a juxtaglomerular apparatus (Lacy and Reale, 1990) and some components of the renin–angiotensin system (Henderson *et al.*, 1981; Hazon *et al.*, 1989; Galli and Kiang, 1990; Uva *et al.*, 1992), suggesting a much finer regulation of renal and perhaps peripheral vascular function than previously believed.

The current chapter synthesizes the available data on the anatomy of the marine and freshwater elasmobranch kidney with an attempt to integrate functional studies where possible.

II. LOBES AND KIDNEY ZONES

The kidneys are paired organs flanking the dorsal aorta in the abdominopelvic region. They are roughly semilunar in most skates, elongated in sharks, but always dorsoventrally flattened (Leydig, 1852). Extensive and sometimes deep sulci extend from the lateral kidney margin toward the medial border, forming irregularly shaped lobes that are further subdivided into lobules. In some cases these indentations are so deep as to separate the tissue into distinct lobes held together only by the connective tissue surrounding the organ.

Internally two different regions (zones) of the parenchyma can be distinguished grossly because of their distinctly different color. This reflects the compartmentalization of tubular and vascular components differing in their size, pathway, arrangement, and density. In both freshwater and marine species, the renal corpuscles are aligned roughly in a plane.

In *marine* species, glomeruli lie along the border separating the two kidney zones. The peripheral zone, which is usually dorsal but sometimes laterally positioned, is thin and reveals tightly packed coils of tubules forming the *bundle zone* (Lacy and Reale, 1985a). The inner or ventral zone is thicker than the bundle zone and is composed of tubules meandering in large blood sinuses, thus forming the *sinus zone* (Lacy and Reale, 1985a). The presence of these two zones initially identified in the little skate has been confirmed in all marine and euryhaline elasmobranchs thus far examined, even though their position may be shifted within the kidney (Lacy *et al.*, 1984, 1985; Hentschel, 1988; Hentschel and Elger, 1989).

In the *freshwater elasmobranch* (*Potamotrygon* sp.) the two zones are called "complex" and "simple" and occur in the medullary and cortical region of each lobule, respectively.

III. CIRCULATION

Blood supply to the kidneys derives from the nearby dorsal aorta via renal arteries and from intercostal as well as iliac arteries. Both the number and arrangement of these arteries vary among individuals as well as species (Ghouse *et al.*, 1968; Deetjen and Antkowiak, 1970; Hentschel, 1988; Lacy and Reale, 1985a).

In the spiny dogfish, segmental arteries of the dorsal aorta branch to give off renal arteries, each of which gives off two branches, one going caudad and one going cephalad (Fig. 1). These anastomose with the counterpart from adjacent renal arteries forming a longitudinal chain that supplies blood to the mesial portion of the kidney. Three further branches of the renal artery give rise to arcuate vessels that reside at three different anatomical levels within the kidney: ventral, dorsal, and horizontal. The afferent arterioles arise from all subdivisions of both renal arteries and the intrarenal arterial chain (Ghouse *et al.*, 1968).

In the little skate, branches from the dorsal aorta enter the kidney substance and course just below the dorsal surface, sending branches

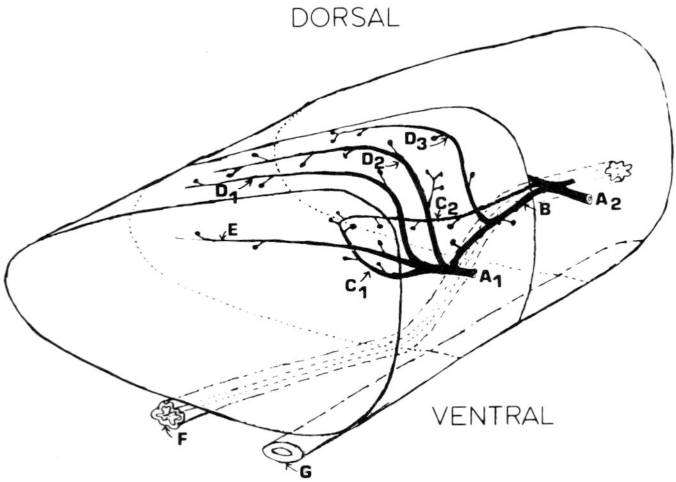

Fig. 1. Schematic drawing of the arterial supply to the kidney of *Squalus acanthias*. A1, A2, renal arteries; B, intrarenal arterial chain; C1, C2, ventral, and D1, D2, D3, dorsal arcuate arteries; E, horizontal artery; F, archinephric duct; G, ureter. (From Ghouse *et al.*, 1968, with permission.)

(intrarenal arteries) to each of the imperfectly formed lobes. These branches subsequently divide, sending arterioles to the glomeruli (Deetjen and Antkowiak, 1970; Hentschel, 1988). Intrarenal arteries also have been reported to give off separate branches to supply the interstitial tissue between the bundles and the collecting duct musculature (Hentschel, 1988). The same branches probably supply the bundles as well. However, the origin of the bundular vessels could not be identified (Hentschel, 1988).

The kidney also receives venous blood from the caudal vein and posterior body wall. The vessels enter the renal parenchyma as part of the renal portal system. Upon entering the kidney, these vessels become continuous with the venous sinuses within the sinus zone (Ghouse *et al.*, 1968; Lacy and Reale, 1985a; Hentschel, 1988). Postglomerular vessels (efferent arterioles) also anastomose with the venous sinus. Blood from these two sources, venous and arterial, mixes freely and leaves the kidney via one to many renal veins that anastomose with the cardinal veins, which lie adjacent to the dorsal aorta and carry blood to the heart (Parker, 1881; Rand, 1905).

In the kidneys of both the *marine and euryhaline species,* the blood sinuses are occasionally surrounded for a short segment by a few large, smooth muscle cells arranged around the vessel's wall like a sphincter (Fig. 2), as described in the renal connective tissue of the renal lobules

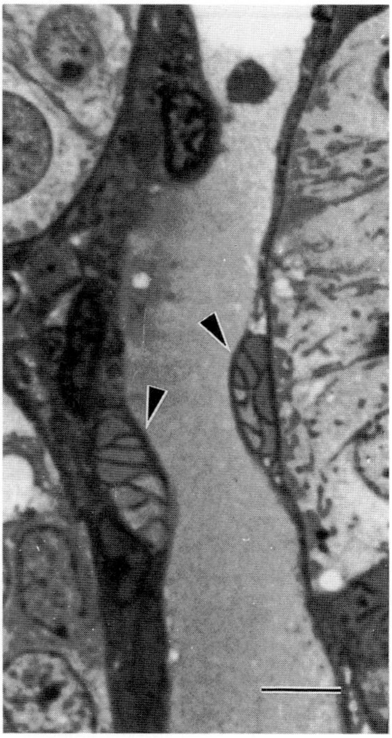

Fig. 2. Minute "Turbanorgane" (the perivascular bundle of smooth muscle cells is indicated by the arrowheads) in the kidney of *Dasyatis sabina,* sinus zone. The small vein has been cut longitudinally. ×960; bar = 10 μm.

of a ray (*Trygon violaceus*) and named turban-organ ("Turbanorgane," Bargmann, 1937). Although this location of a sphincter in a postcapillary position would seemingly appear to be an anomaly since they are not found in mammals, similar venous sphincters have been observed in elasmobranch intestine (Mayer, 1888), gill (Acrivo, 1939), and meninges (Bargmann, 1954), suggesting a common regulatory site for blood flow in these animals (Bargmann, 1954).

Microscopically the small arteries and arterioles in the little skate have an extremely thin tunica media composed of flattened smooth muscle cells (Fig. 3a). The endothelial cells lining the large blood sinuses are extremely flattened with fenestrations closed by a diaphragm about 60 nm in diameter (Fig. 3b). Within the endothelia of small arteries, elongated bodies (about 0.5×0.05 μm) with longitudinal striations are found, which correspond to the bodies observed in endothelia of other vertebrates (Weibel and Palade, 1964) and shown to be the storage site of von Willebrand Factor (Simionescu and Simionescu, 1988).

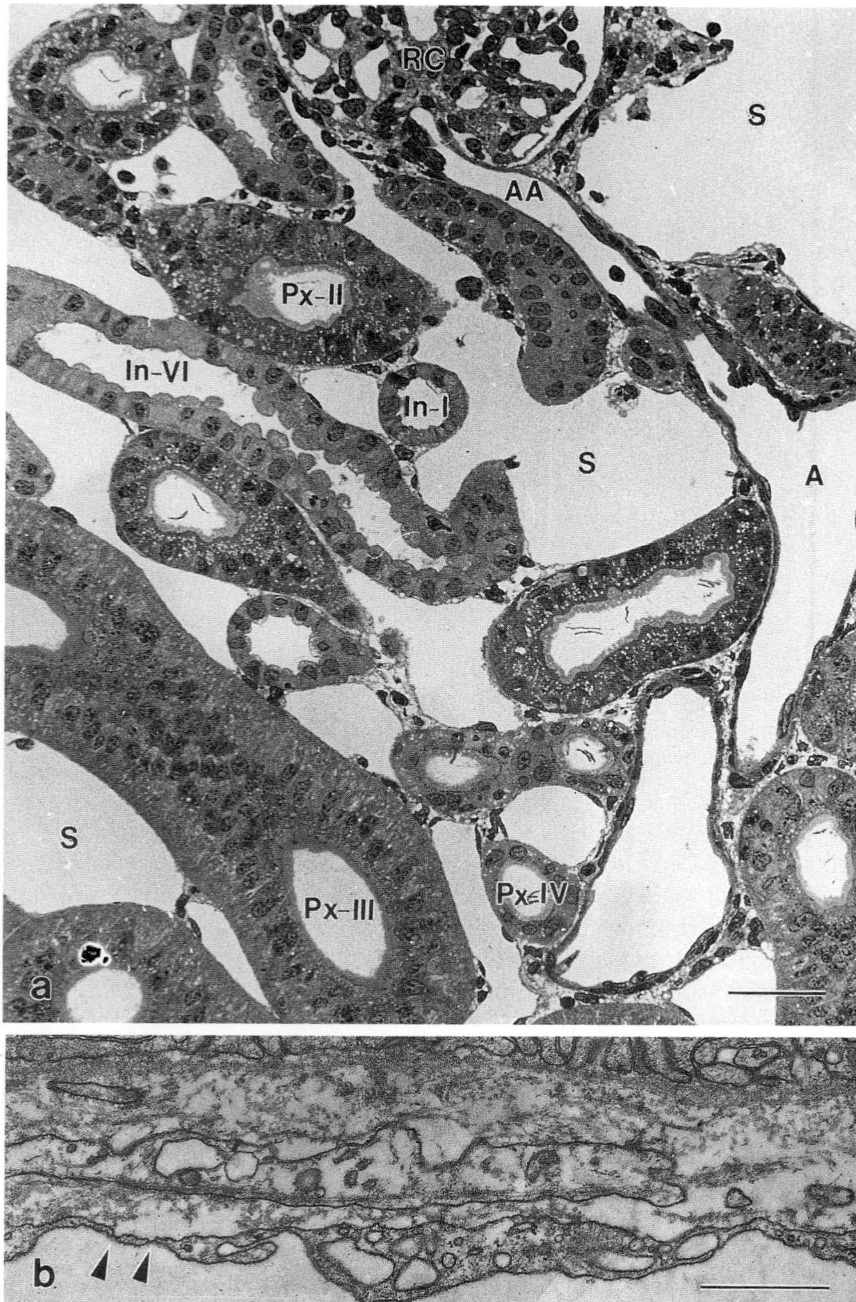
RC
S
AA
Px-II
In-VI
In-I
S
A
S
Px-III
Px-IV
a
b

IV. CONFIGURATION, SEGMENTATION, AND DISTRIBUTION OF THE RENAL TUBULE

A. Marine and Euryhaline Species

The configuration of the tubule is extremely complex. In the kidney of all marine elasmobranchs investigated to date, the tubule leaving the renal corpuscle is arranged in four loops (Fig. 4a). Two loops are directed into the bundle zone, where tubule segments are tightly packed into discrete bundles and the other two loops extend into the sinus zone, where tubules are loosely packed and segregated from each other by large blood sinuses.

The tubule emanating from the renal corpuscle first moves in the general direction of the dorsal kidney surface (bundle zone), then sharply turns back to the parent corpuscle to form the first loop (Fig. 4a). From this location it moves in the opposite direction to the ventrally positioned sinus zone, where it meanders among large blood sinuses and then returns to the same corpuscle, completing the second loop. It passes again into the bundle zone and lies adjacent to the first hairpin loop, and thus forms the third loop. The tubule returns once more to the corpuscle and meanders into the sinus zone, generating the fourth and final loop as it again returns to the parent renal corpuscle, where it is interposed between afferent and efferent glomerular arterioles. From this point it again moves into the bundle zone and proceeds in a straight course adjacent to the other two hairpin tubule loops (numbers I and III). In contrast to the first two loops, the final nephron segment, which enters the bundle zone, does not make a loop but continues to the point where the other two loops make their turn, pierces the peritubular sheath at the distal end of the bundle, and then passes into the interbundular interstitium (Fig. 4a). The two loops and the last single tubule segment form a bundle of tubules, all portions of the same nephron. Tubular fluid travels in one direction in each of three tubular segments and in the opposite direction in the two other segments, thus forming the renal countercurrent system (Boylan, 1972; Lacy *et al.*, 1985).

This arrangement of tubular segments presents the possibility of pas-

---

Fig. 3. Sinus zone, kidney of *Squalus acanthias*. In (a), the renal corpuscle (RC) and its afferent arteriole (AA) originating from a small artery (A), as well as profiles of Px-II, PX-III, PX-IV, In-I, and In-VI, can be seen among the larger blood sinuses (S). In (b), fenestrations (arrowheads) of endothelial cells in an efferent arteriole close to the distal tubule. (a) ×260; bar = 50 μm. (b) ×15,840; bar = 1 μm. (b: from Lacy and Reale, 1990, with permission from Springer-Verlag, New York.)

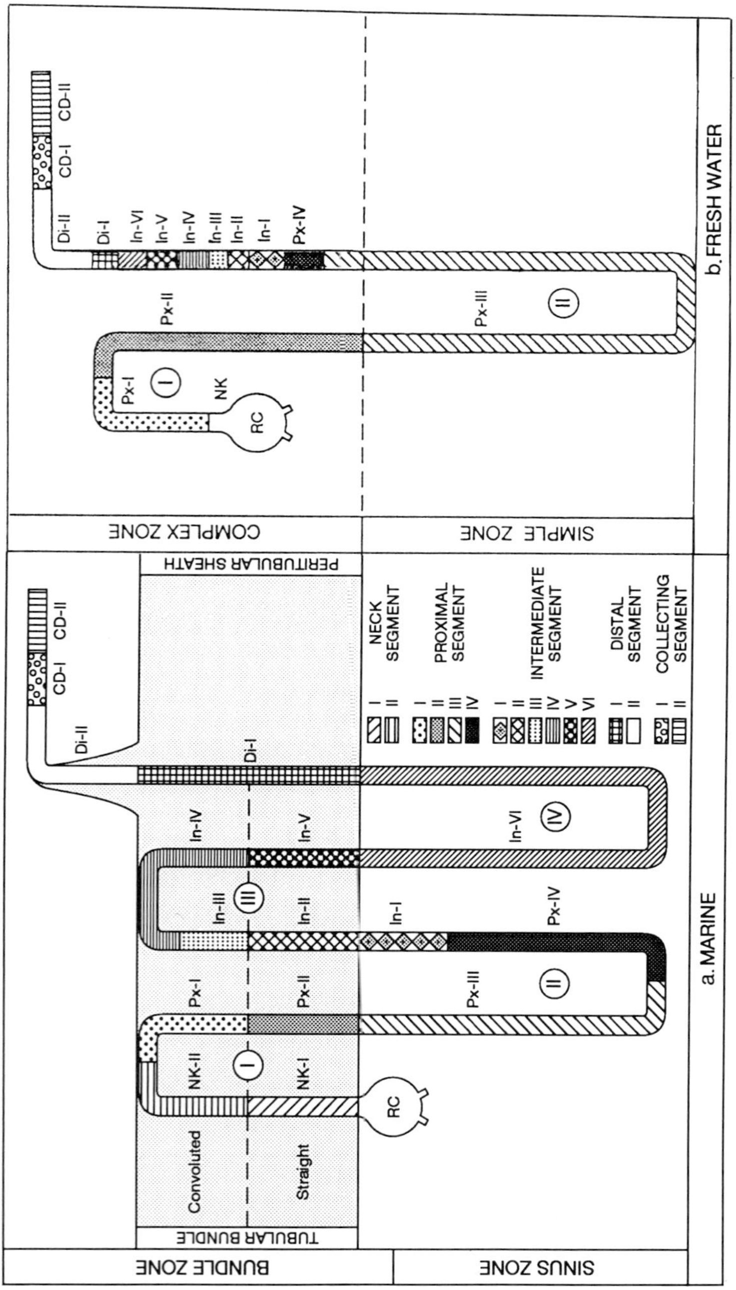

Fig. 4. Schematic drawing of the course of the renal tubule from the renal corpuscle (RC) to the collecting duct (CD) in marine (a) and in freshwater (b) elasmobranchs. In marine fishes the loops I and III as well as the distal tubule segments are wrapped by the peritubular sheath; the segments of the bundle form the countercurrent system. Tubular bundle, peritubu-

sive urea reabsorption (Boylan, 1972), which could account for the extremely small fraction of filtered urea excreted. However, for this anatomical arrangement to function as a countercurrent system similar to that found in other animals, several additional anatomical relationships are necessary. Computer-assisted reconstruction of serial photomicrographs in conjunction with electron microscopy verified that these features were indeed present (Lacy *et al.,* 1985; Lacy and Reale, 1985a,b). First, the two loops and final distal segment of each bundle are very tightly apposed to one another, thus minimizing diffusion distances for solute and water. Second, tubule segments from other nephrons do not intervene in the bundle. This has several implications, the most important of which is that the countercurrent flow pattern is not disrupted and that the tubular transport dynamics are not influenced by that of other nephrons. Third, each bundle has an anastomosing network of capillaries around and among the tubule segments in the bundle (Lacy *et al.,* 1985). There are arteries originating from intrarenal arteries, which probably supply the bundles only. These vessels that enter and exit the bundle at the same end (that nearest the renal corpuscle) give anatomical support to the probability that a functional countercurrent multiplier system is present in this kidney (Lacy *et al.,* 1985). Fourth, each tubular bundle with the surrounding capillary network is wrapped together by overlapping sheets of tightly adherent squamous cells (Lacy and Reale, 1986). The ends of each of these peritubular sheaths are closed, as these cells form a sausagelike casing around the tubular bundle and capillaries, thus creating a barrier between the environment inside this sheath (tubules, capillaries, scant connective tissue) and that outside it (capillaries, connective tissue, late distal tubules) (Lacy *et al.,* 1985; Lacy and Reale, 1986).

Figure 5 schematically shows the course of the renal tubule of the little skate, *Raja erinacea.* The bundle (tubular segments, capillaries, and sheath) that forms the countercurrent system is not straight despite its having a proximal (nearer renal corpuscles) and distal end. At the proximal end, the peritubular sheath is pierced by five tubular segments (forming the two limbs of loops I and III and the final distal segment) and the afferent and efferent vessels of the vasa recta (Fig. 5). The opposite end of the sheath is pierced only by the exiting distal tubule, which joins the collecting duct in the area between bundles. Given this arrangement, it

---

lar sheath, and countercurrent arrangement are not present in freshwater elasmobranchs. After the second loop the tubule meanders freely in the complex zone. The subdivisions of each tubule segment, based on the structure of the epithelium, are indicated by a symbol. (a: From Lacy and Reale, 1985b, with permission from Springer-Verlag, New York; b: from Grabowski *et al.,* 1995).

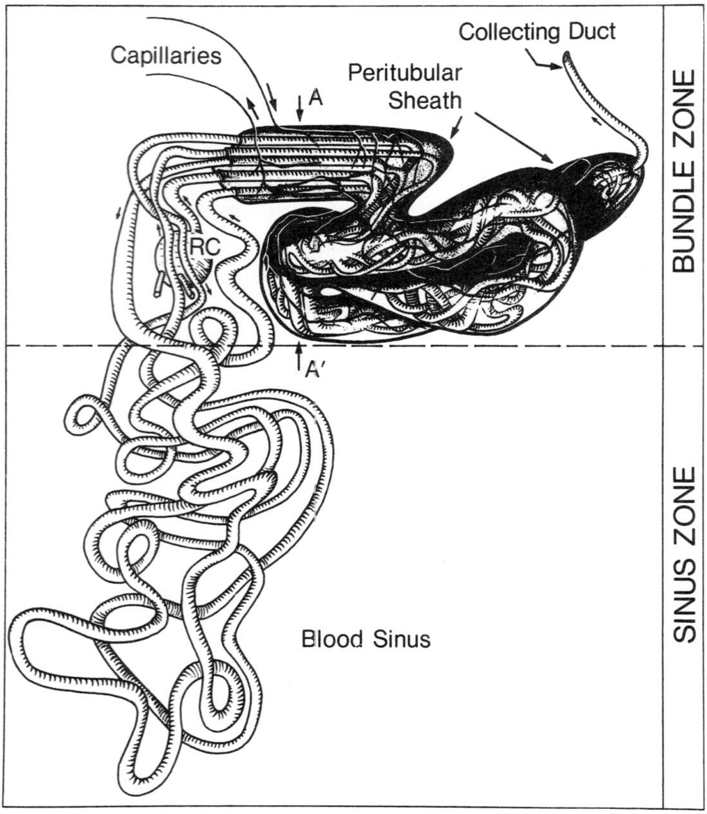

Fig. 5. Drawing of the renal tubule configuration in the bundle and sinus zone of the *Raja erinacea* kidney. Capillaries enter and exit the bundle and form an anastomosing network around the entire length of the bundle. A–A′ indicates a plane orthogonal to the bundle (close to RC, similar to Fig. 12a) and oblique to the bundle (as in Fig. 16a). [Reprinted with permission from Lacy, E. R., Reale, E., Schlusselburg, D. S., Smith, W. K., and Woodward, D. S. (1985). *Science* **227,** 1351–1354. Copyright 1985 American Association for the Advancement of Science.]

could be expected that a linear gradient of solutes would be present within the peritubular sheath. Circumstantial morphological evidence and some physiological studies suggest that these bundles are the sites of urea reabsorption from tubular fluid. Except that an extra loop is present in the bundle, this anatomical arrangement bears remarkable similarity to the mammalian renal papilla in which capillaries and loops of Henle form a countercurrent multiplier system. However, the segregation of "bundular units" from one another in marine elasmobranchs suggests that each unit, which consists of tubular segments from a single nephron and capillaries from a single arteriole, has morphological and perhaps physiological analogies to the entire mammalian papilla in which segments from nearly all nephrons function as a single unit.

Stenohaline *freshwater stingrays,* which do not reabsorb significant filtered urea nor use it as a regulatory osmolyte, do not have tubular bundles (countercurrent system) (Hentschel and Elger, 1982; Lacy *et al.,* 1989a). Although there is a correlation between the renal reabsorption of filtered urea, its high levels in tissue, and the anatomical presence of tubular bundles in species, there are no data indicating that urea is reabsorbed within the bundles themselves. The bundle may simply act as a "priming area" in which the optimal concentration of osmolytes is produced in the tubules and/or in the bundular capillaries so that when this fluid enters the sinus zone, urea is efficiently reabsorbed. Nevertheless, the anatomical presence of a renal countercurrent system appears to "oblige" the fish to maintain a significant hypertonicity compared with the surrounding water. The euryhaline Atlantic stingray, *Dasyatis sabina,* has typically arranged tubular bundles (countercurrent) and can adapt for long periods to extremely dilute environments (DeVlaming and Sage, 1973). Nevertheless, when placed in environments of <100 mOsm, the plasma osmolality and urea concentrations are significantly greater than that of the surrounding water even though they are less than that found when the fish is adapted to full-strength (~950 mOsm) seawater. Until analyses of tubular fluid along the nephron are done, important questions as to the function of the countercurrent system will remain unanswered.

The segregation of nephron segments that occurs in the bundle zone of *marine species* is not present in the sinus zone. Tubules from many different nephrons randomly intermingle in a large vascular sinus that appears to be continuous throughout this zone of the kidney. Tubules are never observed in clusters, but instead each is nearly surrounded by scant interstitial tissue and endothelia (Fig. 3a). This structural arrangement optimizes the tubular surface area exposed to blood, suggesting the importance of this relationship instead of the tubule and tubule to capillary relationship found in the bundle zone. The large and cavernous arrangement of these sinuses and the presence of venous sphincters suggest an environment of slow blood flow that would favor tubular and vascular solutes reaching equilibrium. Another factor that would influence the solute and water movement across the tubules in the sinus zone is the fact that blood within the vascular sinus is a combination of that from the glomerular efferent arteriole and that from the renal portal system as described earlier. The relative contributions of sinus zone blood originating from each source is not known.

B. Freshwater Species

The two kidney zones in the renal lobules of the *freshwater stingray* (*Potamotrygon* sp.) are composed of different nephron segments than in their marine counterparts, and there is an absence of tubular bundles wrapped by a peritubular sheath that form a countercurrent system. Furthermore, the large blood sinuses bathing the tubules of loops II and IV in marine fish are extended to the whole parenchyma in freshwater rays.

The renal tubule emanates from corpuscles and sinuously moves in the direction of the prominent interlobular connective tissue septa (Fig. 4b). The tubule meanders back to the parent renal corpuscle forming the first loop, which is in the complex zone. The tubule then enters the simple zone and as one morphologically distinct subdivision (Px-III), makes several turns, and finally heads back to the complex zone forming loop II. From this point, the tubule meanders extensively in the complex zone forming an extremely ill-defined "loop" as it returns to the vascular pole of its renal corpuscle (see description of juxtaglomerular apparatus in the following). After extensive contact with Bowman's capsule, vessels and extraglomerular mesangium, the tubule moves toward the interlobar septa, where it merges with the collecting ducts (Fig. 4b).

V. THE RENAL CORPUSCLE

As in other vertebrates, Bowman's capsule and the glomerulus with associated vessels form the filtration unit of the kidney, called the renal corpuscle. In *marine species* the glomeruli are generally large (e.g., 150×170 μm in diameter), but there is clearly a wide range of sizes within one kidney (Lacy *et al.,* 1987).

A. Bowman's Capsule

Over most of its extension, Bowman's capsule (Fig. 6) is formed by a single layer of epithelial cells bearing a single cilium near the vascular pole and long *multiple flagella* near the urinary pole (Lacy *et al.,* 1987). These unique flagellar cells continue into the initial segments of the renal tubule, the neck, and proximal segments, where they are interspersed among the tubular epithelial cells. They are also in the intermediate tubular segments depending on the species, but are consistently absent in the distal and collecting duct segments (Lacy and Reale, 1991a,b; Lacy *et al.,* 1989c).

The flagellar ribbons are always bent in the direction of tubular fluid

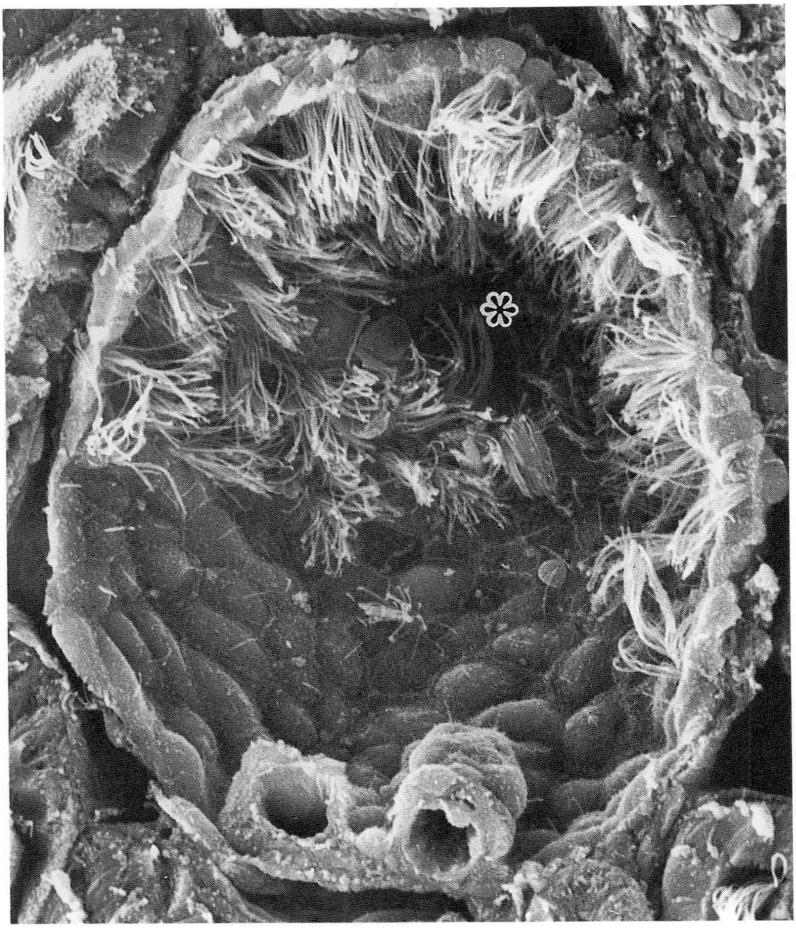

Fig. 6. Bowman's capsule, kidney of *Raja erinacea,* scanning electron microscopy. Cells with a single cilium toward the vascular pole (lower side of the figure) and cells bearing multiple flagella toward the orifice of the neck segment (asterisk). Note the close apposition of the afferent and efferent arterioles at the vascular pole. ×750. (From Lacy, E. R., Castellucci, M., and Reale, E. 1987. The elasmobranch renal corpuscle: Fine structure of Bowman's capsule and the glomerular capillary wall. *Anat. Rec.*, v. 218. Copyright © 1987 by Wiley-Liss, Inc., a subsidiary of John Wiley & Sons, Inc.)

flow. The rapid flagellar beat could generate a negative pressure inside both the tubule as well as Bowman's space, thus increasing glomerular filtration (Kempton, 1943, 1964). In addition, an efficient propulsion of the tubule contents and an improvement of glomerular filtration are probably necessary to compensate for the low blood pressure (glomerular filtration pressure) in marine and presumably freshwater elasmobranchs, the extremely long renal tubule, and the great variations in tubular diameter. Observations of tubular fluid transit times confirm that linear flow velocity is about 3.5× greater in the early tubular segments, where flagellar cells are extremely abundant, than in the later ones, where they are absent (Deetjen and Antkowiak, 1970).

The most distinguishing feature of these cells is a compact ribbon of 10 to 30 tightly packed flagella ordered side by side in one to three parallel rows perpendicular to the direction of tubular fluid flow (Fig. 7a). The length of the flagellar ribbon is up to 50 μm, thus they extend "downstream" beyond many (up to six) adjacent cells. Each flagellum within the ribbon begins deep, with typical structures such as a basal body, rootlets, and a row of intramembranous particles ("necklace") in the basal region. The axoneme has the usual $9 \times 2 + 2$ microtubular pattern with the central microtubule doublet of each axoneme orthogonal to the main tubular axis indicative of the beat direction (Fig. 7b). The axonemes of adjacent flagella are bound together by their surface coat and perhaps by transmembrane units (Lacy *et al.*, 1991a), thus forming a paddlelike structure that beats vigorously as a single entity (Bargmann, 1937; Kempton, 1943, 1962, 1964; Lacy *et al.*, 1989b, 1991a).

As Bowman's capsule reaches the vascular pole of the glomerulus, there is a transition from the simple-squamous cells of the visceral epithelium that line the central and upper region of the capsule to *peripolar cells* that form a collar around the afferent and efferent vessels (Fig. 8). They lie on the basement membrane of Bowman's capsule and border the urinary space. The peripolar cells are flattened to cuboidal and contain distinctive vacuoles and/or granules of varying contents and size. These specialized epithelial cells are found in both marine (Lacy and Reale, 1989) and freshwater (Grabowski *et al.*, 1995) species and probably occur in all vertebrates (Thumwood *et al.*, 1993). The function of these cells is unknown but their strategic position around the afferent and efferent vessels suggests that they may be involved in regulation of glomerular blood flow and thus formation of the ultrafiltrate (Gardiner and Lindop, 1985; Ryan *et al.*, 1979). In any case, they should probably be considered as one of the components of the juxtaglomerular apparatus discussed in Section V,C.

B. Glomerular Capillary Wall

The afferent vessel splits into multiple capillary loops within Bowman's capsule, which anastomose with one another, finally reuniting to form

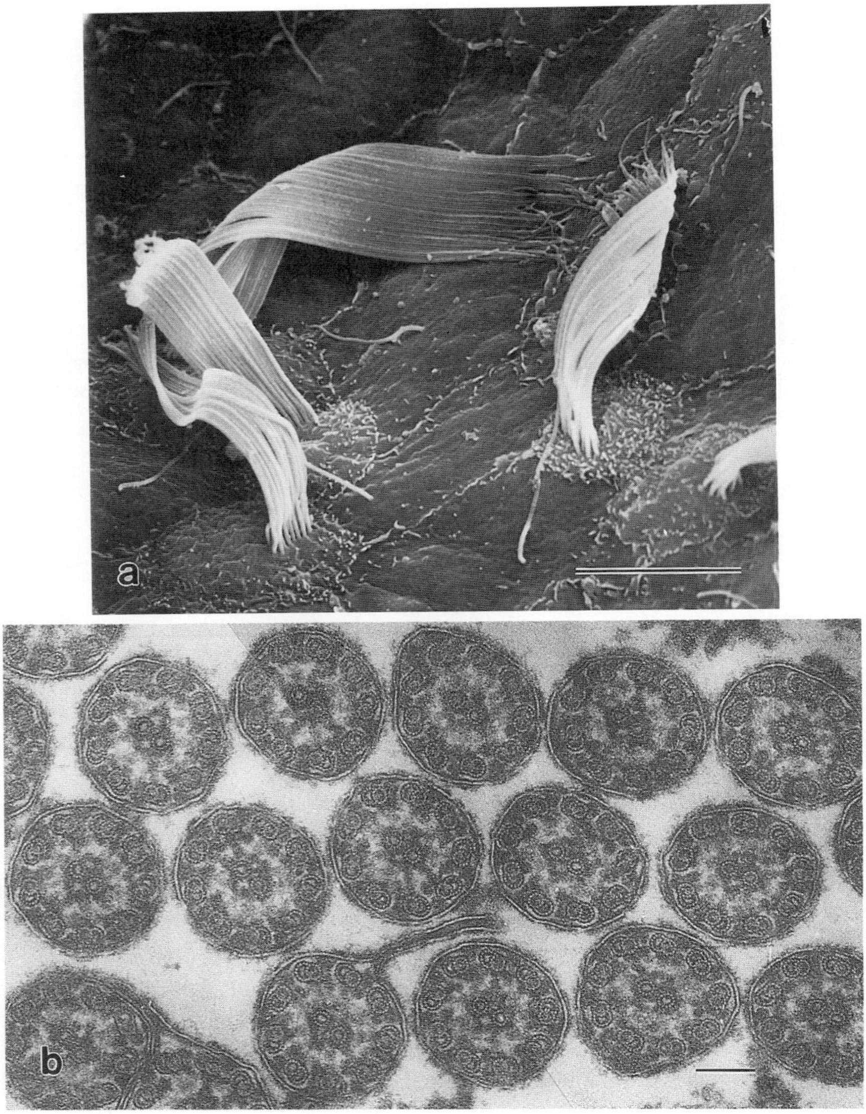

Fig. 7. Bowman's capsule. Arrangement and structure of the flagellar ribbons in *Mustelus canis* (a) and *Potamotrygon* sp. (b). Scanning electron microscopy (a) demonstrates flagella arranged in wide ribbons. Transmission electron microscopy (b) reveals microarchitecture of the axoneme; 9 × 2 peripheral microtubular doublets, central doublet, thick arms, and spokes. (a) ×2155; bar = 10 μm. (b) ×78,400; bar = 0.1 μm. (a: from Lacy *et al.*, 1991a, with permission from Churchill Livingstone, Edinburgh, UK.)

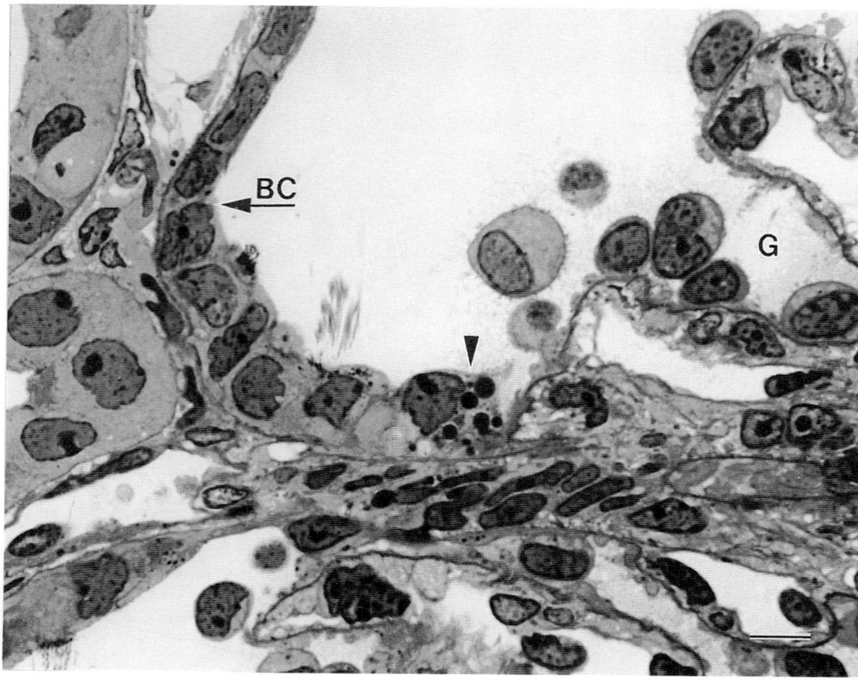

Fig. 8. Vascular pole of a renal corpuscle in *Mustelus canis*. Granular peripolar cell (arrowhead) with deeply stained, homogeneous granules of different size. BC, Bowman's capsule; G, glomerulus. ×800; bar = 10 μm.

one or sometimes multiple vessels. Continuous with the peripolar cells at the vascular pole, the specialized epithelial cells, called podocytes, spread out and cover these capillary loops (Fig. 9a). The podocytes have large, pedunculated, roughly spherical bodies and thin, uniformly wide foot processes (pedicels) surrounding the capillary wall. Occluding junctions as well as gap junctions have been identified on extensions of *R. erinacea* podocytes (Lacy *et al.*, 1987). The glomerular capillary wall consists of pedicels that lie on a thick, continuous basement membrane, a thin and

---

Fig. 9. Glomerular capillary wall in *Squalus acanthias* (a) and *Raja erinacea* (b). In (a), bodies and pedicels of the podocytes, as seen from the inside of the urinary space, by low magnification scanning electron microscopy. In (b), podocytes (P) lying on a continuous basement membrane (arrows) and endothelium (E) on a discontinuous basement membrane (arrowheads). Some endothelial pores are present. Between the two basement membranes lie mesangial cells (M) and mesangial matrix. (a) ×835; bar = 10 μm. (b) ×8310; bar = 1 μm.

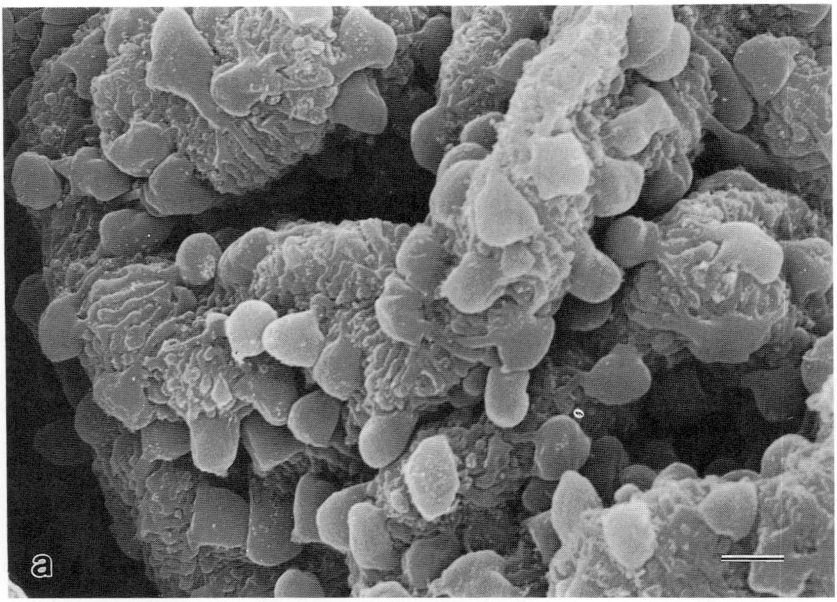
a

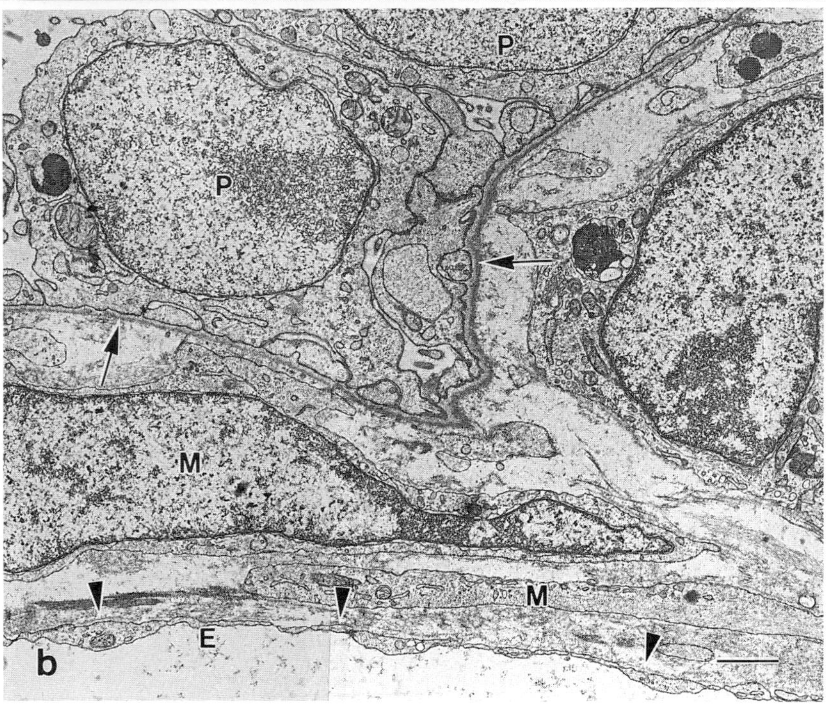
P
P
M
M
E
b

highly discontinuous endothelial basement membrane, and usually but not always a significant layer of mesangium that intervenes between the two layers of basal lamina (Fig. 9b). The epithelial basement membrane is composed of a thick lamina densa and a lamina rara containing glycoconjugates (glycoproteins and proteoglycans) (Boyd and De Vries, 1983; Decker and Reale, 1991). The intervening matrix is composed of mesangial cells with numerous flattened, long processes, scattered collagen fibrils, anchoring fibrils, and microfibrils (oxytalan fibrils) (Fig. 9b). Elaunin and elasticlike fibers have been shown to exist within this mesangium (Lacy *et al.*, 1991b). The endothelial cells are thin, with fenestrations 40 to 70 nm in diameter that are closed by a diaphragm. The presence of two basal laminae and an intervening layer of interstitial matrix forming the filtration barrier is significantly different from the single basal lamina in that position in higher vertebrates.

C. The Juxtaglomerular Apparatus

The presence of a juxtaglomerular apparatus (JGA) in any species belies a complex series of mechanisms for control of both renal tubular function as well as peripheral circulation. Although frequently perceived as an organ without such complexity (see Lacy and Reale, 1990), recent studies have shown that not only are the structural components of the JGA present in elasmobranchs, but that the biochemical components are also present. Angiotensin II-like immunoreactivity is present in elasmobranch tissues (Galli and Kiang, 1990) and partially purified renal tissue extracts from the dogfish (*S. canicula*) elevate blood pressure in rat and dogfish as well as generate angiotensin from porcine and synthetic angiotensinogen (Hazon *et al.*, 1989; Henderson *et al.*, 1981; Uva *et al.*, 1992).

In *marine elasmobranchs* as in higher vertebrates, the JGA lies at the vascular pole of the renal corpuscle and is composed of afferent arteriole(s) with granular smooth muscle cells in its tunica media, efferent arteriole(s), macula densa of the distal tubule, and extraglomerular mesangium (Lacy and Reale, 1990; Fig. 10). In *marine* but more frequently in *freshwater species,* more than one efferent vessel can be present, which makes identification of these structures more difficult as well as expanding the conventional description of the components of the JGA. In the river ray (*Potamotrygon* sp.) kidney, a single afferent arteriole meets the glomerulus at the vascular pole, but several efferent arterioles usually leave the glomerulus, each emerging from separate openings of Bowman's capsule. These openings form a low groove on one side of the renal capsule whose base is formed by the efferent vessels and by the interposed extraglomerular mesangium. The distal segment of the renal tubule lies within this lateral

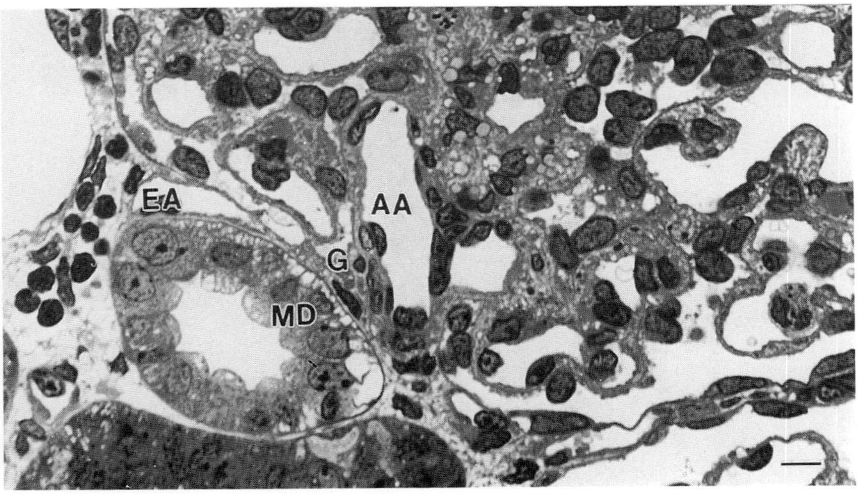

Fig. 10. Juxtaglomerular apparatus, *Squalus acanthias*. AA, afferent arteriole; EA, efferent arteriole; MD, macula densa of the distal tubule; G, extraglomerular mesangium. ×500; bar = 10 μm.

groove, thus forming a long region of the tubule in contact with the extraglomerular mesangium. Although this region has been called the "vascular field" (Bargmann, 1937; Hentschel, 1988) in marine elasmobranchs, it is more precisely the JGA because of a distinct macula densa of the distal tubule in direct contact with the extraglomerular mesangium in addition to the afferent and efferent glomerular vessels.

The afferent arteriole in both freshwater and marine species has a thin wall composed of endothelial cells and a single complete layer of flattened smooth muscle cells (Fig. 11a and 11b). These are positioned close to the glomerulus or scattered in the wall of the vessels at varying distance from the renal corpuscle. These cells contain a few granules that have the same dimension and distribution as those that bind antibodies against mouse submandibular gland renin (Kon *et al.*, 1995). These cells have conspicuous bundles of filaments that can be seen in the cytoplasm closer to the endothelium, whereas granules, glycogen particles, and most of the cell organelles lie on the opposite side (Figs. 11a and 11b). Ultrastructurally the granules are membrane-bound and their contents show a crystallinelike substructure (Figs. 11a and 11b) as described in the JG granules of mammals (for a brief review, see Lacy and Reale, 1990). Therefore, the smooth muscle cells with "renin" granules can be considered analogous to the epithelioid cells of the afferent arterioles of the mammalian kidney.

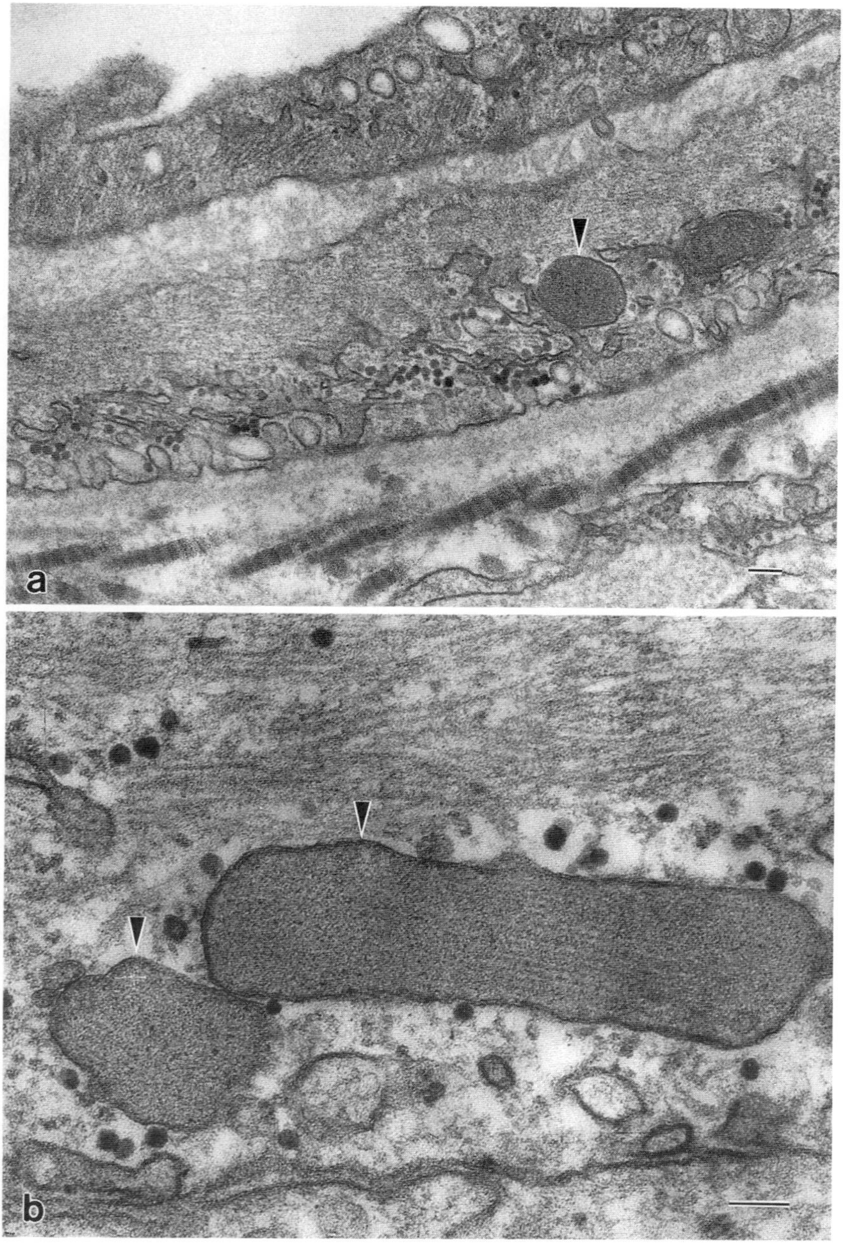

Fig. 11. Juxtaglomerular apparatus in *Potamotrygon* sp. (a) and *Squalus acanthias* (b). Afferent arterioles are shown with granules in the smooth muscle cells of the tunica media (arrowheads). The content of the granules displays a paracrystalline structure (b). Microfilaments in the upper region of the micrographs lie toward the endothelium. The granules lie in areas rich in glycogen particles and vesicles. (a) ×38,000; (b) ×78,850; bar = 0.1 μm. (b: from Lacy and Reale, 1990, with permission from Springer-Verlag, New York.)

Investigations in a number of *marine* sharks, skates, and rays showed that the efferent vessel(s) were thin-walled and devoid of typical smooth muscle cells (Fig. 10). Cell processes from pericytes surrounded by a discontinuous basement membrane were present and the endothelial cells themselves were fenestrated and lay on a very thin and discontinuous basement membrane (Lacy and Reale, 1990).

In both *marine* and *freshwater* species the extraglomerular mesangium is composed of scattered cells and abundant extracellular matrix, which include collagen fibrils and microfibrils as well as particles and granules, presumably proteoglycans and/or glycoproteins (Lacy *et al.,* 1991b).

The final structural component of the JGA, the *macula densa* portion of the distal tubule, is characterized by high columnar cells with convex luminal surfaces. Basolateral intercellular spaces are dilated between distal tubule epithelial cells that face the extraglomerular mesangium and the afferent and efferent arterioles (Fig. 10). The Golgi apparatus is located lateral or basal to the nucleus, and mitochondria are sparse and found throughout the cytoplasm. In *freshwater rays* the macula densa is not found where the distal tubule is adjacent to the afferent arteriole, but instead where the tubule runs in the lateral groove of the renal corpuscle, as described earlier. It is within this long contact area with the extraglomerular mesangium that the numerous epithelial cells of the tubule have large intercellular spaces. At their periphery, tiny, long microvilli- and microplicaelike processes of the bordering cells project into the dilated intercellular spaces. These morphological features are characteristic of the macula densa of higher vertebrates (for a review, see Taugner and Hackenthal, 1989).

Both *marine* and *freshwater species* possess all the components of the JGA. However, their unconventional arrangement both at the vascular pole and along the lateral groove of the renal corpuscle hinders an immediate identification of this complex structure and explains, in part, why the existence of a complete JGA in the elasmobranchs has been denied until recently. Taken together, the morphological immunohistochemical and biochemical data strongly support the notion that elasmobranchs have a functional renin–angiotensin-like system. Its role in osmotic and vascular homeostasis remains to be elucidated.

VI. THE RENAL TUBULE

Numerous investigators since Haller (1902) have described the cellular architecture of various nephron segments, but these reports were unable to identify the position of those segments along the nephron except in the most general terms because of its complexity and length (see Lacy *et al.,*

1985; Lacy and Reale, 1985a,b, 1986). This led to a confusing nomenclature for the renal tubule until the use of computer reconstruction techniques (Smith *et al.,* 1983; Lacy *et al.,* 1984, 1985), which sequenced the different epithelia along the entire nephron. Despite these terminology differences in previous studies, the sequence of different tubular segments remains consistent with other vertebrates and thus the nomenclature used here follows that set forth for all major taxa and is consistent with the earlier terminology (Haller, 1902).

In all elasmobranch kidneys thus far investigated, the renal tubule has five distinctive segments—neck, proximal, intermediate, distal, and collecting duct—each of varying subdivisions (Figs. 4a and 4b). These subdivisions are subtle and the morphological characteristics can often be identified only by transmission electron microscopy or semithin sections of epoxy-resin-embedded tissue. Furthermore, the subdivisions are not exactly equivalent among different species.

A. Neck Segment

The neck segment (Nk) has a similar structure in both *marine* and *freshwater* species even though in the former a major portion of its length lies within the tubular bundle, which does not exist in the freshwater species. This segment is a short portion of the tubule extending from the urinary pole of the renal corpuscle toward the renal surface (Figs. 4a, 4b, 12a, and 16a) and lined by a cuboidal epithelium with flagellar and nonflagellar cells (Leydig, 1852; Lacy and Reale, 1991a). The flagellar cells do not vary in their morphology from those described in Bowman's capsule (see earlier) or any other position along the tubule. The nonflagellar cells bear scanty, if any, microvilli- and microplicaelike apical projections and have the common cytoplasmic organelles and a prominent cytoskeleton (Lacy and Reale, 1991a). These characteristics, along with the fact that the lumen of Nk-I is densely packed with vigorously beating flagella (Figs. 12a and 12b), support the notion that the major function of Nk is to propel urine along the tubule.

B. Proximal Segment

The second segment of the renal tubule, the proximal (Px), is composed of flagellar and nonflagellar cells, the latter of which differ from the Nk-nonflagellar cells. Px is a long segment of the nephron with portions in the bundle zone and in the sinus zone of marine species; it occurs in the simple and complex zones of freshwater species (Figs. 4a and 4b).

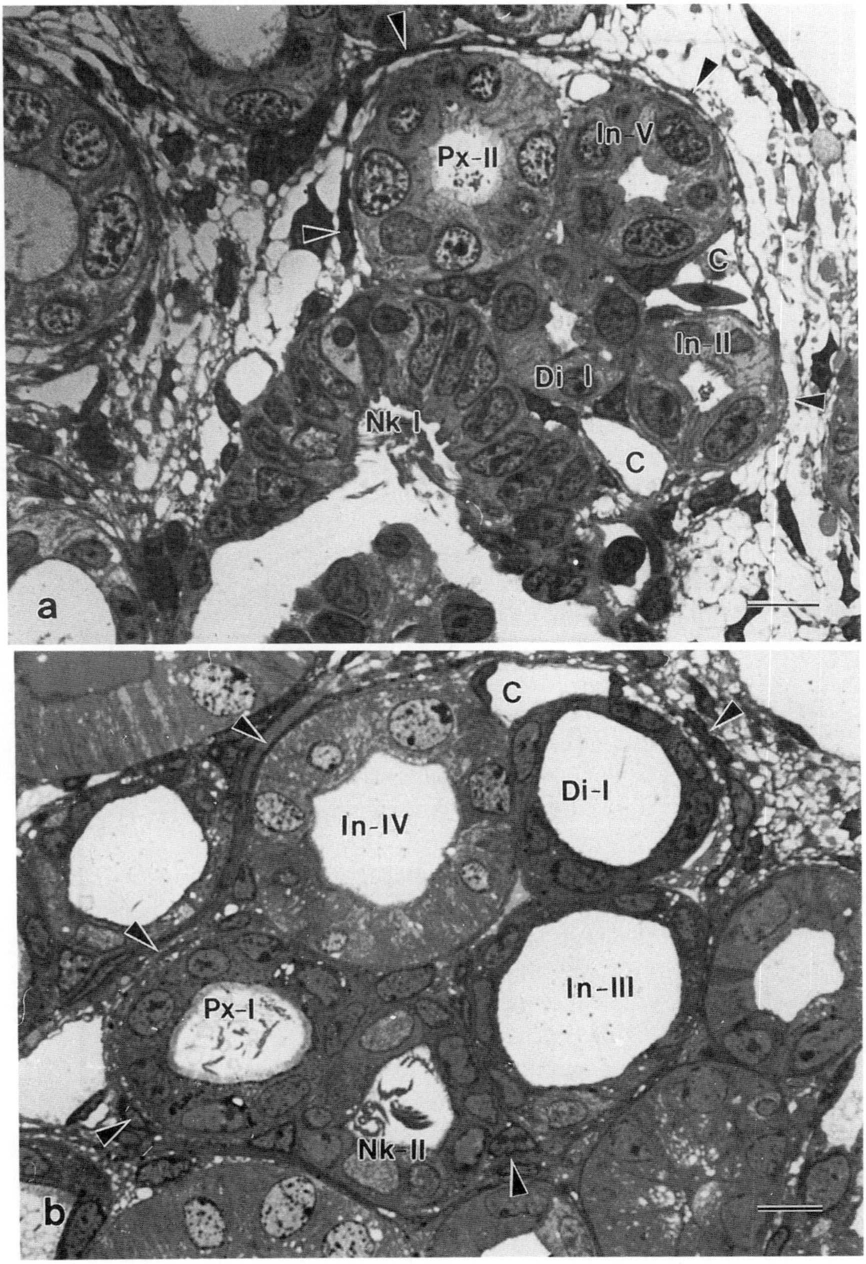

Fig. 12. Cross sections of tubular bundles in the cownose ray, *Rhinoptera bonasus* (a) close to the renal corpuscle and in *Squalus acanthias* (b) distal to the renal corpuscle. Nk, neck segment; Px, proximal segment; In, intermediate segment; Di, distal segment; C, blood capillary. Arrowheads indicate the peritubular sheath. (a) ×910; (b) ×770; bar = 10 μm.

In *marine* species such as the little skate, there are four subdivisions to this segment (Fig. 4a). The first subdivision, Px-I, forms a limb of the first loop. The transition between Nk-II and Px-I occurs at approximately the end of loop I (Fig. 4a). The transition between Px-I and Px-II occurs within the bundle at approximately the same position as the Nk-I to Nk-II transition (Fig. 4a). Thus Nk and Px (I and II) run parallel to each other but the tubular fluid runs in opposite directions. Px returns to the parent renal corpuscle, where it lies adjacent to the Bowman's capsule. The proximal segment is easily recognizable in the tubular bundle (Figs. 12a and 12b) because it has a luminal surface elaborated into a brush border and numerous mitochondria (Lacy and Reale, 1991a). The following characteristics increase from Px-I to Px-II to Px-III but decrease in Px-IV (Figs. 12a and 12b, and 13a–13d): height of the cells and their brush border, apical PAS staining for glycoproteins, extension and number of basolateral plasma membrane invaginations, and outer diameter of the tubule. The mitochondria are extremely diverse in the various subdivisions. In addition, in all the Px subdivisions (Fig. 13a–13d), an extensive tubular network of smooth-surfaced endoplasmic reticulum is present in the apical region of the brush border cells, especially in the Px-II segment (Fig. 13a).

In the little skate and in other elasmobranchs, the transition region from Px-II to Px-III is marked by structures suggesting a prominent receptor-mediated endocytosis (coated pits and vesicles, apical tubules connecting with small vesicles, basally located lysosomelike dense bodies) (Lacy and Reale, 1991a).

Px-IV is markedly smaller in diameter than the preceding Px-III subdivision and is characterized in part by numerous flagellar cells (Fig. 13d). The nonflagellar cells have a low brush border and numerous dense bodies of different size and contents. In the little skate (*R. erinacea*) kidney, Px-IV shows varying inner and outer diameters because it is lined by cells of different heights (Lacy and Reale, 1991a). Freeze-fracture replicas reveal *zonulae occludentes* composed of several superimposed strands in all Px subdivisions. Between the cells of Px-I, II, and III but not Px-IV, gap junctions are present (Lacy and Reale, 1991a).

Fig. 13. *Raja erinacea* kidney, proximal segment. (a) Second subdivision (Px-II); basal (b) and apical (c) parts of cells at the transition zone from Px-II to Px-III; (d) fourth subdivision (Px-IV). Note the different structures of the mitochondria and, in (d), dense bodies in the cytoplasm of the brush border cells and the flagella inside of the tubular lumen. ×5820; bar = 1 μm. (a–c: from Lacy, E. R., and Reale, E. 1991a. Fine structure of the elasmobranch renal tubule: Neck and proximal segments of the little skate. *Am. J. Anat.*, v. 190. Copyright © 1991 by Wiley-Liss, Inc., a subsidiary of John Wiley & Sons, Inc.)

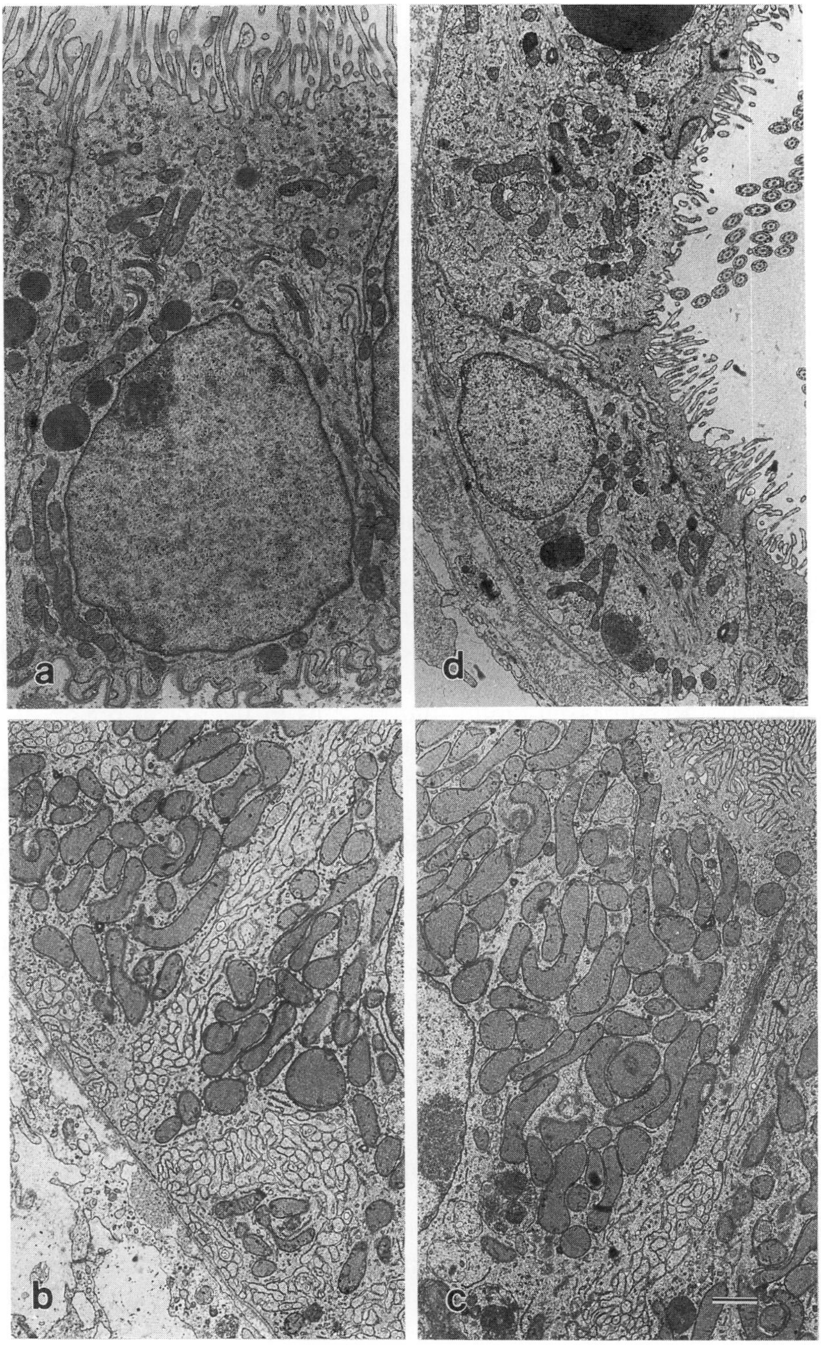
a
d
b
c

Proximal tubule segments in the little skate (Px-III) were micropunctured *in vivo* from the dorsal and ventral renal surfaces by Stolte *et al.*(1977) and found to be the principal sites for Mg^{2+}, phosphate, and sulfate secretion and the main site for reabsorption of sodium and chloride in excess of water. Sawyer and Beyenbach (1985) isolated and perfused Px-III segments from the sinus zone of the shark *S. acanthias* and found that this segment readily secreted fluid and NaCl, providing further evidence for active Cl^- secretion in shark proximal tubules (Beyenbach and Frömter, 1985).

One of the most prominent morphological aspects peculiar to the renal tubule in some but not all *marine* elasmobranchs concerns the occurrence and location of a "granular" segment. In the little skate, the granules are found in the first subdivision of the intermediate tubule. In spiny dogfish (*S. acanthias*) and in smooth dogfish (*M. canis*), this corresponds to the Px-II segment, whose nonflagellar cells are characterized by numerous apical granules of varying size and staining affinities. The reasons for these distribution differences are unknown. Gérard and Cordier (1934), in a comparative study on the occurrence of granular inclusions in the proximal tubule of vertebrates, speculated that they consisted of material reabsorbed from the glomerular ultrafiltrate.

The proximal segment of the *freshwater ray, Potamotrygon* sp., has the general structure similar to that described here for the marine species. The smooth endoplasmic reticulum prevails in the apical region of the cell and, in the transition zone from Px-II to Px-III, it shows a peculiar arrangement in stacks, straight or disposed around a lipid droplet (Fig. 14). These structures are clearly visible by light microscopy, appearing as onionlike bodies up to 9 μm in diameter. Mitochondria with abundant matrix and peroxisomelike bodies are common between the other cytoplasmic components in Px-II and III.

The characteristic thin, last subdivision of the proximal segment (Px-IV) in marine species is also present in the freshwater ray. The lumen of Px-IV is filled with flagella, and the dense bodies are numerous in the cytoplasm of both flagellar and nonflagellar cells.

C. Intermediate Segment

In both *marine* and *freshwater* species, the intermediate segment (In) corresponds to the position of the loop of Henle in mammals in that it is juxtaposed between the end of the brush border segment (proximal tubule) and the distal tubule that begins at the JGA. However, there are few other morphological similarities between these two (In and Henle's loop) renal tubular segments.

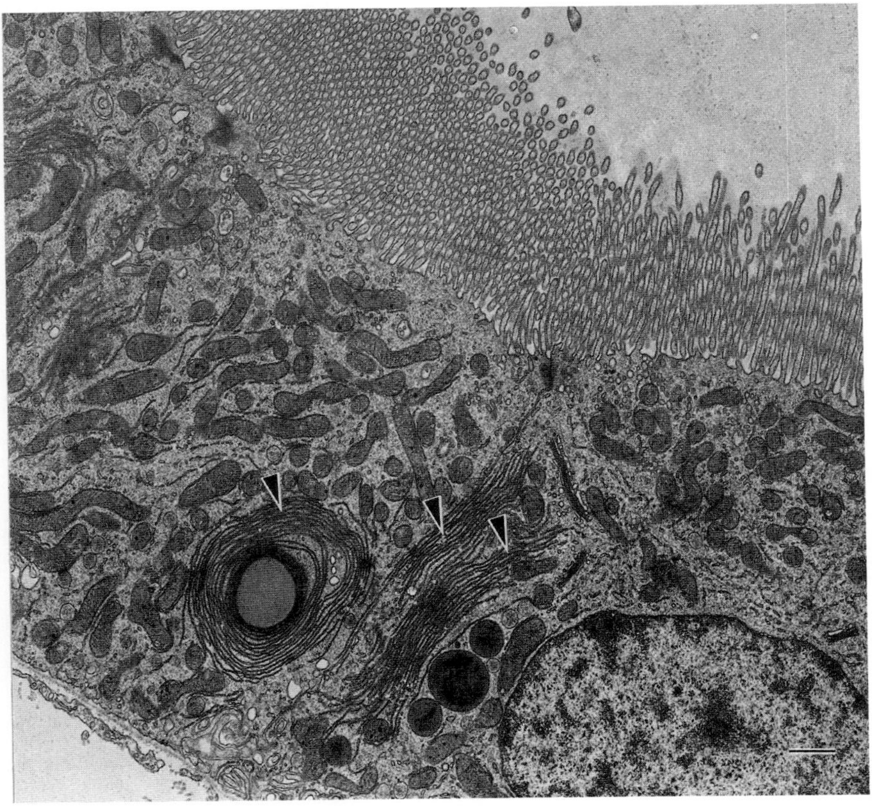

Fig. 14. Proximal tubule in *Potamotrygon* sp. Px-III with remarkable stacks of cisternae of the smooth endoplasmic reticulum (arrowheads) is shown. A stack surrounds a lipid droplet on the left side of the micrograph. ×5820; bar = 1 μm.

The *marine elasmobranch* intermediate segment is the longest segment in the renal tubule, where it forms two of the four loops (Fig. 4a). Its counterpart in freshwater rays is considerably shorter than in marine species, where it begins as the terminal part of the ascending branch of the second loop in the sinus zone. In *marine* species it forms the third loop (in the bundle) and the fourth loop (in the sinus zone), and at the vascular pole of the renal corpuscle (Fig. 4a), it ends by merging into the distal segment. The most prominent characteristic distinguishing this segment in all of its subdivisions is the absence of an easily recognizable brush border. Instead, the apical parts of the epithelial cells bear low, microvillilike projections observed by light and electron microscopy. The transition between proximal segment brush border cells to the intermediate segment is distinct and abrupt (Fig. 15).

Cells in the intermediate segment of both *marine* sharks, skates, and rays and *freshwater* rays show remarkable similarities, suggesting a commonality of function. Light micrographs of some subdivisions are shown in Figs. 12a,12b, 15, and 16a. The microscopic differences in the six subdivisions are extremely subtle and are best observed ultrastructurally. By electron microscopy the basolateral cell borders are striking because of the highly interdigitated infoldings of rather uniform depth (Figs. 17a–17d). They are separated from the junctional complex to the basement membrane by an intercellular space of extremely regular width of about 30 nm. The interdigitating infoldings vary in complexity and intensity depending on the segment subdivision but are always completely devoid of cytoplasmic organelles and arranged roughly parallel to one another (Figs. 17a–17d). The other major characteristic of this segment is the numerous, elongated mitochondria that do not appear to have an organized relationship with the basolateral infoldings. These latter two features generated the appearance of striations at the light microscopic level (Figs. 12b and 16a) described by Haller (1902) as "Streifenabschnitt."

There are major species differences in the morphology of the first subdivision of the intermediate segment (In-I). In the dogfish shark

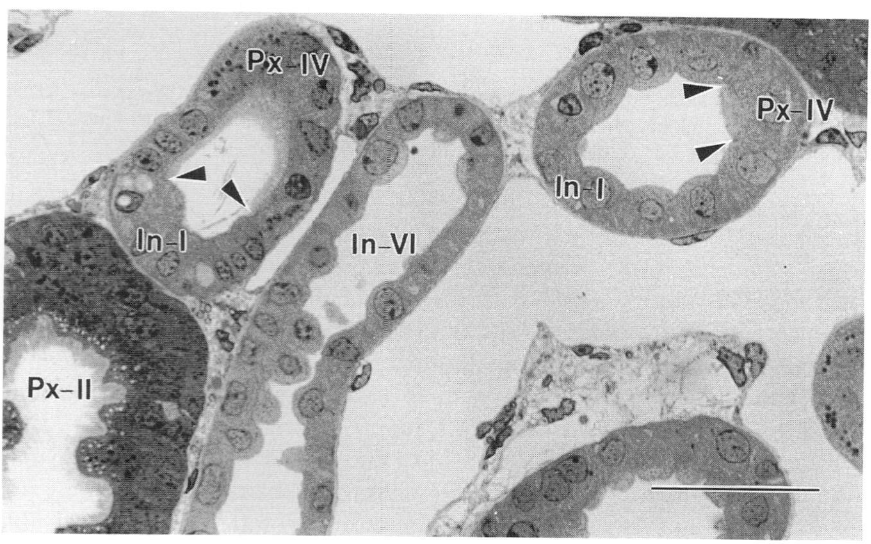

Fig. 15. Sinus zone in *Squalus acanthias*. Transitions from brush border cells (Px-IV) to cells without brush border (In-I) are marked by arrowheads. Parts of Px-II and In-VI subdivisions are recognizable. ×440; bar = 50 μm.

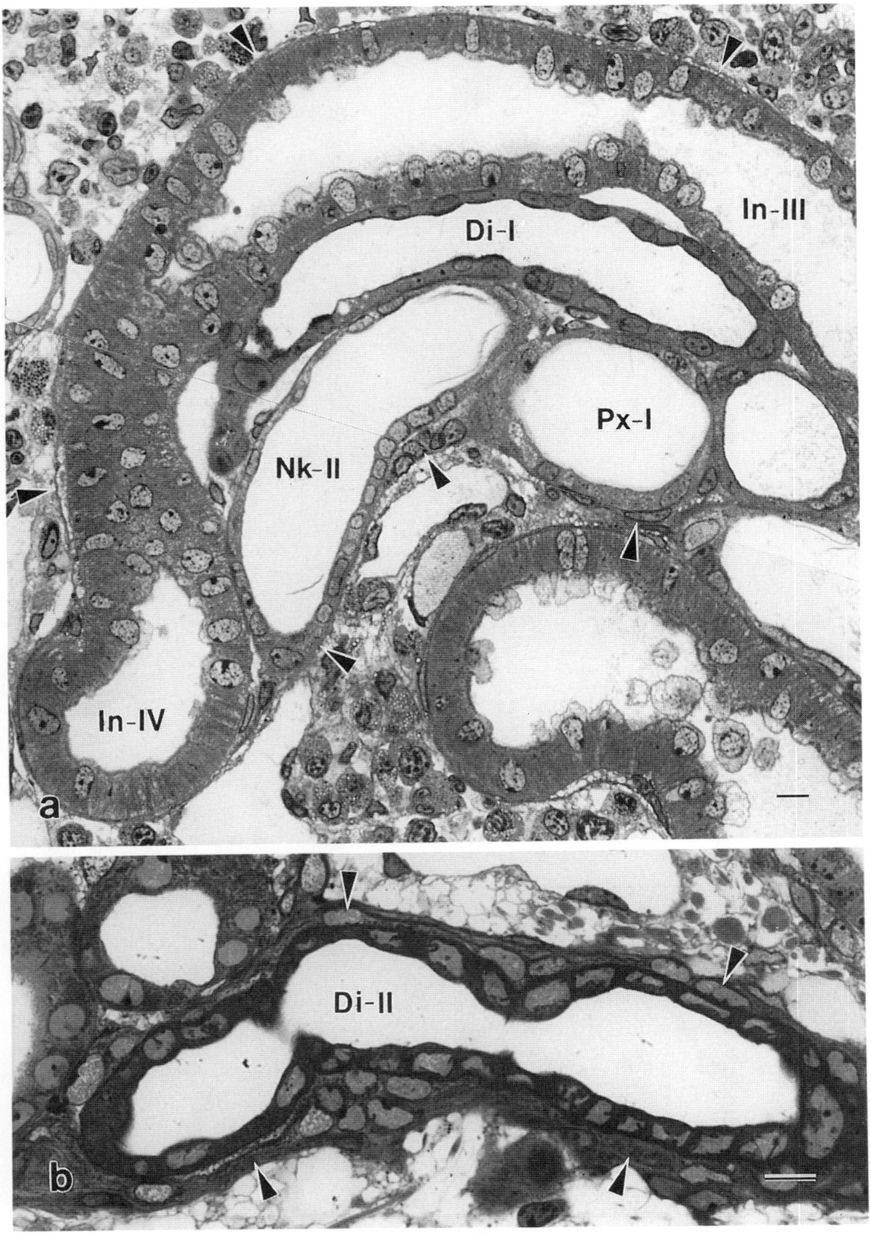

Fig. 16. Bundle zone in *Squalus acanthias* (a) and *Raja erinacea* (b). The tubules have been exposed in longitudinal to oblique sections. Arrowheads indicate the peritubular sheath. In (a), Nk-II, Px-I, In-III, In-IV, and Di-I can be recognized inside the bundle. In (b), Di-II lies outside the bundle still surrounded by the peritubular sheath (arrowheads). (a) ×410; (b) 640; bar = 10 μm.

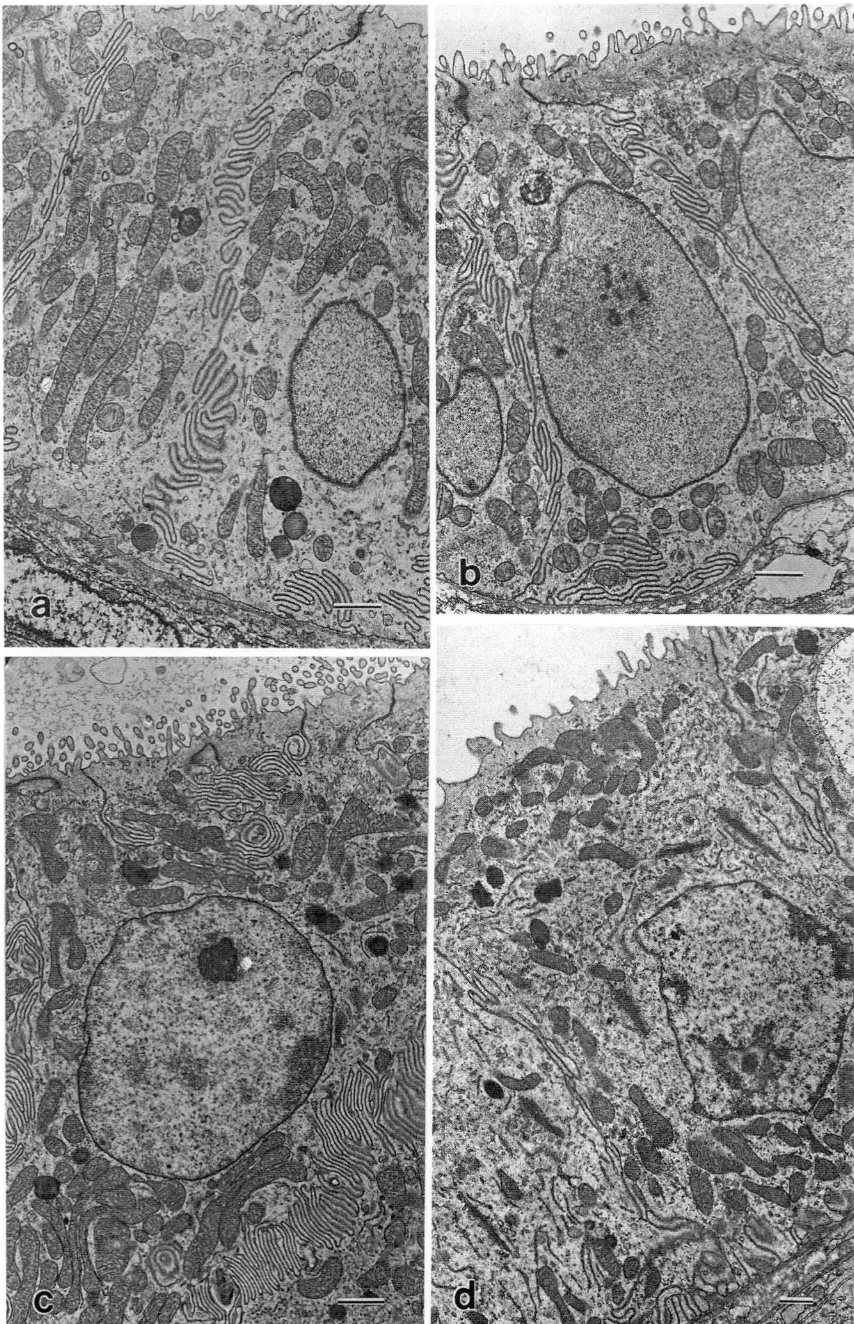
a
b
c
d

(*S. acanthias*) it is thin, with a consistently wide lumen lined by cuboidal cells that bulge into the lumen and, importantly, have no intercellular granules (Fig. 15). In the little skate (*R. erinacea*), the first intermediate subdivision, In-I, is short, lying in the sinus zone extremely close to the renal corpuscle. It is lined by an irregular epithelium of squamous to columnar cells with heterogeneously stained granules up to 3 μm in diameter. Because of this cytoplasmic component, In-I in this elasmobranch species deserves the name "granular" since caveolae, multivesicular bodies, and granules with lipid layers and/or homogeneous, electron-dense material can be found in the same cell. They probably represent successive stages of an endocytoticlike process (Lacy and Reale, 1991b).

In-II and In-V have nearly identical structure and are often difficult to differentiate from each other. The epithelial cells of these subdivisions are columnar-cuboidal, with a large pale nucleus and long mitochondria (Fig. 17a). In the third subdivision (In-III), the nonflagellar cells are cuboidal, with a large, pale nucleus and roundish mitochondria (Fig. 17b). In-IV (Fig. 17c) is the largest tubule within the bundle and at the light microscopic level (Fig. 16a) has a characteristic striated appearance that probably corresponds to the "special segment" reported by Haller (1902). The In-IV cells are highly interdigitating, with complex lateral borders and numerous mitochondria. Isolated perfused tubule studies indicated that this subdivision (In-IV), which is found within the tubular bundle, is a "diluting segment" of the dogfish (*S. acanthias*) nephron (Friedman and Hebert, 1990; Hebert and Friedman, 1990).

In-VI is bordered by columnar cells with stubby microvilli and numerous mitochondria (Figs. 15 and 17d). Lateral cell interdigitations are uncommon but basal infoldings are numerous. In In-VI of some species, such as the Atlantic sharpnose shark, there are apical secretory granules that stain with the PAS reaction indicating the presence of glycoproteins.

In *freshwater species,* the entire intermediate segment is confined to the complex zone. Its subdivisions are characterized by a continuous progressive increase of the cell height and of tightly interdigitating infoldings of the lateral plasma membrane. Along the entire length of this segment the Golgi apparatus is supra- and paranuclear, the rough endoplas-

---

Fig. 17. Ultrastructural aspects of some subdivisions of the intermediate segment in *Raja erinacea*. (a) In-II, (b) In-III, (c) In-IV, (d) In-VI. The intercellular spaces between adjacent epithelial cells are gradually longer from (a) to (c) but extremely uniform in width. (a,b,d) ×5880; (c) ×4410; bar = 1 μm. (a–c: from Lacy, E. R., and Reale, E. 1991b. The fine structure of the elasmobranch renal tubule: Intermediate, distal, and collecting duct segments of the little skate. *Am. J. Anat.*, v. 192. Copyright © 1991 by Wiley-Liss, Inc., a subsidiary of John Wiley & Sons, Inc.)

mic reticulum is scanty, the smooth endoplasmic reticulum is abundant, especially apically, the glycogen particles are numerous, and dense bodies are few. A granular subdivision similar to that found in In-I of the little skate or Px-II of the dogfish (*S. acanthias*) is absent. The cells lining the In segment are usually all nonflagellar. The height of the cells, their mitochondrial complement, and the extension and complexity of their basolateral borders gradually but remarkably increase from In-I to In-VI; with the intercellular space, however, remaining constant. In-IV is characterized by few scattered cells with numerous granules.

In both *marine* and *freshwater* elasmobranchs the transition from the In segment to the Di segment takes place close to the vascular pole of the parent renal corpuscle.

D. Distal Segment

As in all other vertebrate kidneys, the distal tubule is defined as that portion of the nephron that begins at the position where the nephron is in close apposition to the vascular pole (afferent and efferent arterioles) of the renal corpuscle.

There are two subdivisions of the distal segment (Di) in *marine elasmobranchs* (Fig. 4a). The first (Di-I) contacts the extraglomerular mesangium at the vascular pole of the renal corpuscle as described earlier and then enters and extends to the opposite end of the bundle (Figs. 12a, 12b, and 16a). It leaves the other tubules and, still surrounded by an extension of the peritubular sheath, continues as a single tubule, Di-II, lined by a different epithelium (Fig. 16b). After a short path in the interbundular and subcapsular connective tissue, it meets a collecting duct.

In *marine species* a portion of the tubule is specialized to form the macula densa, located at the vascular pole of the renal corpuscle (Fig. 10). Beyond this region the darkly staining tubular cells become low cuboidal, bulging into the tubule lumen, and their lateral limits are sharply defined at the light microscopic level. Near the distal end of the tubular bundle the epithelium becomes flatter, similar to that of Di-II (Figs. 16b, 18a, and 18b).

In thin sections the epithelial cells of Di-I possess irregularly shaped

---

Fig. 18. Distal tubules in *Raja erinacea*. (a) Di-I with basally large intercellular spaces; (b) Di-II enwrapped by the end of the peritubular sheath (PS). Colloidal carbon retrogradely injected filled this segment thus allowing its identification. ×6000; bar = 1 μm. (From Lacy, E. R., and Reale, E., 1991b. The fine structure of the elasmobranch renal tubule: Intermediate, distal, and collecting duct segments of the little skate. *Am. J. Anat.*, v. 192. Copyright © 1991 by Wiley-Liss, Inc., a subsidiary of John Wiley & Sons, Inc.)

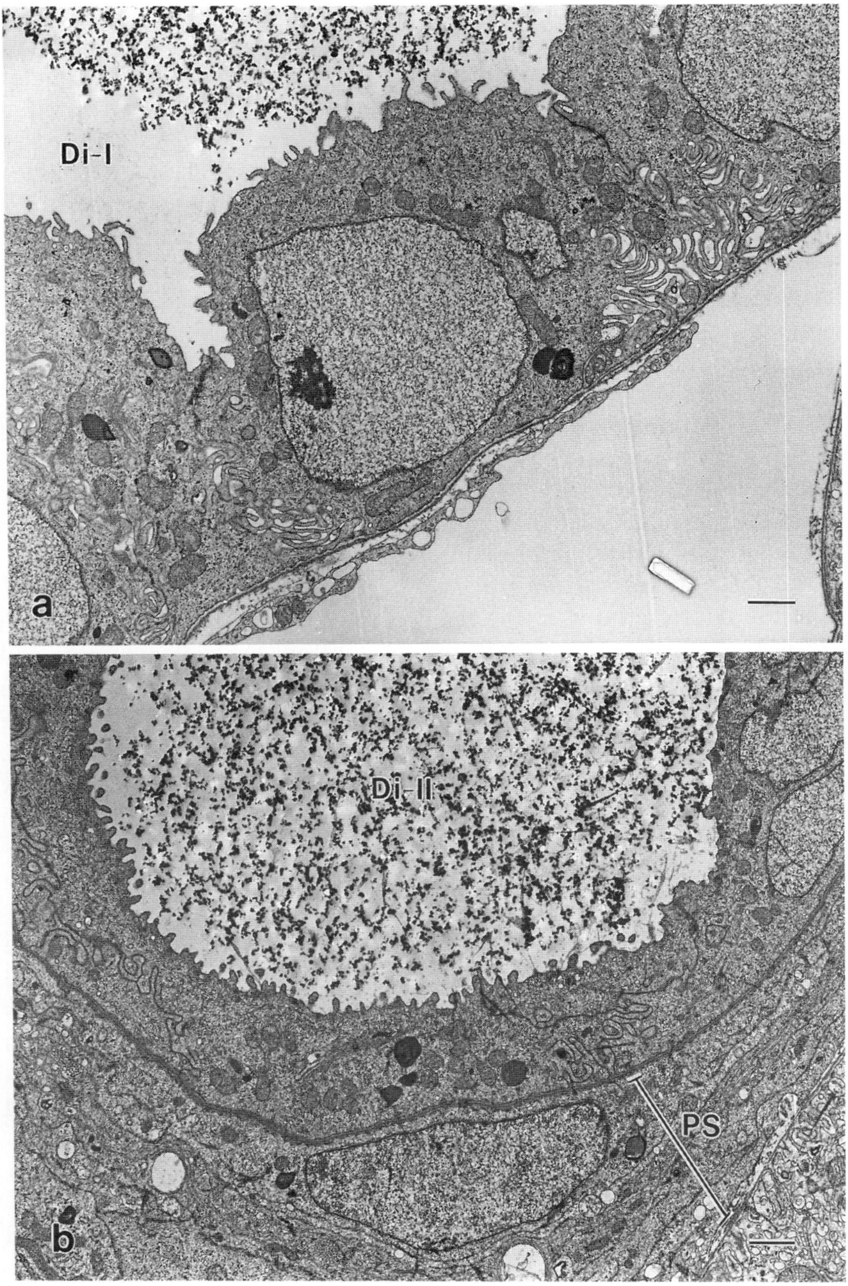
Di-I
a
Di-II
PS
b

microvilli- or microplicaelike apical projections, few mitochondria, a supranuclear Golgi complex, and few cisternae of the rough and numerous cisternae of the smooth endoplasmic reticulum (Fig. 18a). The highly interdigitating lateral cell infoldings separated by a uniformly narrow intercellular space present in the In segment subdivision are substituted by lateral microplicae or folds. These folds are much more elaborated basally than apically (Fig. 18a; Lacy and Reale, 1991b). This configuration of the intercellular borders and spaces is continuous along the length of the distal segment. In Di-I of the little skate and scalloped hammerhead shark, a row of apical secretory granules is stained by PAS and Alcian blue (pH 1.0) reactions, indicating the presence of sulfated glycoconjugates. Similar granules characterize Di-II in all other marine species. The secretory, apical granules of Di-II are round in the early part of the tubule of the little skate, but oblong in the more distal regions similar to those of CD-I (see Section VI,E). Di-II cells, as observed in *R. erinacea,* are low cuboidal to squamous and have an irregularly shaped nucleus generally parallel to the epithelial basement membrane (Fig. 18b). The intercellular space remains narrow and long with numerous interdigitations of irregular form and shape.

The distal tubule in the *freshwater ray* follows a peculiar pathway inside the lateral groove of the renal corpuscle (see JGA description in Section V,C). Beyond the lateral groove, the cells are high columnar and their apical cytoplasm is dome-shaped, with numerous mitochondria, some cisternae of the endoplasmic reticulum, a Golgi complex, and a few dense bodies; the intercellular space is open and filled with numerous long, flattened, interdigitating folds (Fig. 19). After a short path in the complex zone, Di abruptly merges in the collecting duct system (Fig. 20a).

E. Collecting Duct Segment

In *marine species* the collecting duct segment (CD) resides in the connective tissue septa between lobes and in the renal capsule. There are two subdivisions, the first (CD-I) consisting of cuboidal-columnar cells arranged in a single layer followed by the second (CD-II) lined by a pseudostratified epithelium (Lacy and Reale, 1985b; Fig. 4a). Both subdivisions show a characteristic pleomorphic, deeply indented nucleus, and apical assembles of roundish secretory granules (>0.5 to 1.0 μm in diameter). The cells with larger granules prevail in CD-II. By electron microscopy, these granules containing glycoconjugates have a banded substructure (Fig. 20b). These granules appear to be secretory in nature and occasionally exocytotic figures have been observed. Desmosomes are numerous and the lateral cell borders highly interdigitated in both subdivisions. In CD-II, the cells below the luminal cell layer are devoid of granules

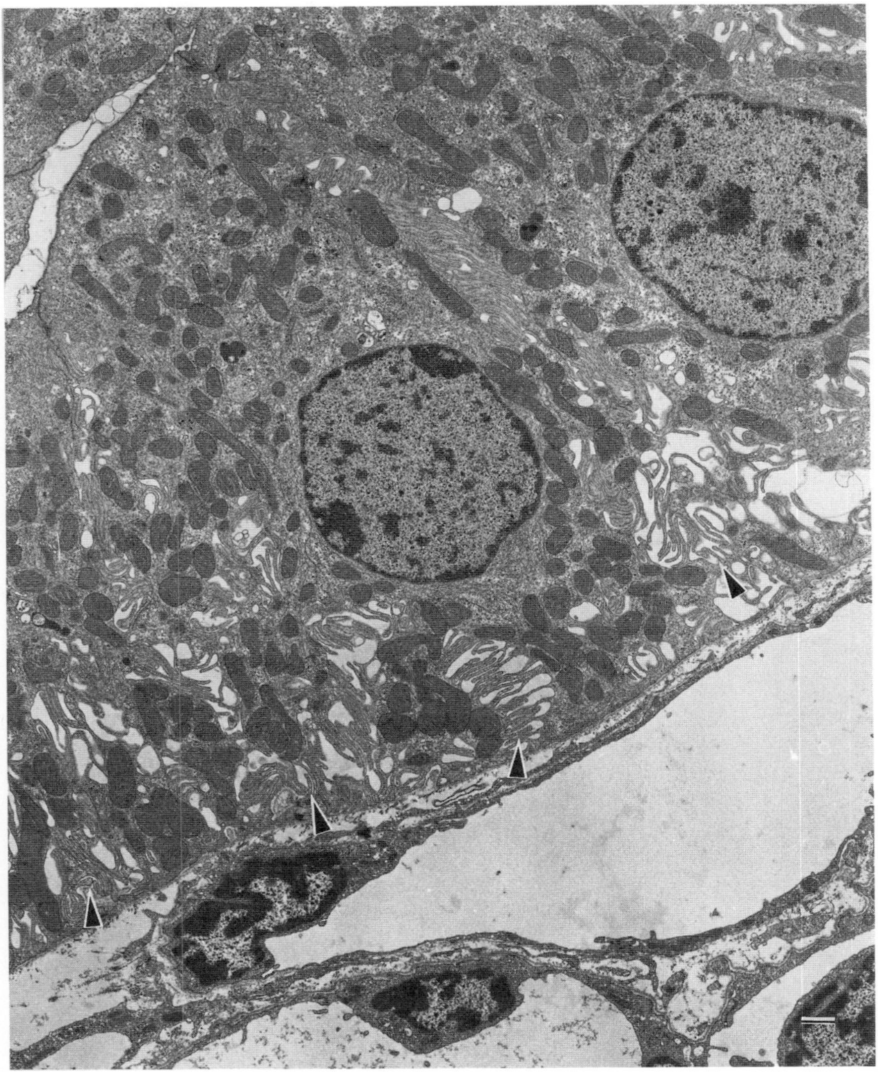

Fig. 19. Distal tubule in *Potamotrygon* sp. The intercellular spaces are basally open and largely occupied by microfolds (arrowheads). ×4400; bar = 1 μm.

but rich in bundles of intermediate filaments directed toward numerous desmosomes, characteristics found in epithelia such as mammalian urinary bladder, which is subjected to stretch. The peritubular sheath surrounding Di-II is not continuous over the collecting duct but is replaced in both

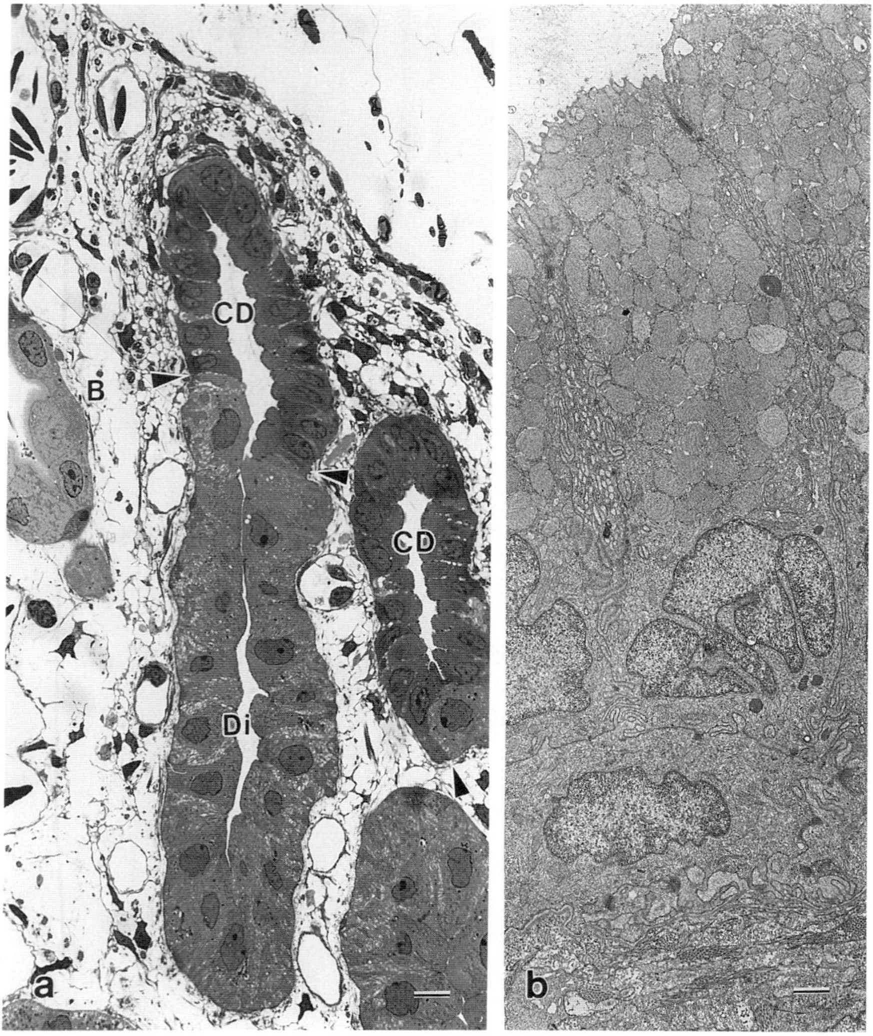

Fig. 20. In (a), transitions (arrowheads) from the distal tubule (Di) to the collecting duct (CD) in *Potamotrygon* sp. In (b), thin sections of the collecting duct epithelium in *Raja erinacea*. Granules are located in the apical region of the cell, which borders the lumen of the duct. (a) ×470; bar = 10 μm. (b) ×4700; bar = 1 μm.

subdivisions by layers of fibrocytes alternating with bundles of collagen fibrils).

In some species, such as the spiny dogfish (*S. acanthias*), ciliary cells, as distinguished from flagellar cells found in the earlier parts of the tubule (described earlier), are intercalated among the "secretory" cells just described.

In *freshwater species* the general structure of the CD subdivisions is similar to that of the marine counterpart. The transition from Di-II to CD-I is abrupt in the river ray (Fig. 20a); after a short path the epithelium becomes pseudostratified and then stratified. The cells are slender, with a basally positioned nucleus, deeply stained cytoplasm, and just a few basolateral invaginations. The apical granules contain glycoconjugates.

The various CD subdivisions repeatedly merge, generating large ducts located on the ventromedial surface of the kidney toward the dorsal aorta. The largest ducts are surrounded by some layers of smooth muscle cells, suggesting a pulsatile propulsion of fluid that has been observed in renal tubules of teleost fish (Townsley and Scott, 1963).

Large collecting ducts of the little skate (*R. erinacea*) were catheterized *in vivo* and found to be the most important site of urinary dilution when the animals were acclimated to full-strength seawater (Stolte *et al.*, 1977). Upon dilution to 75% seawater, fluid entering the collecting duct had already been diluted, suggesting that at least one nephron segment proximal to the end collecting duct was responsible for urinary dilution (Stolte *et al.*, 1977).

VII. OVERVIEW

When the microanatomy of the complete elasmobranch renal tubule was elucidated in the late 1980s, it set the stage for long-overdue physiological investigations of epithelial transport in this organ. Unfortunately, the "wave of modern technology" had passed *in vivo* micropuncture and isolated perfused tubule studies, which are still the most useful tools for elucidating fundamental transepithelial transport. The very powerful techniques of "molecular biology" cannot yet be applied to elasmobranch renal function because of the paucity of basic knowledge about the dynamics of solute and water movement across the nephron and vasculature. The anatomical complexity of the elasmobranch renal tubule no doubt reflects an equally complex array of renal tubular transport mechanisms. However, one cannot help but think that within this daunting complexity there lie unique physiological and molecular mechanisms that probably have applicability to the maintenance of human renal function.

REFERENCES

Acrivo, C. (1939). Über Muskelbildungen längs der Kiemenblattblutgefässe einiger Selachier. *Zool. Anz.* **126,** 125–129.

Bargmann, W. (1937). Untersuchungen über Histologie and Histophysiologie der Fischniere. *Z. Zellforsch. Mikrosk. Anat.* **26,** 765–788.

Bargmann, W. (1954). Über die Endomeninx der Fische (zugleich ein Beitrag zur Kenntnis der Turbanorgane). *Z. Zellforsch. Mikrosk. Anat.* **40,** 88–100.

Beyenbach, K. W., and Frömter, E. (1985). Electrophysiological evidence for Cl-secretion in shark renal proximal tubules. *Am. J. Physiol.* **248,** F282–F295.

Boyd, R. B., and De Vries, A. L. (1983). The seasonal distribution of anionic binding sites in the basement membrane of the kidney glomerulus of the winter flounder. *Pseudopleuronectes americanus. Cell Tissue Res.* **234,** 271–277.

Boylan, J. W. (1967). Gill permeability in *Squalus acanthias. In* "Sharks, Skates, and Rays" (P. W. Gilbert, R. F. Mathewson, and D. P. Rall, eds.), pp. 197–206. Johns Hopkins Univ. Press, Baltimore.

Boylan, J. W. (1972). A model for passive urea reabsorption in the elasmobranch kidney. *Comp. Biochem. Physiol. A* **42,** 27–30.

Decker, B., and Reale, E. (1991). The glomerular filtration barrier of the kidney in seven vertebrates classes. Comparative morphological and histochemical observations. *Basic Appl. Histochem.* **35,** 15–36.

Deetjen, P., and Antkowiak, D. (1970). The nephron of the skate, *Raja erinacea. Bull. Mt. Desert Island Biol. Lab.* **10,** 5–10.

DeVlaming, V. L., and Sage, M. (1973). Osmoregulation in the euryhaline elasmobranch, *Dasyatis sabina. Comp. Biochem. Physiol. A* **45,** 31–44.

Forster, R. P. (1967). Osmoregulatory role of the kidney in cartilagineous fishes (Chondrichthyes). *In* "Sharks, Skates, and Rays" (P. W. Gilbert, R. F. Mathewson, and D. P. Rall, eds.), pp. 187–195. Johns Hopkins Univ. Press, Baltimore.

Friedman, P. A., and Hebert, S. C. (1990). Diluting segment in the kidney of dogfish shark. I. Localization and characterization of chloride absorption. *Am. J. Physiol.* **258,** R398–R408.

Galli, S., and Kiang, K. (1990). Angiotensin II-like immunoreactivity in sharks: Response to hemorrhage and sodium depletion. *FASEB J.* **4**(4), 4000.

Gardiner, D. S., and Lindop, G. B. M. (1985). The granular peripolar cell of the human glomerulus: A new component of the juxtaglomerular apparatus? *Histopathology* **9,** 675–685.

Gérard, M. P., and Cordier, M. R. (1934). Comment interpréter les formations granulaires recontrées dans le tube contourné du rein, chez les Vertébrés? *Bull. Acad. R. Méd. Belg.* [5] **14,** 160–185.

Ghouse, A. M., Parsa, B., Boylan, J. W., and Brennan, J. C. (1968). The anatomy, microanatomy, and ultrastructure of the kidney of the dogfish, *Squalus acanthias. Bull. Mt. Desert Island Biol. Lab.* **8,** 22–29.

Grabowski, G. M., Reale, E., and Lacy, E. R. (1995). Renal morphology of the freshwater elasmobranch, *Potamotrygon* sp. (Submitted for publication.)

Haller, B. (1902). Über die Urniere von *Acanthias vulgaris. Gegenbaurs Morphol. Jahrb.* **29,** 283–316.

Hazon, N., Balment, R. J., Perrott, M., and O'Toole, L. B. (1989). The renin–angiotensin system and vascular and dipsogenic regulation in elasmobranchs. *Gen. Comp. Endocrinol.* **74,** 230–236.

Hebert, S. C., and Friedman, P. A. (1990). Diluting segment in kidney of dogfish shark. II. Electrophysiology of apical membranes and cellular resistances. *Am. J. Physiol.* **258,** R409–R417.

Henderson, I. W., Oliver, J. A., McKeever, A., and Hazon, N. (1981). Phylogenetic aspects of the renin–angiotensin system. *In* "Advances in Animal and Comparative Physiology" (G. Pethes and V. L. Frenyo, eds.), pp. 355–363. Pergamon, Oxford.

Henderson, I. W., O'Toole, L. B., and Hazon, N. (1988). Kidney function. *In* "Physiology of Elasmobranch Fishes" (T. J. Shuttleworth, ed.), pp. 201–214. Springer-Verlag, Berlin.

Hentschel, H. (1988). Renal blood vascular system in the elasmobranch, *Raja erinacea* Mitchill, in relation to kidney zones. *Am. J. Anat.* **183,** 130–147.

Hentschel, H., and Elger, M. (1982). Strukturelle Organisation der Elasmobranchier-Niere. *Verh. Dtsch. Zool. Ges.* **263.**

Hentschel, H., and Elger, M. (1989). Morphology of glomerular and aglomerular kidneys. *In* "Structure and Function of the Kidney" (R. K. H. Kinne, ed.), pp. 1–72. Karger, Basel.

Kempton, R. T. (1943). Studies on the elasmobranch kidney. I. The structure of the renal tubule of the spiny dogfish (*Squalus acanthias*). *J. Morphol.* **73,** 247–263.

Kempton, R. T. (1953). Studies on the elasmobranch kidney. II. Reabsorption of urea by the smooth dogfish, *Mustelus canis. Biol. Bull.* (*Woods Hole, Mass.*) **104,** 45–56.

Kempton, R. T. (1962). Studies on the elasmobranch kidney. III. The kidney of the lesser electric ray, *Narcine brasiliensis. J. Morphol.* **111,** 217–225.

Kempton, R. T. (1964). Some anatomical features of the elasmobranch kidney. *Biol. Bull.* (*Woods Hole, Mass.*) **127,** 377.

Kon, Y., Sugimura, M., Murakami, K., Lacy, E. R., and Reale, E. (1995). The presence of a unique juxtaglomerular apparatus and renin containing smooth muscle-like cells in the freshwater stingray, *Potamotrygon* sp. (Submitted for publication).

Lacy, E. R., and Reale, E. (1985a). The elasmobranch kidney. I. Gross anatomy and general distribution of the nephrons. *Anat. Embryol.* **173,** 23–34.

Lacy, E. R., and Reale, E. (1985b). The elasmobranch kidney. II. Sequence and structure of the nephrons. *Anat. Embryol.* **173,** 163–186.

Lacy, E. R., and Reale, E. (1986) The elasmobranch kidney. III. Fine structure of the peritubular sheath. *Anat. Embryol.* **173,** 299–305.

Lacy, E. R., and Reale, E. (1989). Granulated peripolar epithelial cells in the renal corpuscle of marine elasmobranch fish. *Cell Tissue Res.* **257,** 61–67.

Lacy, E. R., and Reale, E. (1990). The presence of a juxtaglomerular apparatus in elasmobranch fish. *Anat. Embryol.* **182,** 249–262.

Lacy, E. R., and Reale, E. (1991a). Fine structure of the elasmobranch renal tubule: Neck and proximal segments of the little skate. *Am. J. Anat.* **190,** 118–132.

Lacy, E. R., and Reale, E. (1991b). The fine structure of the elasmobranch renal tubule: Intermediate, distal, and collecting duct segments of the little skate. *Am. J. Anat.* **192,** 478–497.

Lacy, E. R., Reale, E., Schlusselburg, D. S., Smith, W. K., and Woodward, D. S. (1984). Computer-assisted reconstruction of the renal countercurrent system in a marine elasmobranch fish. *Anat. Rec.* **208,** 97A.

Lacy, E. R., Reale, E., Schlusselburg, D. S., Smith, W. K., and Woodward, D. S. (1985). A renal countercurrent system in marine elasmobranch fish: A computer-assisted reconstruction. *Science* **227,** 1351–1354.

Lacy, E. R., Castellucci, M., and Reale, E. (1987). The elasmobranch renal corpuscle: Fine structure of Bowman's capsule and the glomerular capillary wall. *Anat. Rec.* **218,** 294–305.

Lacy, E. R., Reale, E., Grabowski, G. M., and Mackanos, L. A. (1989a). Freshwater elasmobranchs do not have a renal counter-current system. *Anat. Rec.* **223,** 64A.

Lacy, E. R., Williams, J. C., Rivers, R., Karnaky, K. J., Jr., Mackanos, L. A., Grabowski, G. M., Luciano, L., and Reale, E. (1989b). Motion analysis of flagellar ribbon and ciliary beat cycles in teleost and elasmobranch fish nephrons. *Eur. J.Cell Biol.* **49**(Suppl. 27), 54.

Lacy, E. R., Luciano, L., and Reale, E. (1989c). Flagellar cells and ciliary cells in the renal tubule of elasmobranchs. *J. Exp. Zool., Suppl.* **2,** 186–192.

Lacy, E. R., Luciano, L., and Reale, E. (1991a). Membrane specializations in flagellar ribbons of elasmobranch fish. *Tissue Cell* **23,** 223–234.

Lacy, E. R., Luciano, L., and Reale, E. (1991b). Elastic-like tissue, a new component of the renal glomerular capillary wall in a cartilagineous fish. *Anat. Embryol.* **183,** 475–481.

Leydig, F. (1852). "Beiträge zur Mikroskopischen Anatomie und Entwicklungsgeschichte der Rochen und Haie." Wilhelm Engelmann, Leipzig.

Marshall, E. K. (1930). A comparison of the function of the glomerular and aglomerular kidney. *Am. J. Physiol.* **94,** 1–10.

Marshall, E. K., Jr. (1934). The comparative physiology of the kidney in relation to theories of renal secretion. *Physiol. Rev.* **14,** 133–159.

Meyer, P. (1888). Über Eigenthuemlichkeiten in den Kreislaufsorganen der Selachier. *Mitt. Zool. Stn. Neapel* **8,** 307–373.

Parker, T. J. (1881). On the venous system of the skate (*Raja nausata*). *Trans. N.Z. Inst.* **13,** 413–418.

Rand, H. W. (1905). The skate as a subject for classes in comparative anatomy; Injection methods. *Am. Nat.* **39,** 364–379.

Ryan, G. B., Coghlan, J. P., and Scoggins, B. A. (1979). The granulated peripolar epithelial cell: A potential secretory component of the renal juxtaglomerular complex. *Nature (London)* **277,** 655–656.

Sawyer, D. B., and Beyenbach, K. W. (1985). Mechanism of fluid secretion in isolated shark renal proximal tubules. *Am. J. Physiol.* **249,** F884–F890.

Shuttleworth, T. J. (1988). Salt and water balance—Extrarenal mechanisms. *In* "Physiology of Elasmobranch Fishes" (T. J. Shuttleworth, ed.), pp. 171–199. Springer-Verlag, Berlin.

Simionescu, N., and Simionescu, M. (1988). The cardiovascular system. In "Cell and Tissue Biology: A Textbook of Histology" (L. Weiss, ed.). Urban & Schwarzenberg, Baltimore and Munich.

Smith, H. W. (1931). The absorption and excretion of water and salts by the elasmobranch fishes. II. Marine elasmobranchs. *Am. J. Physiol.* **98,** 296–310.

Smith, W. K., Schlusselberg, D. S., Culter, B. G., Woodward, D. J., and Lacy, E. R. (1983). Hierarchical database design for biological modelling. *Nat. Comput. Graphics Assoc. Conf. Proc.*, pp. 106–116.

Stolte, H., Galaske, R. G., Eisenbach, G. M., Lechene, C., Schmidt-Nielsen, B., Boylan, J. W. (1977). Renal tubule ion transport and collecting duct function in the elasmobranch little skate, *Raja erinacea*. *J. Exp. Zool.* **199,** 403–410.

Taugner, R., and Hackenthal, E. (1989). "The Juxtaglomerular Apparatus: Structure and Function." Springer-Verlag, Berlin.

Thumwood, C. M., McCausland, J., Alcorn, D., and Ryan, G. B. (1993). Scanning and transmission electron microscopic study of peripolar cells in the newborn lamb kidney. *Cell Tissue Res.* **274,** 597–604.

Townsley, P. M., and Scott, M. A. (1963). Systolic muscular action of the kidney tubules of flounder. *J. Fish. Res. Board Can.* **20,** 243–244.

Uva, B., Masini, M. A., Hazon, N., O'Toole, L. B., Henderson, I. W., and Ghiani, P. (1992). Renin and angiotensin converting enzyme in elasmobranchs. *Gen. Comp. Endocrinol.* **86,** 407–412.

Weibel, E. R., and Palade, G. (1964). New cytoplasmic components in arterial endothelia. *J. Cell Biol.* **23,** 101–102.

6

SOLUTE TRANSPORT BY FLOUNDER RENAL CELLS IN PRIMARY CULTURE

J. LARRY RENFRO

I. INTRODUCTION

The development of a primary culture system for winter flounder renal epithelial cells has provided an alternative means to investigate the mechanisms and regulation of transepithelial transport processes (Dickman and Renfro, 1986). The utility of primary cell cultures for this purpose obviously requires the full expression of differentiated characteristics, which in turn depends on the specific cell selection and culturing process. The advantages and disadvantages associated with various renal cell culture systems have been thoroughly reviewed (Handler, 1986; Stanton and Seifter, 1987; Valentich, 1986; Dickman and Renfro, 1986). Primary cultures of confluent epithelium provide control of all transepithelial chemical and electrical gradients, the effects of which can be assessed in a biophysically precise manner in Ussing flux chambers. Long-term exposures to possible regulatory factors can be examined, and the autonomous responses of the tissue can be assessed independently of the influence of other body systems.

The techniques used to attain this state in flounder renal proximal tubule primary monolayer cultures (PTCs) have been reviewed in detail elsewhere (Dickman and Renfro, 1993) and are only briefly restated here. The primary consideration in this chapter will be the transepithelial transport mechanisms of various solutes by this system and the regulatory processes influencing their transport.

II. SUMMARY OF CULTURE METHODOLOGY

The nephrons of many marine teleosts, including those of the winter flounder (*Pleuronectes americanus*), consist almost exclusively of proximal tubule-like segments. The ultrastructural evidence presented by Hickman and Trump (1969) and Trump and Jones (1977) has shown that the winter flounder nephron appears to lack a distal convoluted-type segment. The latter authors identified proximal segments I, II, and III based primarily on the similarity to mammalian proximal convoluted tubule morphology. The structural identification is supported by the proximal tubule-like transport properties observed in several studies. Organic anion secretion (Forster, 1948; Hickman and Trump, 1969; Kinter, 1975) and calcium (Renfro, 1978; Renfro *et al.*, 1982), magnesium (Renfro and Shustock, 1985; Cliff *et al.*, 1986), hexose (Kleinzeller and McAvoy, 1973), and sulfate (Renfro and Dickman, 1980) transport, as well as electrophysiological properties (Beyenbach *et al.*, 1986), all reflect the proximal tubule nature of this nephron.

In PTCs (Dickman and Renfro, 1986; modifications in Gupta and Renfro, 1989), the kidneys are perfused and teased apart in tissue culture medium (modified M-199). Incubation of tubules in Ca- and Mg-free balanced salt solution with 0.2% trypsin at 22°C is used to remove extrarenal tissues. Tubule fragments are separated from nontubule tissues on 20-μm Nitex nylon mesh. The epithelial cells are released from these fragments by 3 days of cold trypsinization at 5°C (Cole and Paul, 1966). Released cells are collected through 50-μm Nitex mesh, washed, pelleted, suspended in modified M-199 with 10% flounder serum, and plated to confluency. The study of transepithelial transport by cultured epithelia necessitates the use of a removable porous culture support so that cultures can be mounted in Ussing chambers. For this purpose, cells are plated on native rat tail collagen gels, which are released from the plastic culture dishes after cell attachment (Bisbee *et al.*, 1979).

Floating collagen gels allow access of nutrients and hormones to the basal cell surface; they are relatively transparent so that the course of culture development can be monitored, and the "contractability" of collagen gels has been shown to influence differentiation of cultured cells (Burwen and Pitelka, 1980), i.e., through reorganization of the collagen fibrils the cellular monolayer remains confluent and contact inhibited. The collagen is detached from the culture dish 4 days after plating. At 12 days in culture, gel diameter is reduced about 70%, an amount that depends on both cell plating densities and gel collagen concentration (Dickman and Renfro, 1993).

Floating collagen gels also allow the cells to assume their natural columnar shape in culture. Studies with both mammary gland cells (Burwen and Pitelka, 1980) and hepatocytes (Michaelopoulos and Pitot, 1975) indicate that proliferation is minimal when they are cultured on contractable collagen gels. It has been suggested, at least for the mammary gland cells, that the lack of cell growth contributes to the enhanced expression of differentiated function. Following this rearrangement of the collagen fibrils, which is called contraction, 12-day-old cultures form an apical carpet of microvilli comparable to that of flounder renal tubule *in vivo* with a correspondingly appropriate paracellular junctional area. Ultrastructural studies (Dickman and Renfro, 1986) revealed apical tight junctions (zonula occludens), prominent lateral interdigitations, and numerous apical vesicles. This reexpression of normal morphology is directly related to the development of electrophysiological and transport properties in primary monolayer cultures (Dickman and Renfro, 1986), which are extremely similar to those seen in intact, microperfused flounder proximal tubule (Beyenbach *et al.*, 1986).

III. ASSESSMENT OF TRANSPORT PROPERTIES

Reconfiguration of the flounder renal epithelium into the form of an epithelial sheet permits handling it in the same manner that Ussing and coworkers dealt with the frog skin (Ussing and Zerahn, 1951). This straightforward, elegant technique provides perhaps the least ambiguous means of distinguishing active and passive epithelial transport processes. Transepithelial electrical potential difference (PD), resistance (R), total short-circuit current (I_{sc}), and phloridzin-sensitive current (PS-I_{sc}) are measured on PTCs supported by 150-μm nylon mesh and mounted in Ussing chambers after Day 12 in culture. The Ussing chamber methods have been previously described (Gupta and Renfro, 1989) and salient points are reemphasized here. Each Plexiglas chamber with fluid volume of 1.6 ml per hemichamber and separating aperture of 0.332 cm^2 is maintained at 22°C with circulating water. The flounder saline in each hemichamber is stirred with small magnetically driven stir bars and gassed with humidified 99% O_2–1% CO_2.

A. Transepithelial Electrophysiology and Permselectivity

PD is determined with calomel reference electrodes connected to the peritubular and luminal compartments by 3 *M* KCl–2% agar bridges. Tissues are short-circuited with a high input impedance automatic dual-voltage clamp. Transepithelial R is determined from the change in PD produced by a brief 10-μA pulse controlled by the computer through the clamps' remote logic circuits.

Spontaneous transepithelial PD exhibited by the PTCs with flounder saline on both peritubular and luminal sides generally averages less than 1 mV, lumen negative (Table I). Transepithelial R is characteristic of the "leaky" epithelium, ranging from 20 to 45 ohms cm^2 in several different studies (Renfro, 1989). Both PD and R are reduced to zero by 30 min exposure to 5 m*M* EDTA (Dickman and Renfro, 1986), which has been shown to open Ca-sensitive tight junctions (Contreras *et al.*, 1991). Average short-circuit current is equivalent to a net ionic flux of about 900 nEq h^{-1} cm^{-2}. Prior to 11 days in culture, R and PD are variable, reflecting either a lack of confluence or incomplete establishment of polarity.

The selectivity of the paracellular junctional pathway in renal proximal tubule is an important determinant of the functional characteristics. Beyenbach *et al.* (1986) have shown that intact perfused flounder proximal

Table I
Transepithelial Electrical Properties of Flounder Renal Proximal Tubule in Primary Monolayer Culture[a]

Potential difference, mV[b]	−0.6 ± 0.10 (6)
Resistance, Ω cm^2	23 ± 2.3 (6)
Short-circuit current, μA cm^{-2}	24 ± 2.7 (6)
Phlorizin-sensitive current, μA cm^{-2}	4.5 ± 0.97 (5)

[a] Values are means ± SE; 3–7 cultures from each of 5 or 6 different preparations (*n*) were tested. Data taken from Dickman and Renfro (1986).

[b] Peritubular side is ground.

tubule is more permeable to sodium than to chloride. Transepithelial permselectivity was based on NaCl dilutional potentials. In the PTCs, transepithelial polarity is dependent on the direction of the NaCl concentration gradient and indicates that sodium is about four times more permeant than chloride (Dickman and Renfro, 1986).

B. Glucose Transport

Glucose reabsorption by the mammalian proximal tubule occurs through a concentrative process in the apical brush border membranes, which is phloridzin-sensitive and sodium-coupled. The process is rheogenic and accounts for a portion of the tissue net charge transfer. Eveloff *et al.* (1979) showed that a Na-coupled glucose transporter was present in winter flounder brush border membrane vesicles. Early work with the PTCs extended the characterization of glucose transport to transepithelial transport and confirmed the rheogenic nature and phloridzin sensitivity of transport. In Ussing chambers, exposure of the luminal side to glucose stimulates the short-circuit current in a saturable fashion (Dickman and Renfro, 1986). The K_m is 0.16 ± 0.012 mM and V_{max} is 5.3 ± 1.30 μA cm^{-2}. Glucose addition to the peritubular bath has no effect on current. With identical luminal and peritubular solutions, glucose accounts for about one-third of the total charge transfer. Luminal phloridzin (10^{-4} M) decreases short-circuit current by 4.5 ± 0.97 μA cm^{-2} at 0.5 mM glucose. Net fluxes determined from the difference between simultaneous measurements of unidirectional peritubular-to-luminal and luminal-to-peritubular fluxes of ^{3}H-labeled glucose reveal net reabsorption with flux ratios of 3–4. The nonmetabolized glucose analog, [^{14}C]α-methyl glucoside, is a competitive substrate on the apical Na–glucose cotransporter and yields similar flux ratios. In addition, the nonmetabolizeable glucose analog α-methyl glucoside in place of luminal glucose, produces a phloridzin-sensitive current of 4.9 ± 0.37 μA cm^{-2}. These studies also show that net glucose reabsorption by the PTCs remains stable in the Ussing chambers for at least 2 h while tissues are bathed in modified Forster's flounder saline.

Because Na-coupled glucose transport is a direct reflection of the integrity of the plasma membrane electrochemical Na gradient, it is a good indicator of the general metabolic state of the tissues. Presumably, if the cells begin to deteriorate, ATP levels will drop, plasma membrane ion gradients will begin to dissipate, and PS-I_{sc} will decrease. For this reason, in all PTC transport studies, PS-I_{sc} is used as an indicator of nonspecific metabolic effects of pharmacological or hormonal treatments.

C. Sulfate Transport

To compensate for osmotic water loss, marine teleosts must ingest seawater. The sulfate concentration of normal seawater is 25 m*M*. Part of this is absorbed along with the water and is excreted primarily by the kidneys. Flounder plasma inorganic sulfate concentration averages 0.6 m*M* (Renfro and Dickman, 1980) and renal clearance ratios exceed 12 (Hickman and Trump, 1969). It is thus apparent that a well-developed sulfate secretory system is present in the marine teleost renal tubule.

Transepithelial unidirectional sulfate fluxes have been reported in several different studies (Dickman and Renfro, 1986; Renfro, 1989; Brown *et al.*, 1992; Renfro *et al.*, 1993). Typical measurements in PTCs yield secretory fluxes of 98.4 ± 6.67 nmol cm^{-2} h^{-1}, reabsorptive fluxes of 4.8 ± 1.08 nmol cm^{-2} h^{-1}, and net fluxes of 93.6 ± 7.64 nmol cm^{-2} h^{-1} in the presence of flounder saline (1 m*M* SO_4; Dickman and Renfro, 1986) and under short-circuited conditions. This is a flux ratio of almost 20 to 1.

The cellular mechanisms of sulfate secretion have been examined in both basolateral and brush border membrane vesicles isolated from two flounder species (Renfro and Pritchard, 1982, 1983; Renfro, 1989). Sulfate entry, interstitium-to-cell, can be driven by the basolateral membrane pH gradient as either proton symport or hydroxyl exchange. In the intact tubule, this process is Na gradient dependent (Renfro and Dickman, 1980), perhaps secondary to apical membrane $Na^+ : H^+$ exchange. Sulfate exit, cell-to-lumen, is stimulated by counteranion gradients. HCO_3^-, SCN^-, Cl^-, and $S_2O_3^-$ countergradients can drive sulfate transport in flounder brush border membrane vesicles. The latter process is inhibited by disulfonic acid stilbenes but is not sensitive to H^+, Na^+, or K^+ gradients. These basolateral and brush border anion exchangers are depicted in Fig. 1. In this model the Na^+ electrochemical gradient established by Na^+, K^+-ATPase drives an assumed $Na^+ : H^+$ exchanger, which establishes a pH gradient across the basolateral membrane favoring $OH^- : SO_4^-$ exchange. SO_4^- exits apically in an electroneutral exchange for luminal atoms, mainly Cl^- and HCO_3^-.

D. Glucocorticoid Effects on SO_4 Secretion

Most of the sulfate excreted by the winter flounder and other marine fishes is actively secreted into the urine by the renal epithelium (Renfro and Dickman, 1980). Apparently, net Na-dependent reabsorption predominates in almost all other vertebrates even though a secretory machinery may be present (Brazy and Dennis, 1981). There may not be a comparable

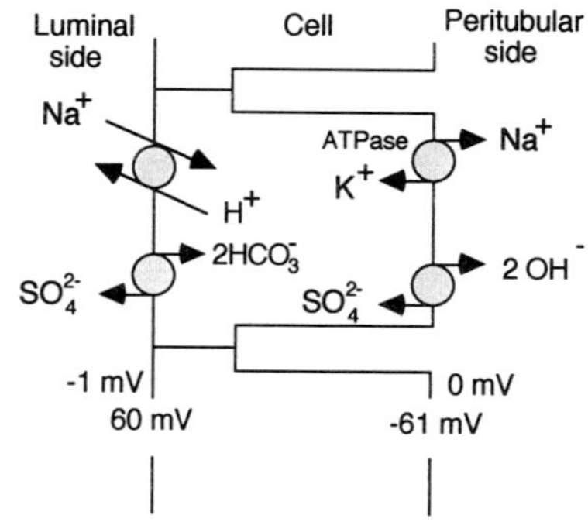

Fig. 1. Suggested mechanism of inorganic sulfate secretion by flounder renal proximal tubule in primary monolayer culture. Sulfate enters the cell from the peritubular interstitium on an anion exchanger that functions electroneutrally and can be driven by a pH gradient (Renfro and Pritchard, 1982). Sulfate exits the cell into the luminal urine on another electrically neutral anion exchanger that operates best with a bicarbonate or chloride countergradient (Renfro and Pritchard, 1983). No active reabsorptive mechanism has been found.

reabsorptive process in marine teleosts since none can be demonstrated in brush border membrane vesicles (Renfro and Pritchard, 1983; Renfro, 1989). Acclimation of winter flounder to 10% SW (SO_4-free) reveals that net sulfate secretion decreases (Renfro, 1989). This corresponds to a great decrease in bicarbonate : sulfate exchange in brush border membrane vesicles isolated from 10% SW-acclimated animals compared to 100% SW-acclimated animals. The glucocorticoid cortisol has been repeatedly implicated in teleost salinity adaptation (Loretz and Bern, 1983). Tests of a slowly metabolized glucocorticoid analog in the intact animals show that the decreased $HCO_3 : SO_4$ exchange in 10% SW-acclimated animals is restored by daily treatment of intact animals with 60 μg dexamethasone per 100 g body weight for 5 days. To determine whether the effect of glucocorticoid treatment acted directly on the kidney or indirectly through other induced changes, the effect of cortisol treatment on flounder PTCs was examined.

Cortisol (hydrocortisone phosphate) is an important ingredient in most animal cell culture media (Freshney, 1987) and is normally included in the flounder PTC medium at 5 mg liter^{-1}. The effect of this steroid on SO_4 transport was tested by deleting it from the culture medium on Day 4 after plating (Renfro, 1989). Following 8–10 days in medium containing only the small amount of cortisol present due to the use of 10% flounder serum, net transepithelial sulfate secretion decreased to one-third of that in normal culture medium. A three-fold decrease in secretory flux accounted for the decrease in net secretion; reabsorption flux was unchanged by the treatment. Thus, cortisol may be important for full expression of SO_4 transport capacity.

E. Phosphate Transport

The teleost kidney is capable of either net reabsorption or net secretion of inorganic phosphate (P_i), but our understanding of the mechanisms and regulation of P_i clearance remains incomplete. Marine species frequently exhibit net tubular secretion (Hickman and Trump, 1969). One micropuncture study in skates (*Raja*) indicated that phosphorous (form unknown) secretion occurred in the second proximal segment (Stolte *et al.*, 1977). Cliff and Beyenbach (1988) showed net phosphorous secretion in the second proximal segment of a bony fish. Initial studies of transepithelial P_i transport by flounder PTCs showed that both secretory and reabsorptive transepithelial P_i fluxes in this system can be active and sensitive to extracellular P_i concentration (Gupta and Renfro, 1989). Unidirectional fluxes of $^{32}P_i$ by PTCs mounted in Ussing chambers and short-circuited indicate that, following usual culturing procedures, tissues invariably display net phosphate reabsorption (secretory flux = 2.3 ± 0.52; reabsorptive flux = 7.1 ± 1.77; net reabsorptive flux = 4.9 ± 1.45 (SE) nmol cm^{-2} h^{-1}) (P_i = 0.4 m*M*). Raising the bathing flounder saline P_i concentration in both luminal and peritubular chambers from 0.4 m*M* to greater than 0.5 m*M* stimulates net secretion of P_i. The apparent $K_{1/2}$'s for reabsorption and secretory fluxes are about 2 m*M*, and the relationship between secretory flux and P_i concentration is sigmoidal. Between 0.5 and 2.0 m*M* (which is the physiological range in most fishes) (Hickman and Trump, 1969), small increases in P_i concentration result in larger increases in secretory flux than in reabsorptive flux. Even though reabsorptive flux also rises in response to increasing extracellular P_i concentration, the response is hyperbolic. This may represent an important physiological response. At low plasma P_i levels, the sigmoid shape of the secretory flux curve indicates that there will be little response to changes in extracellular P_i by the renal tubule. If the intact animal behaves similarly, during periods of P_i deprivation, the reabsorptive flux would predominate and conserve filtered P_i. During periods of P_i surfeit, plasma P_i would rise and renal secretion would predominate.

The mechanisms of apical and basolateral membrane P_i transport are still somewhat speculative. A working model of reabsorption, based on preliminary flounder brush border membrane vesicle studies and published data on bird and mammalian P_i transport (Hoffman *et al.*, 1976; Renfro and Clark, 1984), consists of Na : P_i cotransport process in the apical membrane and a voltage-sensitive, Na-independent exit pathway in the basolateral membranes (Schwab and Hammerman, 1986) (Fig. 2).

Net P_i secretion is not apparent in mammals and much less is known of this mechanism. Work with brush border membrane vesicles from chickens induced to secrete P_i by parathyroid hormone injection has revealed a likely membrane mechanism for P_i secretion (Barber *et al.*, 1993). The model of transepithelial P_i secretion derived from these studies (Fig. 2) consists of a Na-coupled P_i entry step from interstitium to cell coupled with a K^+-dependent and voltage-dependent rheogenic transfer of monovalent P_i from the cell to the lumen.

Although the voltage-dependent P_i secretory transport process was found in chicks, flounder PTCs were very well suited to testing several aspects of the vesicle-derived model in an intact epithelium. Both chicks and flounder perform net tubular P_i secretion and evolution of substantially different mechanisms seems unlikely. In addition, the PTCs can be readily stimulated to secrete P_i, thus providing a stable state with which to study the process.

Exposure to 100 μM 1,2-dihexanoyl-*sn*-glycerol (DAG), 10 μM

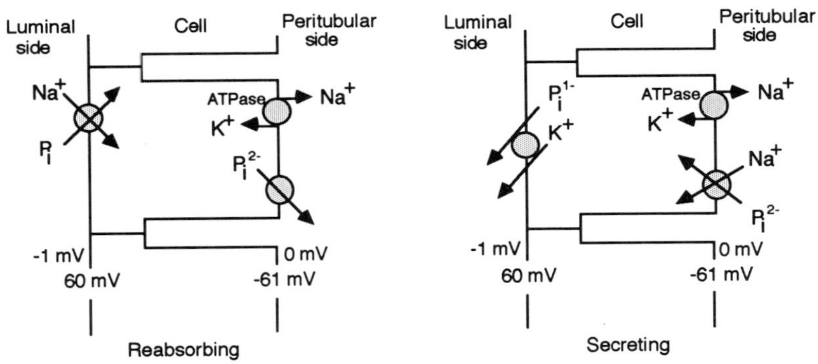

Fig. 2. A working model of the mechanism of inorganic phosphate (P_i) reabsorption by flounder renal proximal tubule in primary monolayer culture. P_i enters the cell from the luminal medium on a Na : P_i cotransporter and exits to the peritubular side down its electrochemical gradient. The model is based on preliminary data from flounder renal brush border membranes and mechanisms present in bird and mammalian renal cortical brush borders (Renfro and Clark, 1984; Hoffman *et al.*, 1976). The exit step is based on data from mammalian renal cortical basolateral membrane vesicles (Schwab and Hammerman, 1986). The suggested mechanism of inorganic phosphate (P_i) secretion by flounder renal tubule in primary monolayer cultures is shown on the right. P_i entry from the peritubular interstitium is sodium dependent and may be on a Na : P_i cotransporter (Schwab and Hammerman, 1986). P_i exit, cell-to-lumen, is voltage gradient (electrically negative) and potassium dependent (Barber *et al.*, 1993; Lu *et al.*, 1994a).

phorbol-12,13-didecanoate, 10 μM of the calcium ionophore A23187, or 200 nM thapsigargin stimulates net transepithelial P_i secretion within 30 min (Gupta and Renfro, 1989; Lu *et al.*, 1994a). Thapsigargin (200 nM) acts by inhibiting the Ca^{2+} reuptake process in the endoplasmic reticulum and secondarily causes plasma membrane Ca^{2+} channels to open. Elevation of extracellular [Ca^{2+}] from 2 to 5 mM apparently overwhelms the calcium efflux pumps, elevates $[Ca^{2+}]_{inside}$, and stimulates P_i secretion. Verapamil, a plasma membrane calcium channel blocker, prevents the effect of thapsigargin, which indicates that release of intracellular Ca^{2+} alone is insufficient to sustain maximal P_i secretion. Forskolin (10 μM) stimulates an increase in net reabsorption of P_i after a long latency (2 h) and increases cAMP production 30-fold (Gupta and Renfro, 1989). These data implicate intracellular messengers in the regulation of renal transepithelial P_i secretion and reabsorption. In addition, lowering the peritubular pH from 7.5 to 6.5 strongly stimulates net P_i secretion, but lowering the pH of the luminal medium has no effect on secretion (Gupta and Renfro, 1991). H-7 (a protein kinase inhibitor) inhibits both the phorbol ester-stimulated and acidosis-induced P_i secretion. Because H-7 has no effect following thapsigargin treatment, Ca^{2+} effects in addition to those on protein kinase C are likely (Lu *et al.*, 1994b). Luminal amiloride (2×10^{-4} M) or peritubular ouabain (10^{-4} M) or isosmotic replacement of all luminal Na with mannitol or LiCl significantly decreases P_i secretion. Thus, simulation of metabolic acidosis stimulates P_i secretion in flounder PTCs in part through an amiloride-sensitive protein kinase C-activated process.

The apical plasma membrane potential of PTCs is depolarized from −60 to −23 mV when 100 mM KCl is substituted for 100 mM NaCl in the luminal bath (Lu *et al.*, 1994a). This shift probably reflects the presence of K^+ conductance pathways in this pole of the cells. Stimulation of PTCs to secrete with 200 mM thapsigargin and subsequent depolarization of the luminal membranes with 100 mM KCl produce a significant inhibition of secretion with no effect on reabsorption (Lu *et al.*, 1994a). $^{32}P_i$ loading and washout studies on the PTCs mounted in the Ussing chambers and preequilibrated with $^{32}P_i$ show that high K^+ in the luminal side significantly decreases cell-to-lumen $^{32}P_i$ efflux and has no effect on cell-to-peritubular side efflux. The most likely cause of this effect is the depolarization of the apical membranes by high K^+. This possibility is further supported by the fact that 2 mM $BaCl_2$ and 20 nM charybdotoxin added to the luminal solution also inhibit P_i secretion. The primary action of both of these substances in the lumen of the proximal tubule has been reported to be blockade of apical K^+ channels (Miller *et al.*, 1985; Zweifach *et al.*, 1991). Thus, the response of the P_i secretory process described in the flounder

PTCs is consistent with the characteristics of P_i transport by chick renal cortical brush border vesicles (Barber *et al.*, 1993).

F. Effect of Stanniocalcin (STC) on Renal P_i Transport

Control of renal P_i transport in fishes is not well studied. Because protein kinase C activators stimulate secretion and forskolin (an adenylate cyclase activator) stimulates net reabsorption (Gupta and Renfro, 1989), candidate regulatory hormones were sought. In seawater the Ca^{2+} concentration is high (around 10 m*M*), and the ion can be taken up directly across the gills and gut of marine teleosts as needed (Fenwick and So, 1974; Sundell *et al.*, 1992). There is no P_i in seawater, however, and fishes must balance dietary P_i intake and renal excretion. Fish tissues such as bone and scales represent a P_i store that probably participates in P_i homeostasis (Hickman and Trump, 1969).

Because the fish kidney can either reabsorb or secrete P_i (Hickman and Trump, 1969), interdependence of plasma P_i and Ca^{2+} concentration requires hormonal coordination in a manner similar to that exerted by parathyroid hormone in terrestrial vertebrates. The corpuscles of Stannius (Wendelaar Bonga *et al.*, 1986) are unique glands found in teleostean and holostean fishes. They produce stanniocalcin (STC), an antihypercalcemic hormone released in response to elevation of extracellular Ca^{2+} concentration (Wagner *et al.*, 1989; Wendelaar Bonga and Pang, 1991). STC reduces plasma [Ca^{2+}] by inhibition of Ca^{2+} absorption across the intestine (Felder *et al.*, 1990; Sundell *et al.*, 1992) and gills (Lafeber *et al.*, 1988; Wagner *et al.*, 1988a).

Fenwick and Brasseur (1991) showed that stanniectomy of freshwater eels induces hypercalcemia and increases calcium influx. Injection of STC decreases plasma calcium, and they suggested that STC both decreases the influx of calcium and increases the rate of calcium exit from the blood since the reduction in influx alone could not account for the amount of reduction in blood calcium. Wagner *et al.* (1988b) showed that STCs from two different species of salmon are potent inhibitors of gill calcium uptake in rainbow trout and American eel. More recently, Wagner *et al.* (1991) reported that the *in vivo* response of the corpuscles of Stannius is rapid and Ca^{2+}-specific. They have obtained similar results in primary cultures of corpuscle cells with a specific antiserum and radioimmunoassay (Gellersen *et al.*, 1988; Wagner *et al.*, 1988b).

Stanniocalcin was extracted from coho salmon (*Oncorhynchus kisutch*) corpuscles of Stannius by the method of Wagner *et al.* (1988a). Table II

Table II
Effect of Stanniocalcin (STC) on Transepithelial Transport of P_i by Flounder Renal Proximal Tubule Primary Monolayer Cultures[a]

Treatment	Flux (nmol cm^{-2} h^{-1}) Reabsorptive	Secretory	Net
Control	−1.6 ± 0.20	1.9 ± 0.56	0.3 ± 0.74
STC	−3.2 ± 0.68	1.3 ± 0.37	−2.0 ± 0.73**
STC	−3.1 ± 0.56	1.8 ± 0.26	−1.3 ± 0.61
STC + H-89	−1.6 ± 0.31**	3.0 ± 0.79	1.6 ± 1.02*
Forskolin	−3.0 ± 0.29	1.8 ± 0.23	−1.3 ± 0.45
Forskolin + H-89	−1.3 ± 0.29*	3.1 ± 0.71	1.8 ± 0.98*

[a] Values are mean ± SE of determinations made 3 h following the indicated treatments. Secretory, reabsorptive, and net fluxes are shown. Negative sign denotes reabsorptive flux. STC was used at 200 ng ml^{-1}; H-89 and forskolin were used at 10 μM. Significantly different from paired tissues at (*) $P < 0.05$ and (**) $P < 0.01$, respectively. Data are from Lu *et al.* (1994b).

shows the effect of 200 ng ml^{-1} STC following 3 h of treatment of PTCs inside Ussing chambers. Net reabsorption of P_i increased from near zero to 2.0 nmol cm^{-2} h^{-1} (Lu *et al.*, 1994b). The response to STC was apparent within 30 min of initial exposure. The effect appeared to be specific in that no changes were produced in PD, R, or PS-I_{sc}. Significant stimulation of P_i reabsorption was seen with as little as 12.5 ng ml^{-1}. The effect was concentration dependent up to 50 nm ml^{-1}.

STC and forskolin have nearly identical effects on P_i transport. The production and release of cAMP increases more than twofold within 1 h of exposure to STC. The likelihood that STC works through protein kinase A was further supported by the results of treatment with a very specific protein kinase A inhibitor, H-89. This compound not only reduces reabsorptive flux and stops STC-induced P_i reabsorption, it stimulates net P_i secretion. H-89 has a similar inhibitory effect on forskolin-induced P_i reabsorption (Lu *et al.*, 1994b).

Of fourteen different hormones tested, only STC, ovine prolactin (oPRL), and bovine parathyroid hormone (bPTH) stimulate net reabsorption by PTCs (Lu *et al.*, 1994b). No hormones have been found that stimulate P_i secretion. bPTH has no effect at 10 n*M* but slightly stimulates at 100 n*M*. oPRL was tested only at 100 n*M* but is as effective at this dose as the highest effective dose of STC (1 n*M*).

If STC is equally effective *in vivo,* the heightened reabsorption of P_i could contribute to the overall hypocalcemic effect of STC by enhancing Ca^{2+} and P_i deposition as hydroxyapatite into bone and scales. This effect would help to explain the rapidity with which STC lowers plasma Ca^{2+} concentration.

STC has no effect on Ca^{2+} transport by PTCs; however, no net Ca^{2+} transport was observed under standard control conditions (Lu *et al.,* 1994b). Forskolin (10 μM) also has no effect on calcium transport by this tissue.

G. Organic Anion Secretion

Organic anion transport by the renal proximal tubule has been extensively studied in a variety of vertebrates. Ullrich and coworkers have defined the properties of putative substrates for this process through stopped-flow microperfusion in rat kidney (Fritzsch *et al.,* 1989). Their studies show that initial interaction of a substrate with this secretory process requires only a partial negative charge (a carboxyl group) and a hydrophobic moiety. Together with these *in situ* studies, isolated basolateral membrane vesicles from mammalian renal cortex have provided the current models of organic anion secretion (see Pritchard and Miller, 1993). The teleost renal tubule has been historically important in defining the mechanisms of secretion (Forster, 1948). The relationship of the transport properties determined in vesicle studies to transepithelial transport was further examined in the flounder PTCs (Dawson and Renfro, 1990, 1993). There appear to have been few studies of this type on electrically and chemically short-circuited primary renal epithelia.

The transport of *p*-aminohippuric acid (PAH) is widely used as a marker for renal proximal tubule organic anion secretion. As would be predicted from the extensive work of Ullrich and others (see Fritzsch *et al.,* 1989), the kinetics of transport are complex and probably reflect the contribution of several systems to net secretion. Measurement of unidirectional secretory, reabsorptive, and net PAH transport by flounder PTCs, in conditions of no transepithelial chemical or electrical gradients, over the range of 0.6 μM to 1.5 mM revealed that reabsorptive flux is not saturable and behaves as a simple diffusive process. The $K_{1/2}$ for secretion is 0.4 mM with maximum net secretion of 80 nmol/cm^{-2} h^{-1}. Probenecid (1 mM) completely inhibits net PAH (10 μM) secretion within 60 min. This is apparently due to specific interaction with PAH transport since there are no effects on Na-dependent glucose transport (Dawson and Renfro, 1990).

At 10 μM PAH, maximal secretory transport is about 1.6 nmol cm^{-2} h^{-1} while simultaneously determined reabsorptive flux is about 0.1 nmol

cm^{-2} h^{-1}. Substitution of choline chloride for sodium chloride reduces PAH net secretion by 70% within 1 h and to near zero by 2 h. Similarly, 1 m*M* ouabain reduces net secretion to zero within 90 min of initial exposure (Dawson and Renfro, 1990).

Eveloff *et al.* (1979) showed that an electrical gradient is capable of driving concentrative uptake of PAH by flounder kidney brush border membrane vesicles. The effect of transepithelial electrical gradients on PAH secretion by PTCs was examined by voltage clamping the tissues in Ussing chambers to $\pm$10 mV for 60 min (Dawson and Renfro, 1990). These gradients have no effect on either secretory or reabsorptive PAH transport. If reabsorption is a nonfacilitated leak flux through the paracellular pathway it should have been influenced by the transepithelial electrical gradient. Because imposed transepithelial electrical gradients have no effect on PAH reabsorptive flux, a low-affinity cellular route was inferred.

Both Shimada *et al.* (1987) and Pritchard (1988) used mammalian renal cortical tissue to show that the initial step in secretion, transport from blood to cell through the basolateral membrane, occurs by an anion exchanger utilizing glutarate (α-ketoglutaric acid, α-KG) as counteranion. α-KG from cellular metabolism exits the cell in exchange for the extracellular organic anion. The α-KG is then returned to the cellular interior on a Na : dicarboxylate cotransporter. This is probably the basis for the apparent Na dependence of organic acid secretion. In all of the intact renal proximal tubule preparations studied, including those of teleosts, addition of extracellular α-KG or glutarate, at concentrations between 10 and 50 μM, stimulates PAH secretion (Pritchard and Miller, 1993) presumably by entering the cell and accelerating the entry of PAH by exchange. Higher extracellular concentrations are inhibitory. The Na dependence of PAH secretion is readily apparent in PTCs; however, the stimulatory effect of glutarate is not, even though the dicarboxylate strongly inhibited PAH secretion at high concentrations (1 m*M*) in the extracellular medium. The model shown in Fig. 3 for control conditions will likely account for PAH secretion by flounder PTCs. The culture conditions may be responsible for the inability to show α-KG stimulation since cellular metabolism may be altered by placing cells in culture. However, the stimulation of PAH secretion by cytosolic counteranions may be demonstrable in this system with other anions (see the following).

A variety of organic anionic toxins are excreted by the proximal tubule PAH system. The interaction of competitive anions with this system may alter their renal clearance. Using PAH as a measure of the degree of interactions of pesticides with the organic anion transport system, the structural differences in several closely related herbicides are sufficient to alter their interaction with the system. Several herbicides thought to

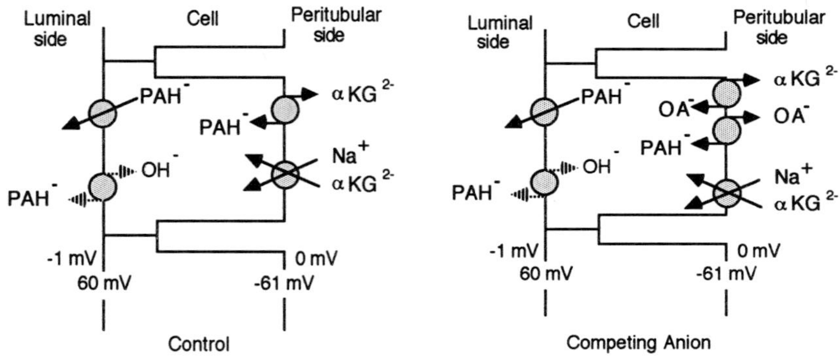

Fig. 3. Hypothetical model of organic anion secretion by flounder renal proximal tubule in primary monolayer culture. Model is adapted from Pritchard and Miller (1993). *p*-Aminohippuric acid (PAH) enters from the peritubular interstitium in exchange for α-ketoglutaric acid (α-KG). Although this initial step is likely to be present, it has not been demonstrable in flounder primary cultures (Dawson and Renfro, 1990). The α-KG is recycled back into the cell on a Na-coupled transporter. PAH exit is probably down its electrochemical gradient (Eveloff *et al.*, 1979), although other mechanisms are possible and remain to be tested. An adaptation of the organic anion (OA) model of secretion is shown on the right. This model could explain the biphasic effects of certain competitive substrates on PAH secretion (Dawson and Renfro, 1993). Accumulated OA may exchange for PAH in a manner identical to that proposed for α-KG.

competitively interact with PAH transport were tested for their ability to block PAH transport (Dawson and Renfro, 1993). 2,4-Dichlorophenoxyacetic acid (2,4-D), 2-methyl-4-chlorophenoxyacetic acid (MCPA), 2-(2-methyl-4-dichlorophenoxy)propionic acid (dichlorprop), and 2-(2,4-chlorophenoxy)propionic acid (mecoprop) vary only in the number of chlorides attached to the ring and the carbon side chain length. Dose–response curves based on the ability of these compounds to inhibit PAH secretion show that the increase in side chain length increases the inhibitory effectiveness.

Several herbicides and insecticides have biphasic effects on PAH transport, that is, at high concentrations they inhibit, but at low concentrations they stimulate net transepithelial PAH secretion (Dawson and Renfro, 1993). This effect is not unlike that of glutarate (Shimada *et al.*, 1987; Pritchard, 1988) and is both dose and time dependent. Mecoprop at 10^{-7} to 10^{-6} *M* stimulates PAH secretion only after 2-h exposure. At 10^{-5} to 10^{-4} *M*, mecoprop inhibits during the first hour of exposure but has no effect beyond that time. At higher concentrations, mecoprop strongly inhibits. The explanation for these biphasic effects is speculative, but perhaps the most probable is interaction on an exchange process in the basolateral membranes as illustrated in Fig. 3 (competing anion). The proposed endogenous metabolite that exchanges for PAH, α-KG or glutarate, inhibits PAH uptake in similar biphasic manner when it is present in the extracellular medium at concentrations above 50–100 μM. Thus, it is possible that certain xenobiotic compounds, when accumulated to sufficient intracellular concentrations, may behave similarly.

H. Taurine Transport

Taurine has been implicated in a variety of important physiological processes in mammals as well as other vertebrates and invertebrates (Wright *et al.*, 1986). Goldstein and coworkers have shown that taurine is released from cells of marine fish in response to environmental dilution (King and Goldstein, 1983). This is part of a volume regulatory decrease. The fluctuations in plasma taurine, which is apparently not metabolized, are dealt with exclusively by the kidneys.

In an examination of taurine transport by flounder renal epithelium, Perlman *et al.* (1991) showed that under control conditions (0.1 mM taurine) flounder PTCs perform net taurine secretion (3.6 ± 0.75 nmol cm^{-2} h^{-1}). The average flux ratio under short-circuited conditions was as high as 19. Thus, the taurine transport pumps are quite potent.

Taurine accumulates to very high levels in the intact renal epithelium (32–64 mM). The flounder PTCs accumulate taurine to 11 ± 1.87 nmol per mg protein (Perlman *et al.*, 1991). Because of this, transepithelial fluxes are slow to reach steady state (90–120 min). Perlman *et al.* (1991) compared the accumulation and secretion of taurine to that of the nonsecreted β-amino acid, β-alanine, and showed that a specific difference in handling of these two compounds occurs at the cell-to-lumen step in transepithelial transport. Probenecid (1 mM), a good inhibitor of PAH secretion, effectively blocks taurine secretion; however, PAH (10 mM) does not, suggesting that taurine may be a better competitor for the organic anion transporter than is PAH. Bromocresol green, a powerful blocker of PAH transport, but very poorly transported itself, blocked taurine secretion from the luminal side.

Asymmetrical addition of 10 mM taurine to the luminal side only does not stimulate secretory ^{14}C-labeled taurine flux from the cell to the lumen. The conclusion was that no taurine–taurine exchange is present in the brush border membranes. The high luminal taurine concentration does significantly inhibit reabsortive flux of ^{14}C-labeled taurine.

Work with intact isolated and perfused flounder renal tubules showed that the basolateral entry step is sensitive to both Na^+ and Cl^- (King *et al.*, 1982), whereas the luminal transport appears to be a Na^+ cotransporter

(King *et al.*, 1985). The data obtained by Perlman *et al.* 1991) on PTCs are consistent with basolateral entry of β-amino acids on a Na^+ cotransporter. However, this concentrative uptake leads to a gradient for taurine to exit cell-to-lumen on a bromocresol green- and probenecid-sensitive transporter, which is apparently distinct from the apical Na cotransporter.

I. Effects of Physicochemical Stress on Transport

PTCs have been important for the study of renal responses to physicochemical challenge with temperature and xenobiotics and subsequent injury. Heat stress and the induction of the heat shock proteins (hsp's) have been examined with an eye toward understanding the response of the tissue to mild injury. At the cellular level, sublethal thermal stress results in an accumulation of denatured proteins. This begins the process of heat shock gene activation (Hightower, 1991). Several families of hsp's are produced in PTCs (Brown *et al.*, 1992). The hsp's have been categorized into several groups based on their molecular masses and similarity in amino acid sequences. The hsp90, hsp70, and hsp30 families are well expressed in heat-shocked PTCs. Hsp90 and hsp70 are among the most highly conserved proteins known (Lindquist, 1986). Their accumulation is associated with the development of resistance to subsequent more severe thermal stress. At the level of the individual cell the hsp's have a chaperoning function that may prevent further abnormal protein folding and allow cell survival. The PTCs have been important in determining the effect of thermal stress on tissue-level function rather than the more extreme criterion of survival.

PTCs have been used as a model system to study the heat shock response. These studies revealed a correlation of the induction of hsp synthesis with effects on the normal transport functions by kidney tissue. Cells exposed to a mild heat stress are able transiently to withstand a normally lethal temperature. This state is known as thermotolerance. Though most studies measure thermotolerance using clonogenic assays, that is, survival assays, use of PTCs has allowed the observation of protection of a differentiated tissue function, namely, sulfate transport. PTCs normally maintained at 22°C lose sulfate transport ability and surface structures such as microvilli and cilia upon exposure to 32°C (severe heat stress, SHS) (Fig. 4A) (Brown *et al.*, 1992). However, if the cultures are given a priming heat exposure of 27°C for 6 h, then returned to 22°C for 1.5 h (mild heat stress, MHS), prior to the severe temperature exposure, they do not show this loss. Incubation of the cultures with the protein synthesis inhibitor cycloheximide during the stresses prevents the acquisition of protection (Fig. 4B). Because members of three families of the

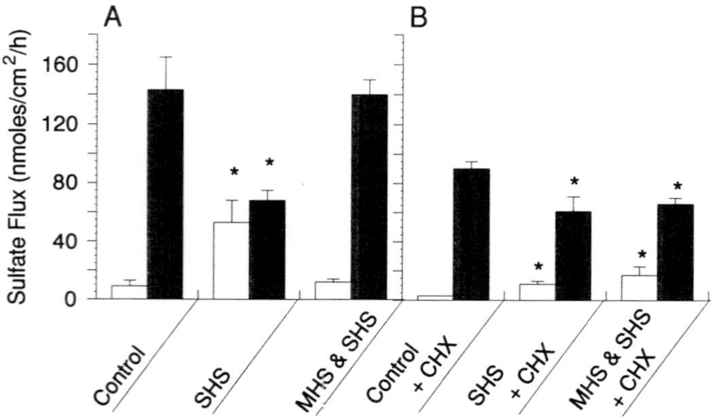

Fig. 4. Unidirectional secretory (shaded bars) and reabsorptive (open bars) sulfate fluxes in flounder renal proximal tubule primary monolayer cultures exposed to varying temperature stresses. (A) Control: tissues were maintained continuously at 22°C. SHS (severe heat stress): tissues were raised from 22 to 32°C for 1.5 h followed by 1.5 h at 22°C. MHS (mild heat stress) & SHS: tissues were exposed to 27°C for 6 h followed by 32°C for 1.5 h then 22°C for 1.5 h. (B) Same as in (A) except cycloheximide (CHX, 50 $\mu g\ ml^{-1}$) was added to all tissues at the time heat exposures began. (*) Significantly different from controls at $P < 0.05$. Data taken from Brown *et al.* (1992).

hsp's show the greatest increase in synthesis at 27°C, it is possible that these proteins play a role in the observed protection (Brown *et al.*, 1992).

Similar to thermotolerance is cross-tolerance, a state in which cells exposed to an inducer of hsp's other than heat will show increased survival after a subsequent exposure to SHS. Using zinc ions as the primary stressor and inducer of hsp's (mild zinc stress, MZS), Renfro *et al.* (1993) have determined that sulfate transport can be protected against the damaging effects caused by exposure to SHS. Sulfate transport is even enhanced after an exposure to the mild stress (Fig. 5A). Incubation of the cultures with cycloheximide also abolishes this form of protection (Fig. 5B).

Exposure of these cultures to the xenobiotic herbicide 2,4-dichlorophenoxyacetic acid (2,4-D) causes a decrease in sulfate transport (Figure 6A). The protection provided by MHS and MZS against severe temperature damage also extends to the inhibitory effect of 2,4-D (Figs. 6A and 6B). The increase in sulfate transport caused by the primary stressors appears to partially compensate for the 2,4-D-induced loss.

These studies show that the induction of stress proteins such as hsp70 not only serves as an indicator of cellular protein damage caused by exposure to xenobiotics, that is, a molecular biomarker of exposure and response, but also marks the presence of an altered cellular physiology,

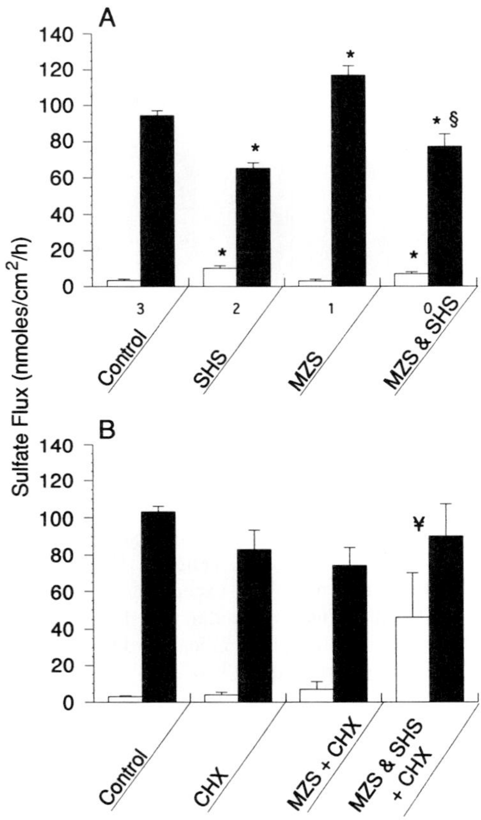

Fig. 5. Demonstration of cross-tolerance of unidirectional reabsorptive (open bars) and secretory (shaded bars) fluxes of sulfate in flounder renal proximal tubule primary monolayer cultures. (A) Control: tissues were exposed to 22°C continuously. SHS (severe heat stress): tissues were raised from 22 to 32°C for 1.5 h then returned to 22°C for 1.5 h. MZS (mild zinc stress): tissues were maintained continuously at 22°C and exposed to 100 μM zinc chloride for 6 h then zinc-free medium for 1.5 h. MZS & SHS: tissues were exposed to zinc prior to thermal stress. (B) Same as in (A) except cycloheximide (CHX, 50 μg ml^{-1}) was added at the beginning of MZS. (*,§) Significantly different from controls and SHS, respectively, at $P < 0.05$. (¥) Net flux is significantly less than controls. Data taken from Renfro *et al.* (1993).

a protected state. Tissues in a protected state are more resistant to the toxic effects of some xenobiotics.

The discovery of one of the primary causes of the development of resistance to cell-killing drugs in cancer cells during the course of chemotherapy may have relevance as well to a normal renal function (Thiebaut

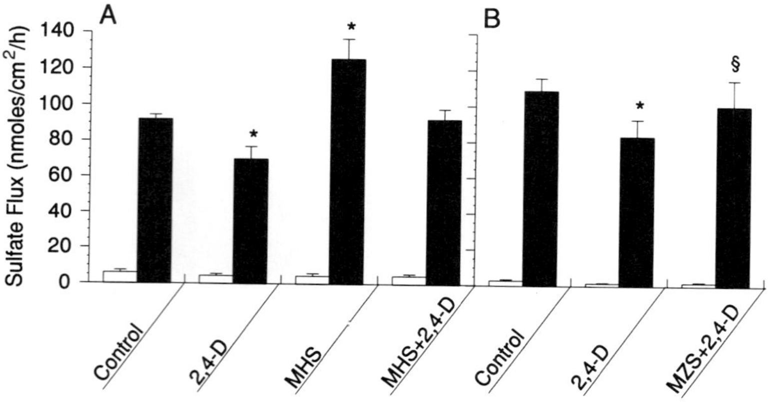

Fig. 6. The effect of mild heat stress (MHS) on unidirectional secretory (shaded bars) and reabsorptive (open bars) sulfate fluxes in flounder renal proximal tubule in primary monolayer culture. (A) Control: tissues were exposed continuously to 22°C. 2,4-D: 2,4-dichlorophenoxyacetic acid, 0.5 m*M*, present in the Ussing chambers during sulfate flux determinations. MHS (mild heat stress): tissues were raised from 22 to 27°C for 6 h then returned to 22°C for 1.5 h prior to flux measurements. MHS & 2,4-D: tissues receiving both aforementioned treatments. (B) Same as (A) except MZS (mild zinc stress: exposure to 100 μM zinc chloride for 6 h then zinc-free medium for 1.5 h prior to flux determinations) treatment instead of MHS was used. (*,§) Significantly different from controls and 2,4-D, respectively. Data taken from Renfro *et al.* (1993).

et al., 1987). The failure of chemotherapy in cancer treatment is frequently due to the amplified expression in these cells of the "multidrug resistance" gene. The gene product is a membrane transport protein dubbed P-glycoprotein (P-gp). All evidence suggests that this is an efflux pump. Substrate specificity is quite broad in that substances of widely varying chemical structure are interactive. However, there are compounds that can be used as specific indicators of the P-gp presence. Since its discovery in tumor cells, P-gp has been found to be a normal constituent of cells in organs such as kidney, liver, and intestine (Thiebaut *et al.*, 1987). Based on pharmacological evidence gained with specific inhibitors of P-gp, all animals, both vertebrate and invertebrate, may use P-gp to protect themselves from naturally occurring cytotoxic agents (Kurelec, 1992).

One of the most commonly used markers of P-gp transport is daunomycin. In renal cells, this transport process is very sensitive to vinblastine, vincristine, and verapamil. Table III shows that a typical substrate for P-gp, daunomycin, is avidly secreted by flounder PTCs under short-circuited conditions (Sussman-Turner and Renfro, 1995). At 5 μM, ^{3}H-labeled daunomycin secretory transport was greater than the reverse flux (Table III). The unidirectional fluxes were quite high, reflecting the relatively high

Table III
Transepithelial Transport of Daunomycin by Flounder Proximal Tubule Primary Monolayer Cultures[a]

Treatment	Daunomycin transport rate (nmol cm^{-2} h^{-1})		
	Reabsorptive	Secretory	Net
Control	−0.10 ± 0.016	0.16 ± 0.025	0.06 ± 0.027
Control + VBL	−0.11 ± 0.006	0.13 ± 0.011	0.02 ± 0.013
Heat-shocked	−0.11 ± 0.014	0.23 ± 0.030	0.11 ± 0.026*
Heat-shocked + VBL	−0.10 ± 0.012	0.12 ± 0.016	0.03 ± 0.011**

[a] Values are mean ± SE (n = 5). Daunomycin concentration was 5 μM and was not toxic during the 2-h time period of flux determination. Vinblastine (VBL) was used at 20 μM. Heat-shocked tissues were raised from 22 to 27°C for 6 h followed by 1.5 h at 22°C. Fluxes were measured at 22°C. Negative sign denotes reabsorptive flux. Significantly different from (*) control and (**) heat-shocked, respectively. Data are taken from Sussman-Turner and Renfro (1995).

lipid solubility. Previously, we had assumed that substances with such high partition coefficients could not be efficiently excreted by the kidney. Net transport is relatively low, about 60 pmol cm^{-2} h^{-1}, and this is reflected in the low ratio of reabsorptive to secretory flux (1:1.5). Vinblastine (VBL, 20 μM) totally inhibits this active flux. However, because the controls secreted inconsistently, the VBL effect was not significant. Note that inhibition was due solely to a drop in secretory transport. Backflux (reabsorptive flux) was unchanged.

Table III also shows the effects of MHS. This treatment increased daunomycin secretory flux almost 2-fold (Table III) (Sussman-Turner and Renfro, 1995). Again, the reabsorptive flux was unaffected, and VBL significantly inhibits the stimulated transport. The increased transport by P-gp following heat shock is at least partially dependent on protein synthesis since this stimulation is prevented by cycloheximide (Sussman-Turner and Renfro, 1995).

IV. CONCLUSIONS

There are obviously complex interactions, both structural and biochemical, that influence the apparent mechanisms of solute handling by flounder proximal tubule *in vivo*. Whereas tissue culture has provided a means to simplify examination of these mechanisms, the relationship of *in vitro* to *in vivo* processes must be directly determined where possible.

Meanwhile, tissue-level function in culture will continue to provide a basis for predictions concerning function of the intact kidneys.

REFERENCES

Barber, L. E., Corić, V., Clark, N. B., and Renfro, J. L. (1993). PTH-sensitive K^+- and voltage-dependent P_i transport by chick renal brush-border membranes. *Am. J. Physiol.* **265,** F822–F829.

Beyenbach, K. W., Petzel, D. H., and Cliff, W. H. (1986). The renal proximal tubule of the flounder. I. Physiological properties. *Am. J. Physiol.* **250,** R608–R615.

Bisbee, C., Machen, T., and Bern, H. (1979). Mouse mammary epithelial cells on floating collagen gels: Transepithelial ion transport and effects of prolactin. *Proc. Natl. Acad. Sci. U.S.A.* **76,** 536–540.

Brazy, P. C., and Dennis, V. W. (1981). Sulfate transport in rabbit proximal convoluted tubules: Presence of anion exchange. *Am. J. Physiol.* **241,** F300–F307.

Brown, M. A., Upender, R. P., Hightower, L. E., and Renfro, J. L. (1992). Thermoprotection of a functional epithelium: Heat stress effects on transepithelial transport by flounder renal tubule in primary monolayer culture. *Proc. Natl. Acad. Sci. U.S.A.* **89,** 3246–3250.

Burwen, S., and Pitelka, D. (1980). Secretory function of lactating mouse mammary epithelial cells cultured on collagen gels. *Exp. Cell Res.* **126,** 249–262.

Cliff, W. H., and Beyenbach, K. W. (1988). Fluid secretion in glomerular renal proximal tubules of freshwater-adapted fish. *Am. J. Physiol.* **254,** R154–R158.

Cliff, W. H., Sawyer, D. B., and Beyenbach, K. W. (1986). The renal proximal tubule of the flounder. II. Transepithelial Mg transport. *Am. J. Physiol.* **250,** R616–R624.

Cole, R. J., and Paul, J. (1966). The effects of erythropoietin on haem synthesis in mouse yolk sac and cultured foetal liver cells. *J. Embryol. Exp. Morphol.* **15,** 245–260.

Contreras, R. G., Ponce, A., and Bolivar, J. J. (1991). Calcium and tight junctions. *In* "Tight Junctions" (M. Cereijido, ed.), pp. 139–150. CRC Press, Boca Raton, FL.

Dawson, M. A., and Renfro, J. L. (1990). Organic anion secretion by winter flounder renal proximal tubule primary monolayer cultures. *J. Pharmacol. Exp. Ther.* **254,** 39–44.

Dawson, M. A., and Renfro, J. L. (1993). Interaction of structurally similar pesticides with organic anion transport by primary cultures of winter flounder renal proximal tubule. *J. Pharmacol. Exp. Ther.* **266,** 673–677.

Dickman, K. G., and Renfro, J. L. (1986). Primary culture of flounder renal tubule cells: Transepithelial transport. *Am. J. Physiol.* **251,** F424–F432.

Dickman, K. G., and Renfro, J. L. (1993). Tissue culture techniques for study of transepithelial transport by teleost renal tubule. *In* "New Insights in Vertebrate Kidney Function" (J. A. Brown, R. J. Balment, and J. C. Rankin, eds.), pp. 65–85. Cambridge Univ. Press, Cambridge, UK.

Eveloff, J., Kinne, R., and Kinter, W. B. (1979). *p*-Aminohippuric acid transport into brush border vesicles isolated from flounder kidneys. *Am. J. Physiol.* **237,** F291–F298.

Felder, C. C., Campbell, T., Albrecht, F., and Jose, P. A. (1990). Dopamine inhibits Na–H exchanger activity in renal BBMV by stimulation of adenylate cyclase. *Am. J. Physiol.* **259,** F297–F303.

Fenwick, J. C., and Brasseur, J. G. (1991). Effects of stanniectomy and experimental hypercalcemia on plasma calcium levels and calium influx in American eels, *Anguilla rostrata,* leSueur. *Gen. Comp. Endocrinol.* **82,** 459–465.

Fenwick, J. C., and So, Y. P. (1974). A perfusion study of the effect of stanniectomy on the net influx of calcium-45 across an isolated eel gill. *J. Exp. Zool.* **188,** 125–131.
Forster, R. P. (1948). Use of thin kidney slices and isolated renal tubules for study of cellular transport kinetics. *Science* **108,** 65–67.
Freshney, R. I. (1987). "Culture of Animal Cells." Alan R. Liss, New York.
Fritzsch, G., Rumrich, G., and Ullrich, K. J. (1989). Anion transport through the contraluminal cell membrane of renal proximal tubule. The influence of hydrophobicity and molecular charge distribution on the inhibitory activity of organic anions. *Biochim. Biophys. Acta* **978,** 249–256.
Gellersen, B., Wagner, G. F., Copp, D. H., and Friesen, H. G. (1988). Development of a primary culture system for rainbow trout corpuscles of Stannius and characterization of secreted teleocalcin. *Endocrinology (Baltimore)* **123,** 913–921.
Gupta, A., and Renfro, J. L. (1989). Control of phosphate transport in flounder renal proximal tubule primary cultures. *Am. J. Physiol.* **256,** R850–R857.
Gupta, A., and Renfro, J. L. (1991). Effects of pH on phosphate transport by flounder renal tubule primary cultures. *Am. J. Physiol.* **260,** R704–R711.
Handler, J. S. (1986). Studies on kidney cells in culture. *Kidney Int.* **30,** 208–215.
Hickman, C. P., and Trump, B. F. (1969). The kidney. *In* "Fish Physiology" (W. S. Hoar and D. J. Randall, eds.), pp. 91–239. Academic Press, New York.
Hightower, L. E. (1991). Heat shock, stress proteins, chaperoning and proteotoxicity. *Cell (Cambridge, Mass.)* **66,** 191–197.
Hoffman, N., Thees, M., and Kinne, R. (1976). Phosphate transport by isolated renal brush border vesicles. *Pfluegers Arch.* **362,** 147–156.
King, P. A., and Goldstein, L. (1983). Organic osmolytes and cell volume regulation in fish. *Mol. Physiol.* **4,** 53–66.
King, P. A., Beyenbach, K. W., and Goldstein, L. (1982). Taurine transport by isolated flounder renal tubules. *J. Exp. Zool.* **233,** 103–114.
King, P. A., Kinne, R., and Goldstein, L. (1985). Taurine transport by brush border membrane vesicles isolated from the flounder kidney. *J. Comp. Physiol., B* **155,** 185–193.
Kinter, W. B. (1975). Structure and function of renal tubules isolated from fish kidneys. *Fortschr. Zool.* **23,** 223–231.
Kleinzellar, A., and McAvoy, E. (1973). Sugar transport across the peritubular face of renal cells of the flounder. *J. Gen. Physiol.* **62,** 169–184.
Kurelec, B. (1992). The multixenobiotic resistance mechanism in aquatic organisms. *CRC Crit. Rev. Toxicol.* **22,** 23–43.
Lafeber, F. P. J. G., Flik, G., Wendelaar Bonga, S. E., and Perry S. F. (1988). Hypocalcin from Stannius corpuscles inhibits gill calcium uptake in trout. *Am. J. Physiol.* **254,** R891–R896.
Lindquist, S. (1986). The heat-shock response. *Annu. Rev. Biochem.* **55,** 1151–1183.
Loretz, C. A., and Bern, H. A. (1983). Control of ion transport by *Gillichthys mirabilis* urinary bladder. *Am. J. Physiol.* **245,** R45–R52.
Lu, M., Barber, L. E., and Renfro, J. L. (1994a). Renal transepithelial phosphate secretion: Luminal membrane voltage and Ca^{2+} dependence. *Am. J. Physiol.* **267,** F624–F631.
Lu, M., Wagner, G. F., and Renfro, J. L. (1994b). Stanniocalcin stimulates phosphate reabsorption by flounder renal proximal tubule in primary culture. *Am. J. Physiol.* **267,** R1356–R1362.
Michaelopoulos, G., and Pitot, H. (1975). Primary culture of parenchymal liver cells on collagen membranes. *Exp. Cell Res.* **90,** 70–78.
Miller, C., Moczydlowski, E., Latorre, R., and Phillips, M. (1985). Charybdotoxin, a protein inhibitor of single Ca^{2+}-activated K^+ channels from mammalian skeletal muscle. *Nature (London)* **313,** 316–318.

Perlman, D. F., Renfro, J. L., and Goldstein, L. (1991). Taurine secretion in cultured winter flounder renal epithelium. *Am. J. Physiol.* **261,** R1155–R1163.

Pritchard, J. B. (1988). Coupled transport of *p*-aminohippurate by rat basolateral membrane vesicles. *Am. J. Physiol.* **255,** F597–F604.

Pritchard, J. B., and Miller, D. S. (1993). Mechanisms mediating renal secretion of organic anions and cations. *Physiol. Rev.* **73,** 765–796.

Renfro, J. L. (1978). Calcium transport across peritubular surface of the marine teleost renal tubule. *Am. J. Physiol.* **234,** F522–F531.

Renfro, J. L. (1989). Adaptability of marine teleost renal inorganic sulfate excretion: Evidence for glucocorticoid involvement. *Am. J. Physiol.* **257,** R511–R516.

Renfro, J. L., and Clark, N. B. (1984). Parathyroid hormone effect on chicken renal brush-border membrane phosphate transport. *Am. J. Physiol.* **247,** R302–R307.

Renfro, J. L., and Dickman, K. G. (1980). Sulfate transport across the peritubular surface of the marine teleost renal tubule. *Am. J. Physiol.* **239,** F143–F148.

Renfro, J. L., and Pritchard, J. B. (1982). H^+-dependent sulfate secretion in the marine teleost renal tubule. *Am. J. Physiol.* **243,** F150–F159.

Renfro, J. L., and Pritchard, J. B. (1983). Sulfate transport by flounder renal tubule brush border: Presence of anion exchange. *Am. J. Physiol.* **244,** F488–F496.

Renfro, J. L., and Shustock, E. (1985). Peritubular uptake and brush border transport of ^{28}Mg by flounder renal tubules. *Am. J. Physiol.* **249,** F497–F506.

Renfro, J. L., Dickman, K. G., and Miller, D. S. (1982). Effect of Na and ATP on peritubular Ca transport by the marine teleost renal tubule. *Am. J. Physiol.* **243,** R34–R41.

Renfro, J. L., Brown, M. A., Parker, S. L., and Hightower, L. E. (1993). Relationship of thermal and chemical tolerance to transepithelial transport by cultured flounder renal epithelium. *J. Pharmacol. Exp. Ther.* **265,** 992–1000.

Schwab, S. J., and Hammerman, M. R. (1986). Electrogenic Na^+-independent P_i transport in canine renal basolateral membrane vesicles. *Am. J. Physiol.* **250,** F419–F424.

Shimada, H., Moewes, B., and Burckhardt, G. (1987). Indirect coupling to Na of *p*-aminohippuric acid uptake into rat renal basolateral membrane vesicles. *Am. J. Physiol.* **253,** F795–F801.

Stanton, R. C., and Seifter, J. L. (1987). Transport and metabolism in renal cells in culture. *Semin. Nephrol.* **7,** 29–36.

Stolte, H., Galaske, R. G., Eisenbach, G. M., Lechene, C., Schmidt-Nielsen, B., and Boylan, J. W. (1977). Renal tubule ion transport and collecting duct function in the elasmobranch little skate, *Raja erinacea*. *J. Exp. Zool.* **199,** 403–410.

Sundell, K., Bjornsson, B. T., Itoh, H., and Kawauchi, H. (1992). Chum salmon (*Oncorhynchus keta*) stanniocalcin inhibits *in vitro* intestinal calcium uptake in Atlantic cod (*Gadus morhua*). *J. Comp. Physiol., B* **162,** 489–495.

Sussman-Turner, C., and Renfro, J. L. (1995). Heat shock-stimulated transepithelial daunomycin secretion by renal proximal tubule primary cultures. *Am. J. Physiol.* **267,** F135–F144.

Thiebaut, F., Tsuruo, T., Hamada, H., Gottesman, M. M., Pastan, I., and Willingham, M. C. (1987). Cellular localization of the multidrug-resistance gene product P-glycoprotein in normal human tissues. *Proc. Natl. Acad. Sci. U.S.A.* **84,** 7735–7738.

Trump, B. F., and Jones, R. T. (1977). Correlation of structure and active transport in the teleost nephron. *J. Exp. Zool.* **199,** 365–382.

Ussing, H. H., and Zerahn, K. (1951). Active transport of sodium as the source of electric current in the short-circuited isolated frog skin. *Acta Physiol. Scand.* **23,** 110–127.

Valentich, J. D. (1986). Innovative approaches for the study of cultured renal epithelia. *Min. Electrolyte Metab.* **12,** 6–13.

Wagner, G. F., Copp, D. H., and Friesen, H. G. (1988a). Immunological studies on teleocalcin and salmon corpuscles of Stannius. *Endocrinology* (*Baltimore*) **122,** 2064–2070.

Wagner, G. F., Fenwick, J. C., Park, C. M., Milliken, C., Copp, D. H., and Friesen, H. G. (1988b). Comparative biochemistry and physiology of teleocalcin from sockeye and coho salmon. *Gen. Comp. Endocrinol.* **72,** 237–246.

Wagner, G. F., Gellersen, B., and Friesen, H. G. (1989). Primary culture of teleocalcin cells from rainbow trout corpuscles of Stannius: Regulation of teleocalcin secretion by calcium. *Mol. Cell. Endocrinol.* **62,** 31–39.

Wagner, G. F., Milliken, C., Friesen, H. G., and Copp, D. H. (1991). Studies on the regulation and characterization of plasma stanniocalcin in rainbow trout. *Mol. Cell. Endocrinol.* **79,** 129–138.

Wendelaar Bonga, S. E., and Pang, P. K. (1991). Control of calcium regulating hormones in the vertebrates: Parathyroid hormone, calcitonin, prolactin, and stanniocalcin. *Int Rev. Cytol.* **128,** 139–213.

Wendelaar Bonga, S. E., Pang, R. K., and Pang, P. K. T. (1986). Hypocalcemic effects of bovine parathyroid hormone (1-34) and Stannius corpuscle homogenates in teleost fish adapted to low-calcium water. *J. Exp. Zool.* **240,** 363–367.

Wright, C. E., Tallan, H. H., and Lin, Y. Y. (1986). Taurine: Biological update. *Annu. Rev. Biochem.* **55,** 427–453.

Zweifach, A., Desir, G. V., Aronson, P. S., and Giebisch, G. H. (1991). A Ca-activated K channel from rabbit renal brush-border membrane vesicles in planar lipid bilayers. *Am. J. Physiol.* **261,** F187–F196.

7

PHENOTYPIC EXPRESSION AND NATRIURETIC PEPTIDE-ACTIVATED CHLORIDE SECRETION IN CULTURED SHARK *(SQUALUS ACANTHIAS)* RECTAL GLAND EPITHELIAL CELLS

JOHN D. VALENTICH[1], *KARL J. KARNAKY, JR., AND WILLIAM M. MORAN*

[1]To whom correspondence should be sent. E-mail: shark@sam.neosoft.com or valentich @beach.utmb.edu; Phone: (409)772-9887.

I. INTRODUCTION

Ion-transporting epithelia play a significant role in modulating physiologically important homeostatic processes such as plasma salt, pH, and water balance. They are also vital for regulating the ionic and fluid composition of the lumenal surfaces of the gut, airways, and reproductive tract. To achieve this level of control, transporting epithelia have evolved regulatory mechanisms utilizing neural, endocrine, paracrine, and autocrine mediators. These mediators function in poorly characterized combinatorial networks. Their integrated effects play a crucial role in determining and defining an organism's adaptive response to environmental change and pathologic or toxicologic insult.

An emerging paradigm in epithelial regulatory biology is the importance of local control mechanisms operating in series or in parallel with the endocrine and nervous systems. Both soluble mediators and their cognate receptors are expressed in specific tissue compartments, creating microenvironments where specialized responses can be elicited by generic signaling molecules. Local control may involve paracrine-mediated heterologous cell interactions between epithelial cells and subepithelial connective tissue, neuronal, and inflammatory cells. Paracrine or juxtacrine interactions between specialized subpopulations of intraepithelial secretory cells, such as gut enteroendocrine cells, are also possible (Grube, 1986). Finally, local control via secretion of autocrine mediators by epithelia may also be physiologically important.

Natriuretic peptides (NPs) and other peptides coupled to the activation of plasma membrane guanylate cyclase are emerging as important players in autocrine/paracrine signaling networks in the kidney (Valentin and Humphreys, 1993; Suzuki *et al.,* 1993; Ritter *et al.,* 1991; Greenwald *et al.,* 1991, 1992), salivary gland (Vollmar *et al.,* 1991), olfactory bulb (Gutkowska *et al.,* 1991), adrenal medulla (Morel *et al.,* 1988b), anterior pituitary (Morel *et al.,* 1988a, 1989), thyroid (Hughes and Sellitti, 1991), and intestine (Gerbes *et al.,* 1991; Forte *et al.,* 1993). Since their first identification and isolation from extracts of mammalian atrial tissue, insight into the physiologic importance of NPs has expanded enormously. The diversity of cell types where NPs are synthesized and act suggests that they have additional physiologic roles beyond modulating blood pressure and extracellular fluid homeostasis.

NP immunoreactivity and responsiveness are found in all vertebrates, including hagfish, elasmobranchs, and teleosts, where they are believed to have important osmoregulatory roles (Evans, 1990). A major part of the natriuretic and diuretic effects of these peptides in mammals are mediated through their action on ion transport processes in epithelia such as the renal distal tubule and intestine. In fish, atrial natriuretic peptide (ANP) modulates NaCl transport in flounder intestine, killifish opercular epithelium, and the isolated perfused dogfish shark rectal gland (O'Grady *et al.*, 1985; Scheide and Zadunaisky, 1988; Solomon *et al.*, 1985). However, owing to the complexity of these systems, cellular and molecular analyses of the mechanism of action of NPs on ion transport in fish epithelia are severely limited.

To elucidate the potential role of endogenous NPs as autocrine/paracrine mediators in the SRG and to define the signaling mechanism through which they activate Cl^- secretion, we have developed methods for primary tissue culture of the salt-secreting epithelial cells from the dogfish shark (*Squalus acanthias*) rectal gland (SRG). This chapter summarizes the techniques we have developed for studying SRG epithelial cells in culture, describes the functional properties of these cultures, and reviews what we have learned in using this *in vitro* system about the regulation of secondary active Cl^- secretion by NPs.

II. OSMOREGULATORY SIGNIFICANCE, ION TRANSPORT FUNCTION, AND STRUCTURE OF THE SHARK RECTAL GLAND

Elasmobranch osmoregulation has been reviewed extensively by Shuttleworth (1988) and is only summarized here. Elasmobranchs living in seawater have plasma solute concentrations that are isoosmotic or somewhat hyperosmotic to their marine environment. However, the NaCl concentration is higher in seawater, resulting in the need to eliminate excess plasma NaCl. By elaborating a secretory fluid containing $[Na^+]$ and $[Cl^-]$ almost twice that of plasma, the SRG is responsible for eliminating NaCl. *In vitro* studies employing the perfused SRG (Silva *et al.*, 1977) and isolated microperfused SRG tubules (Greger and Schlatter, 1984) demonstrate that the secondary active Cl^- secretory mechanism conforms to that observed throughout the animal kingdom (Frizzell *et al.*, 1979). Briefly, Cl^-, K^+ and Na^+ enter the cells via basolateral cotransporters driven by the inward transmembrane Na^+ gradient established by the Na^+ pump. Cl^- exits cells via apical plasma membrane Cl^- channels and Na^+ diffuses into the electrically negative tubular lumen via the paracellular pathway. K^+ exits cells via basolateral channels, hyperpolarizing the cell and maintaining the electrical driving force for Cl^- secretion.

The SRG consists largely of simple and branched, blind-ended secretory tubules that radiate from a central duct. The central duct presumably serves as a simple excretory conduit and is lined by a stratified epithelium 4 to 6 cells thick. The simple columnar cells of the secretory tubules exhibit extensive basolateral plasma membrane interdigitations and a rich population of mitochondria. The extensive amplification of the basolateral plasma membrane places a high density of Na^+/K^+ pumps in intimate contact with ATP-producing mitochondria (Eveloff *et al.*, 1979). This cellular architecture is the structural basis for generating the inwardly directed transmembrane Na^+ gradient that provides the driving force to support the exceptionally high rates of secondary active Cl^- secretion by SRG epithelial cells. At the apical pole, SRG secretory cell borders interlock like pieces of a jigsaw puzzle. This results in an extensive increase in the linear distance of the tight junction, which consists of only one to three junctional strands (Forrest *et al.*, 1982; Ernst *et al.*, 1981). Consequently, SRG tubules exhibit a low transepithelial electrical resistance ($27\ \Omega \cdot cm^2$) and voltage (-11 mV) despite extremely high rates of net NaCl secretion ($400\ \mu A/cm^2$) following activation by cyclic AMP (Greger and Schlatter, 1984).

III. CULTURED SHARK RECTAL GLAND CELLS ARE A UNIQUE MODEL FOR ANALYZING SECONDARY ACTIVE Cl^- SECRETION IN EPITHELIA

Although transepithelial secondary active Cl^- secretion occurs in a variety of vertebrate tissue and cell types, the SRG offers several significant technical and theoretical advantages for analyzing this process at the whole-animal, organ, cellular, and molecular levels (Silva *et al.*, 1990; Valentich, 1991a). This is chiefly due to the SRG's extraordinary specialization for Cl^- secretion. The gland is composed principally of densely packed epithelial cells whose only known ion transport function is secretion of an isosmotic NaCl solution into the tubular lumen (Fig. 1D). Because of this functional homogeneity, the SRG is uniquely amenable for analyses of Cl^- secretion in preparations ranging from catheterized glands in free-swimming sharks (Silva *et al.*, 1990) to purified plasma membrane fractions (Dubinsky and Monti, 1986).

Despite these benefits of native SRG preparations, investigating transport mechanisms and their regulation at the subcellular level requires a model system where transport events and their associated regulatory molecules can be analyzed in a highly manipulable and controllable environment. For this reason we developed methods for analyzing SRG epithelial cell function in tissue culture. The experimental advantages of cultured

SRG cells over native SRG tissue are outlined in Table I. Because the *Squalus acanthias* rectal gland is composed principally of a single parenchymal cell type and contains minimal supporting connective tissue matrix, it is ideal for initiating highly homogeneous primary cultures of Cl^- secreting epithelial cells. In addition, because the phenomenology of hormone-regulated Cl^- secretion is well defined for the native SRG, the fidelity of phenotypic expression by cultured SRG cells can be critically evaluated.

Numerous mammalian cell culture systems for analyzing Cl^- transporting epithelia exist. These include primary cultures as well as established cell lines. The latter have been derived from (1) spontaneously occuring human carcinomas, (2) immortalization of normal human and rodent cells in primary culture using retroviral or plasmid vectors carrying transforming oncogenes, and (3) establishing cultures from tissue of animals expressing SV40 transgenes (McRoberts and Barrett, 1989; Jat *et al.*, 1991; Welsh, 1985; Jefferson *et al.*, 1990). However, certain deficien-

Table I
Advantages of Cultured SRG Cells over Native SRG Preparations

1. Cultures comprised only of Cl^- secretory epithelial cells without associated neural, vascular, and connective tissue elements.
2. Cells are given an opportunity to recover from physical, chemical, and anoxic trauma inherent in tissue, tubule, and cell isolation.
3. Using the short-circuit current technique, changes in Cl^- secretory rate can be determined with greater sensitivity and time resolution than by measuring transepithelial voltage, oxygen, consumption, and Cl^- content of secretory fluid.
4. A greater change in fractional resistance of the apical plasma membrane occurs in response to secretagogues.
5. Ease of direct access to apical plasma membranes for manipulating the ionic and secretagogue milieu, patch clamping, microelectrode impalement, measurement of unidirectional tracer fluxes, addition of chemical probes, and collection of secretory products.
6. Ease of voltage clamping transepithelial electrical potential for direct measurement of short-circuit current and active ion flux.
7. Ability to perform correlative morphologic, electrophysiologic, and biochemical analyses on cells of known and uniform transport state.
8. Capability for long-term, stable control of nutrient, hormonal, and toxicant milieu.
9. Maintenance of optimal cell viability throughout the duration of long-term experiments.
10. Exhibit higher transepithelial electrical resistance.
11. Cl^- secretion can be activated by single rather than a cocktail of secretagogues.

cies are associated with these culture systems: (1) lack of definitive knowledge of the cell of origin when primary cultures or cell lines are derived from a heterogeneous epithelium, such as the kidney, airway, or intestinal mucosa; (2) labor-intensive procedures for isolating, purifying, and identifying the desired cell type when it comprises only a minor fraction in the tissue of interest; (3) phenotypic instability and poor or complete lack of expression of transepithelial Cl^- secretion *in vitro;* and (4) uncertainties about the potentially confounding effects of neoplastic transformation on intracellular regulatory networks.

IV. HOW ARE SHARK RECTAL GLAND CELLS CULTURED?

Methods for culturing SRG cells as monolayers on a variety of solid and microporous substrata, as suspensions of isolated tubules or as isolated tubules embedded in collagen gels, have been described previously in detail (Valentich, 1991a). The steps most critical to the preparation of functional monolayer cultures are: (1) avoiding dissociation of SRG tubules to single cells, (2) plating tubules at a sufficient concentration to maintain a high cell density following migration of cells from tubules onto the culture substratum, and (3) achieving an even distribution of the tubule inoculum across the culture substratum. Suboptimal tubule inoculum densities may result in confluent monolayer formation, but cells assume a highly flattened squamous morphology and secretagogue-stimulated Cl^- secretory activity is minimal (Valentich and Forrest, 1991).

Because of the importance of continually maintaining high cell density for retaining transport activity in SRG cultures, cell dissociation, subculture, or immortalization of cultured SRG cells may not be feasible or desired. Although continuous passage of SRG cells is not currently possible, primary cultures can be maintained for 2 to 3 months with only a slow decrement in the magnitude of maximally stimulated Cl^- secretion.

V. DIFFERENTIATED PROPERTIES OF CULTURED SHARK RECTAL GLAND CELLS

A. Cell Morphology

Cultured SRG cells maintained on collagen-coated microporous membranes or collagen gels retain the cuboidal-columnar shape and polarized differentiation of highly specialized transporting epithelia (Figs. 1 and 2).

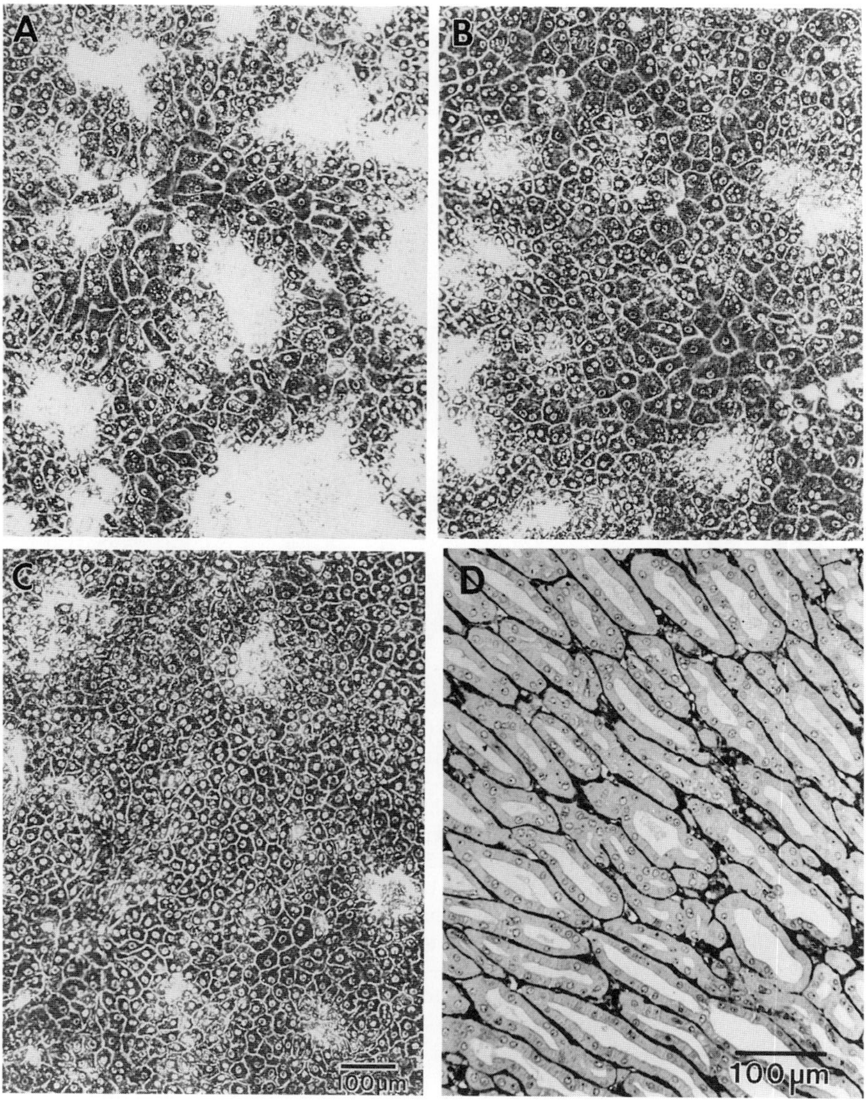

Fig. 1. Phase-contrast micrographs illustrating confluent monolayers of SRG tubular epithelial cells at 5 (A), 10 (B), and 15 days (C) after plating collagenase-dissociated tubules on collagen gel substrata. Phase-bright densities are tubule remnants that gradually disappear over time. (From Valentich and Forrest (1991). Copyright 1991 by The American Physiological Association, Bethesda, MD. Reprinted with permission.) (D) Histologic section through a shark rectal gland showing densely packed secretory tubules composed of a uniform population of cuboidal-columnar epithelial cells. The interstitial connective tissue matrix (stained black) is sparse, allowing large numbers of tubules to be isolated by gentle collagenase digestion with minimal contamination by nonepithelial cell types. (From Valentich, J. D. *Journal of Tissue Culture Methods* **13,** 149–162, 1991. Copyright © 1991 by the Tissue Culture Association, Columbia, MD. Reprinted with permission.)

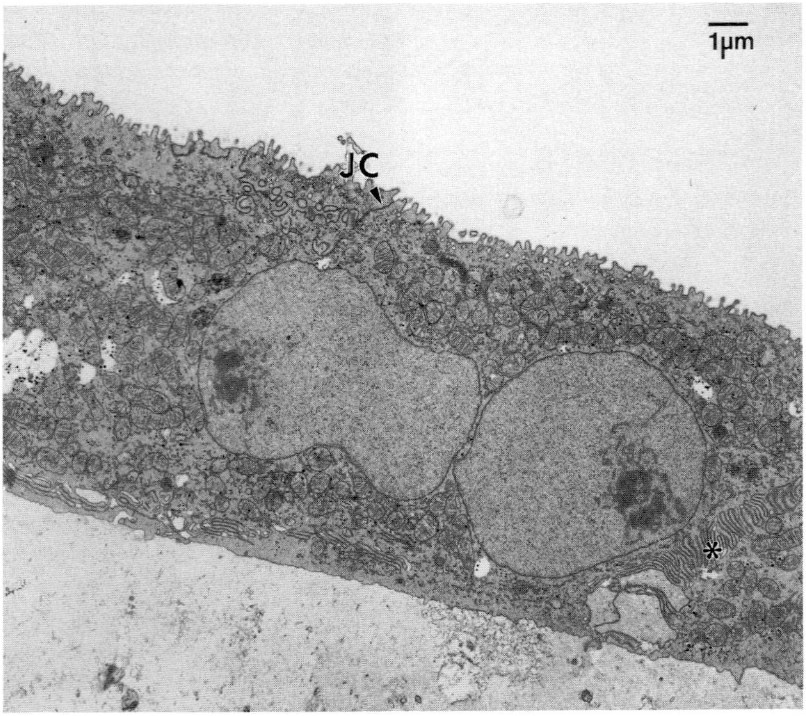

Fig. 2. Transmission electron micrograph illustrating the ultrastructural morphology of SRG cells after 14 days in culture on a collagen gel substratum. Cells contain a rich complement of mitochondria and the lateral cell borders are highly interdigitated (asterisk). There are a large number of tubulo-vesicles in the cytoplasm near the apical plasma membrane. Adjacent cells are linked together by junctional complexes (JC). These ultrastructural features are characteristic of highly specialized transporting epithelia. (From Valentich, J. D. *Journal of Tissue Culture Methods* **13,** 149–162, 1991. Copyright © 1991 by the Tissue Culture Association, Columbia, MD. Reprinted with permission.)

Prominent morphologic features include apical microvilli and junctional complexes, a subapical tubular-vesicular system, supranuclear Golgi complexes, numerous mitochondria, and extensive lateral plasma membrane interdigitation. No basal lamina is observed in cultures grown on collagen or Matrigel substrates (Valentich and Garretson, 1987). Rhodamine-phalloidin staining demonstrates actin microfilaments bounding the perimeter of each polygonal epithelial cell. In whole-mounts of rhodamine-phalloidin-stained cultures the absence of subepithelial fusiform or stellate-shaped cells demonstrates the lack of culture contamination by fibroblasts or neurons. Studies using fluorescent lectin probes specific for tubular epithelial cells (*Helix pomatia* lectin, concanavalin-A, wheat germ agglutinin) or central duct cells (peanut lectin) show that by 7–10 days SRG cultures are composed exclusively of tubular epithelial cells (J. D. Valentich, unpublished observations). These data confirm that the electrophysiologic and biochemical responses of SRG cultures to secretagogues reflect the activity of tubular epithelial cells and not other tissue elements.

B. Ion Transport Processes

1. Transepithelial Electrogenic Cl^- Secretion

The Cl^- secretory properties of cultured SRG cells maintained on permeable supports have been investigated in detail using electrophysiologic methods. Cl^- secretion is determined by calculating the equivalent short-circuit current after measuring transepithelial voltage (V_{ab}) and resistance (Valentich, 1991a) or directly using the Ussing short-circuit current technique (Valentich and Forrest, 1991). Transepithelial ^{36}Cl flux measurements demonstrate that essentially all of the I_{sc} can be accounted for by active Cl^- secretion (Valentich and Forrest, 1991). In the absence of secretagogue stimulation, SRG cultures display an I_{sc} of only 1–10 $\mu A/cm^2$. However, following activation, I_{sc} as high as 300 $\mu A/cm^2$ is rapidly achieved with typical stimulated currents falling in the range of 100–150 $\mu A/cm^2$ (Valentich and Forrest, 1991). Consistent with this high level of secondary active Cl^- secretion, selective addition of specific transport inhibitors to apical or basolateral baths rapidly inhibits I_{sc} and demonstrates the highly polarized plasma membrane distribution of apical Cl^- channels, basolateral K^+ channels, Na^+/K^+ pumps, and $Na^+/K^+/2Cl^-$ cotransporters (Valentich and Forrest, 1991; Valentich, 1991a).

2. Analysis of Electrogenic Cl^- Secretion Using Intracellular Microelectrodes

a. Effect of Cyclic AMP on Apical Membrane Cl^- Conductance. In monolayer SRG cultures, apical membrane electrical potential difference (V_a) and fractional resistance of the apical membrane (fR_a) can be assessed directly because microelectrodes can be inserted across the apical membrane (Moran and Valentich, 1991). In native SRG tubules perfused *in vitro,* microelectrodes can only be inserted across the basolateral membrane. Because this barrier has a dominant K^+ conductance, changes in electrical properties of the apical membrane are difficult to assess. Although cultured SRG cells are not unusually large, stable, long-term microelectrode impalements are relatively easy to obtain, in contrast to most native and cultured epithelia.

In the absence of Cl^- secretagogues, V_a ranges from −55 to −70 mV (cell interior negative) and fR_a is high, usually >0.90 (Moran and Valentich, 1991). Elevation of intracellular cyclic AMP levels with forskolin, a potent adenylate cyclase activator in SRG cells (Valentich and Forrest, 1991; Yancey *et al.*, 1991), rapidly causes dramatic and sustained alterations in the electrophysiologic properties of the apical membrane (Fig. 3). These changes are characterized by depolarization of V_a and V_b, a large fall in fR_a, and a small increase in V_{ab} (Moran and Valentich, 1991). These alterations in membrane properties are consistent with increased electrogenic Cl^- secretion. Following forskolin stimulation, superfusing the apical surface with low-Cl^- shark Ringer causes abrupt depolarization of V_a and an increase in fR_a, both of which are indicative of a cyclic AMP-activated apical membrane Cl^- conductance (G_a^{Cl}) (Moran and Valentich, 1991). In the absence of forskolin, low-Cl^- shark Ringer has little effect on either V_a or fR_a because G_a^{Cl} is very small. The large decrease in fR_a from 0.97 to 0.1 in response to forskolin shows that the apical plasma membrane undergoes a major change in Cl^- permeability following cell activation.

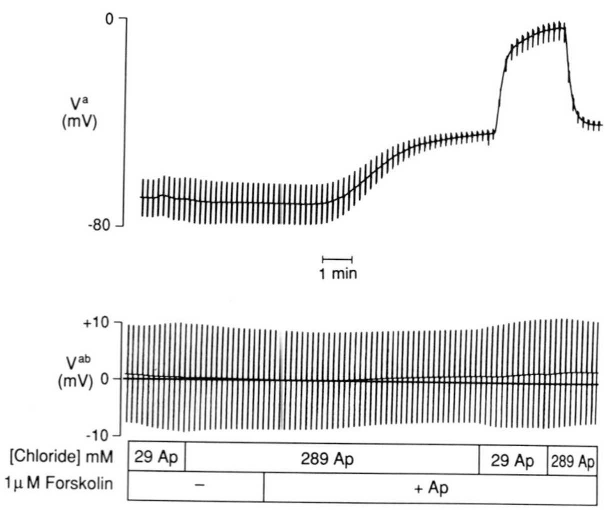

Fig. 3. Effect of 10^{-6} *M* forskolin on apical membrane electrical potential difference (V_a) and apical membrane fractional resistance (fR_a) in SRG primary cultures. Deflections of V_a are due to periodic displacement of transmonolayer electrical potential difference (V_{ab}) by transepithelial constant-current pulses to monitor transmonolayer resistance and fR_a. Ap is the apical superfusate. In absence of forskolin, apical low-Cl^- shark Ringer solution appears to have no effect on V_a; however, after correction for liquid junction potential, V_a actually depolarized by ~6 mV. (From Moran and Valentich (1991). Copyright 1991 by The American Physiological Association, Bethesda, MD. Reprinted with permission.)

b. Effect of Ca^{2+} on Apical Membrane Cl^- Conductance. The response of cultured SRG cells to Ca^{2+} ionophores suggests that rises in intracellular Ca^{2+} also activate G_a^{Cl} and stimulate Cl^- secretion. The temporal and quantitative changes in V_a and fR_a induced by ionomycin or A23187 when added to the apical superfusate are very similar to those observed in forskolin-stimulated cells, indicating that ionomycin activates G_a^{Cl} (Moran and Valentich, 1993). The effect of ionophores on Cl^- secretion and membrane Cl^- conductive properties is not due to the formation of cycloxygenase products since indomethacin has no effect on ionophore-activated I_{sc} and G_a^{Cl}. Although Ca^{2+} ionophores increase G_a^{Cl} to nearly the same extent as forskolin, stimulation of I_{sc} is significantly less (Moran and Valentich, 1993). This disparity may be due to differences in the activation of basolateral membrane transport processes by cyclic AMP and Ca^{2+}. High rates of sustained Cl^- secretion require coordinate activation of G_a^{Cl}, basolateral K^+ channels, $Na^+/K^+/2Cl^-$ cotransporters, and the Na^+/K^+ pump.

In contrast to apically applied ionophores, ionophores added to the basolateral bath fail to activate SRG Cl^- secretion or G_a^{Cl}, even after extended periods of superfusion (Moran and Valentich, 1993). This suggests that the Ca^{2+}-dependent signaling system is localized near the apical plasma membrane. It is possible that the extensive tubular-vesicular complex found near the apical membrane of SRG cells (Fig. 2) is part of such a system. If Ca^{2+} sequestration is a function of the tubular-vesicular elements, they may be intracellular targets of apically applied Ca^{2+} ionophores. Modulation of juxtaposed ion channels by local increases in cytosolic Ca^{2+} near plasma membranes has been described in colonocytes and pancreatic and salivary acinar cells (Morris *et al.*, 1990; Kasai, 1990; Foskett *et al.*, 1989).

One mechanism to account for the massive increase in G_a^{Cl} in cyclic AMP- or Ca^{2+}-activated SRG cells involves secretagogue-stimulated exocytotic delivery to this barrier of Cl^- channels that previously were sequestered in cytoplasmic vesicles. The large number of vesicles in the apical cytoplasm of SRG cells supports this idea. In addition, data from impedance analysis of cultured SRG cells are consistent with an increase in apical plasma membrane capacitance following forskolin stimulation (Kushman *et al.*, 1991). Such a capacitance change is expected if vesicular exocytosis elicits an increase in the area of the apical plasma membrane.

3. Patch-Clamp Analysis of Single Apical Membrane Cl^- Channels

In native, cyclic AMP-stimulated microperfused SRG tubules, ''small''- and ''large''-conductance Cl^- channels have been detected with the patch-clamp technique. In excised patches, a 10-pS ("small") channel in the apical membrane was detected that had a linear I-V relation (Gogelein *et al.*, 1987b). The "large" channel also exhibited a linear I-V relation, but its single-channel conductance was 30–60 pS (Greger *et al.*, 1987b, 1988). Unlike the 10-pS channel, the large-conductance Cl^- channel is activated by cyclic AMP in cell-attached patches and by addition of ATP and the catalytic subunit of protein kinase A to the cytosolic side of excised, inside-out patches (Greger *et al.*, 1985, 1988). Using the whole-cell mode of the patch-clamp technique, Devor *et al.* (1995) have described a cyclic AMP-stimulated Cl^- current in cultured SRG cells. In cell-attached recordings, a cyclic AMP-activated 6-pS channel with a linear I-V relation was observed. In contrast to native SRG cells, no large 50-pS Cl^- channel was detected. However, La *et al.* (1991) have detected a 50-pS channel in the apical membranes of cultured SRG cells stimulated with either forskolin or okadaic acid, an inhibitor of protein phosphatases 1 and 2A.

Which of these Cl^- channels accounts for the increase in G_a^{Cl} in cyclic AMP-stimulated SRG tubules or monolayers is unresolved. The work of Hanrahan *et al.* (1993) may help clarify this issue. They assessed whether the shark cystic fibrosis transmembrane conductance regulator protein (sCFTR) is involved in cyclic AMP-stimulated Cl^- secretion. sCFTR is the shark homolog of human CFTR, the membrane protein in which mutations elicit the Cl^- secretory defect of cystic fibrosis. sCFTR exhibits 72% amino acid identity to human CFTR and retains the characteristic domain organization of CFTR (Marshall *et al.*, 1991; Grzelczak *et al.*, 1990). sCFTR cDNAs were expressed in both *Xenopus* oocytes and Sf9 insect cells and the resulting Cl^- channels were compared with those in native SRG tubules. Cell-attached patch-clamp studies identified linear 4-pS and 46-pS Cl^- channels on the apical membranes of cells in microdissected tubules. The 4-pS channels were recorded more frequently than the 46-pS channel when intracellular cyclic AMP was elevated. Furthermore, the 4-pS channels were the only channels found in excised patches activated by protein kinase A. 4-pS channels exhibiting properties identical to those in native SRG cells were also detected in both sCFTR heterologous expression systems. These data indicate that sCFTR is directly responsible for the cyclic AMP-stimulated G_a^{Cl} in SRG cells or regulates the activation of this conductance (Schwiebert *et al.*, 1994).

4. Analysis of Basolateral Membrane Transporters

a. K^+ Channels. Microelectrode techniques demonstrate a Ba^{2+}-sensitive basolateral membrane K^+ conductance (G_b^K) in both cultured SRG cells and native tubules microperfused *in vitro* (Moran and Valentich,

1991; Gogelein *et al.*, 1987a). Basolateral membrane K^+ channels have also been detected in native SRG tubules with the patch-clamp technique (Greger *et al.*, 1987a). As with $G_b{}^K$, these channels are blocked by Ba^{2+} but open probability was not affected by reducing cytosolic Ca^{2+} (Gogelein *et al.*, 1987a).

The primary function of $G_b{}^K$ appears to be the recycling of K^+ across this barrier. K^+ enters SRG cells via both the basolateral membrane Na^+/K^+ pump and the $Na^+/K^+/2Cl^-$ cotransporter (Greger and Schlatter, 1984; Silva *et al.*, 1977). During sustained Cl^- secretion, $G_b{}^K$ may increase to prevent intracellular K^+ accumulation and cell swelling. Indirect evidence from both cultured SRG cells and native SRG tubules suggests that $G_b{}^K$ increases during cyclic AMP-stimulated Cl^- secretion (Moran and Valentich, 1991; Greger and Schlatter, 1984).

b. Na^+/K^+ Pump. Elevating intracellular cyclic AMP levels increases both the activity and apparent number of basolateral membrane Na^+/K^+ pump units in cultured SRG cells. Forskolin stimulates ouabain-sensitive ^{86}Rb influx threefold. In addition, forskolin or dibutyryl cyclic AMP/theophylline enhances specific equilibrium [^{3}H]ouabain binding by more than twofold in SRG cells grown on permeable substrata (J. D. Valentich, unpublished data). Increases of 50–100% in [^{3}H]ouabain binding are observed in SRG slices or freshly isolated cells following cyclic AMP/theophylline stimulation (Shuttleworth and Thompson, 1980; Silva *et al.*, 1983). Whether increased [^{3}H]ouabain binding reflects direct cyclic AMP-mediated activation or externalization of the Na^+/K^+ pump or occurs secondary to increased Na^+ entry is controversial. Electron microprobe measurements of intracellular Na^+ levels in cultured SRG cells demonstrate that intracellular Na^+ falls rather than rises following secretagogue stimulation (Lear *et al.*, 1992). However, in isolated microperfused SRG tubules, Na^+ activity, measured with ion-selective microelectrodes, increases following activation with a cocktail of secretagogues (Greger *et al.*, 1984).

c. $Na^+/K^+/2Cl^-$ Cotransport. Cotransport activity in SRG cells grown in tissue culture dishes or on microporous membranes is measured by determining the magnitude of bumetanide-sensitive ^{86}Rb influx. Bumetanide is a well-characterized inhibitor of $Na^+/K^+/2Cl^-$ cotransport and rapidly blocks forskolin-stimulated I_{sc} in SRG monolayers with an IC_{50} of approximately 2×10^{-7} *M*; this is more than 10-fold less than the IC_{50} for inhibition of Cl^- secretion in the cyclic AMP-stimulated perfused gland (Palfrey *et al.*, 1984). Activating Cl^- secretion in SRG monolayers with forskolin or 2-chloroadenosine increases cotransport activity 2- to 3-fold

(Fig. 4). ANP also significantly stimulates cotransport activity in SRG cells, consistent with its activation of transepithelial Cl^- secretion (Fig. 4) (Karnaky *et al.*, 1991). This contrasts with the inhibitory effects of ANP on cotransport in flounder intestine and bovine aortic endothelial cells (O'Grady, 1989; O'Donnell, 1989). However, ANP-mediated cotransporter activation is observed in mammalian vascular smooth muscle and medullary collecting duct cells (O'Donnell and Owen, 1986; Rocha and Kudo, 1990).

C. Secretagogues and Second Messengers That Modulate Transepithelial Cl^- Secretion by Cultured Shark Rectal Gland Cells

Figure 5 summarizes the sensitivity of SRG monolayer I_{sc} to different activators and inhibitors of Cl^- secretion and indicates the polarized action of these compounds. All known stimulatory and inhibitory secretagogues described for the perfused SRG or isolated microperfused SRG tubules are also active in cultured SRG cells, including responsiveness to the four major shark osmoregulatory mediators, vasoactive intestinal peptide (VIP), adenosine, somatostatin, and natriuretic peptides (Valentich and Forrest, 1991; Karnaky *et al.*, 1991; Feero and Valentich, 1991).

Many Cl^- secretagogues exhibit characteristic electrophysiologic sig-

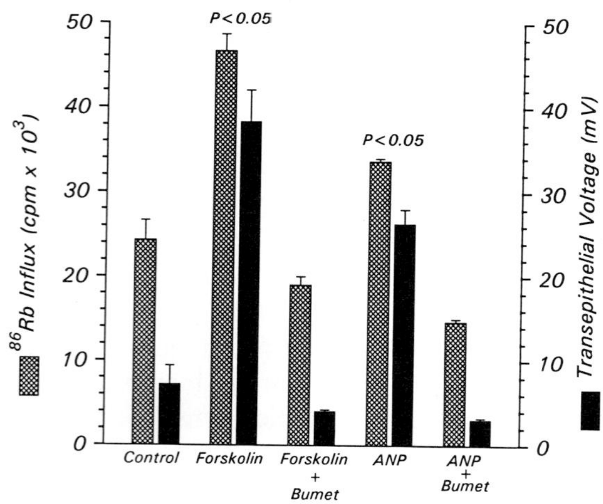

Fig. 4. Activation of $Na^+/K^+/2Cl^-$ cotransport and transepithelial voltage by 5×10^{-8} *M* forskolin and 10^{-6} *M* ANP in cultured SRG cells grown on microporous membranes. Cotransport activity is depicted as the fraction of $^{86}Rb^+$ influx sensitive to 10^{-4} *M* bumetanide. Statistical comparisons are relative to control influx and voltage values.

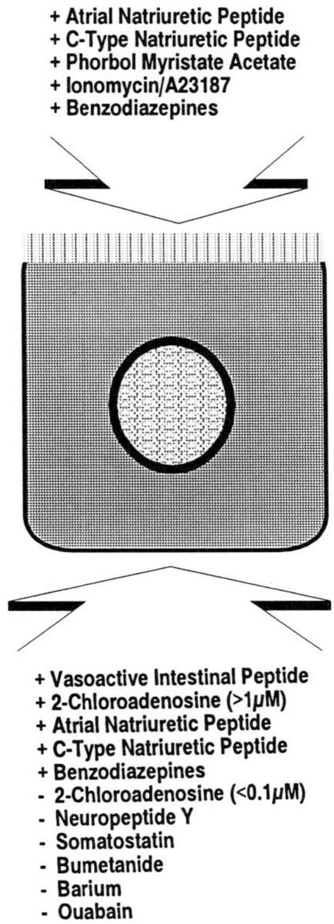

Fig. 5. Polarized effects of Cl^- secretagogues and transporter inhibitors on I_{sc} of SRG cultures. +, activates transport; −, inhibits transport.

natures when the time course of their effects on I_{sc} is analyzed (Fig. 6). Compounds that elevate intracellular cyclic AMP, such as forskolin, cyclic AMP analogues, or 2-chloroadenosine, elicit large, stable increases in I_{sc} that are maintained for prolonged periods. Diacylglycerol analogs also produce stable increases in I_{sc} but maximal stimulation is significantly less than that produced by elevations in intracellular cyclic AMP. In contrast, VIP and Ca^{2+} ionophores only activate I_{sc} transiently. Finally, ANP produces oscillations in I_{sc} that eventually stabilize at a level near the midpoint of the oscillation.

Biochemical analyses of secretagogue-stimulated second messenger formation and addition of exogenous mediators to solutions bathing SRG monolayers demonstrate that cyclic AMP, cyclic GMP, Ca^{2+}, and diacylglycerol are important intracellular messengers involved in activating Cl^-

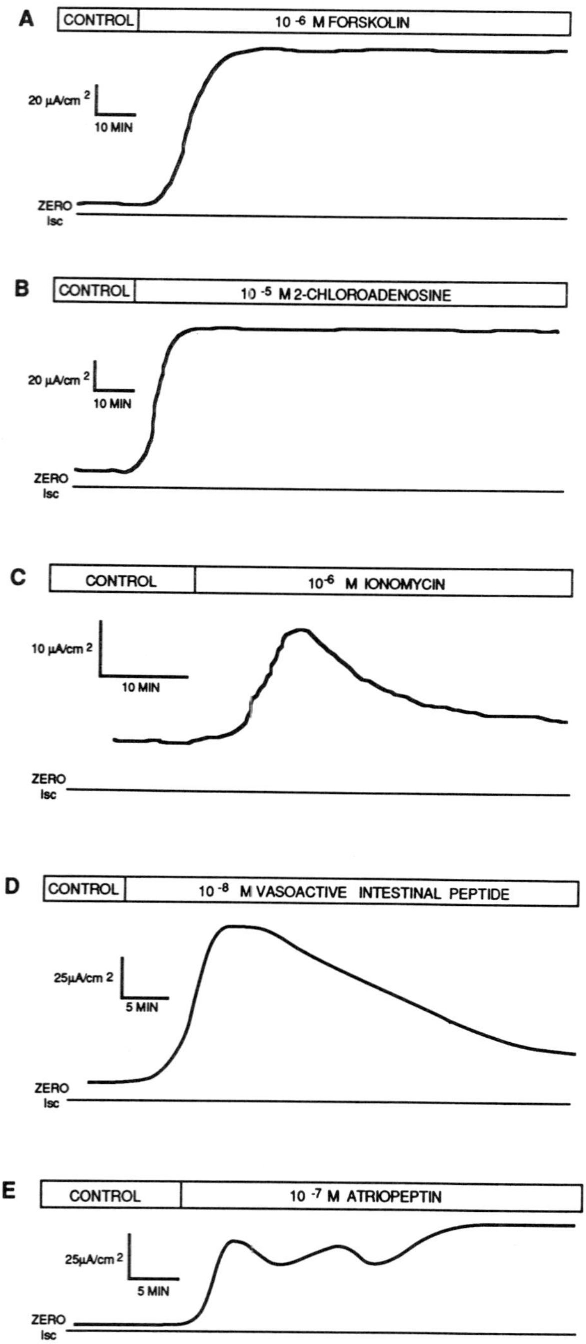
A
CONTROL
10 -6 M FORSKOLIN
20 μA/cm 2
10 MIN
ZERO
Isc
B
CONTROL
10 -5 M 2-CHLOROADENOSINE
20 μA/cm 2
10 MIN
ZERO
Isc
C
CONTROL
10-6 M IONOMYCIN
10 μA/cm 2
10 MIN
ZERO
Isc
D
CONTROL
10 -8 M VASOACTIVE INTESTINAL PEPTIDE
25μA/cm 2
5 MIN
ZERO
Isc
E
CONTROL
10 -7 M ATRIOPEPTIN
25μA/cm 2
5 MIN
ZERO
Isc

secretion (Feero and Valentich, 1991; Yancey *et al.*, 1991; Karnaky *et al.*, 1991; Valentich and Forrest, 1991; Ecay and Valentich, 1991). The unexpected polarized effects of Ca^{2+} ionophores and the diacylglycerol analog phorbol 12,13-dibutyrate on Cl^- secretion suggest that at least a subset of the target molecules for these messengers are localized in or near the apical plasma membranes of SRG cells.

The majority of studies of SRG Cl^- secretion have focused on transport activation. However, Cl^- secretion is also under inhibitory control by both neuroendocrine and autocrine mediators. In the perfused SRG, somatostatin and neuropeptide Y nearly completely inhibit forskolin-stimulated Cl^- secretion (Silva *et al.*, 1985, 1993). Similarly, through a transport-stimulated autocrine feedback mechanism, adenosine, acting at A_1-type receptors, completely inhibits forskolin-activated Cl^- secretion (Kelley *et al.*, 1990, 1991). These results contrast with data from experiments using cultured SRG cells, where inhibitory antagonists elicit only 20–30% reductions in Cl^- secretion or cyclic AMP formation (Valentich and Forrest, 1991; Epstein *et al.*, 1992; Silva *et al.*, 1993). This difference in responsiveness suggests that in the intact SRG, maximum inhibition of Cl^- secretion by somatostatin, neuropeptide Y, and adenosine may require parallel effects on both epithelial cells and nonepithelial tissue elements such as neurons and vascular smooth muscle.

D. P-Glycoprotein-like Transport Activity

P-Glycoprotein is a member of a superfamily of ATP-binding cassette (ABC) membrane transport proteins. These proteins mediate ATP-

---

Fig. 6. Chart recorder tracings showing the short-circuit current (I_{sc}) responses of voltage-clamped shark rectal gland cultures mounted in Ussing chambers. Once the basal I_{sc} stabilized, cultures were stimulated with the indicated compounds. Forskolin (A) and 2-chloroadenosine (B) are believed to stimulate Cl^- secretion by increasing the intracellular levels of cyclic AMP. Both compounds elicit prolonged and sustained increases in I_{sc}. In contrast, ionomycin, a Ca^{2+} ionophore, causes a small and transient activation of I_{sc} (C). Vasoactive intestinal peptide strongly stimulates I_{sc} (D). However, this effect is transient and the I_{sc} spontaneously decays to a steady state, which is approximately 20% of the peak I_{sc} but still above the initial baseline. Atriopeptin I or III stimulation (E) is characterized by a series of I_{sc} oscillations followed by a steady-state current four to five times lower than the maximal I_{sc} elicited by forskolin or 2-chloroadenosine. Analyses of second messenger formation suggest that increases in both cytoplasmic cyclic AMP and inositol phosphates occur in response to vasoactive intestinal peptide (Yancey *et al.*, 1991; Ecay and Valentich, 1991). In contrast, atriopeptin stimulates increases in cyclic GMP and inositol phosphates (Ecay and Valentich, 1991; Karnaky *et al.*, 1991). Interactions between these second messenger systems may be responsible for the complex I_{sc} responses to vasoactive intestinal peptide and atriopeptin. (From Valentich, J. D. *Journal of Tissue Culture Methods* **13,** 149–162, 1991. Copyright © 1991 by the Tissue Culture Association, Columbia, MD. Reprinted with permission.)

dependent efflux of a variety of hydrophobic molecules from both prokaryotic and eukaryotic cells (Higgins, 1992). An intensively studied member of this family is CFTR. SRG cells strongly express the shark homolog of this protein in their apical plasma membranes (Marshall *et al.*, 1991). Cl^- channel and ATP-dependent transport functions have been associated with CFTR and P-glycoprotein (Fuller and Benos, 1992; Higgins, 1992). In mammals, P-glycoprotein is found principally in the apical plasma membranes of transporting epithelia, such as renal tubules, airways, hepatocytes, pancreatic ducts, enterocytes, and endometrium (Gottesman and Pastan, 1993). P-Glycoprotein is also present in lower organisms, including sponges, mussels, oysters, and fish (Chan *et al.*, 1992; Kurelec, 1992). The physiologic function and endogenous substrates for P-glycoprotein are not clearly defined. However, ATP-dependent extrusion of harmful toxic metabolites, xenobiotics, and glutathione conjugates is frequently attributed to P-glycoprotein-like transporters.

Rhodamine 123 is a substrate for P-glycoprotein as well as a vital fluorescent stain for mitochondria. Selective addition of rhodamine 123 to the basolateral side of SRG monolayers grown on microporous membranes rapidly results in intense mitochondrial staining. However, when rhodamine 123 is added to the apical bath, no mitochondrial staining is observed unless cells are pretreated with the P-glycoprotein inhibitor reserpine (Valentich, 1991b). These results are consistent with the presence of a highly active P-glycoprotein-like transport moiety in the apical plasma membranes of SRG cells. In preliminary studies with SRG cultures mounted in Ussing chambers, we have measured a net basolateral to apical secretory flux of [^{3}H]vinblastine, a P-glycoprotein substrate. These data suggest that the SRG may be an important site of excretion of certain xenobiotics in elasmobranchs.

VI. NATRIURETIC PEPTIDE REGULATION OF Cl^- SECRETION IN SHARK RECTAL GLAND CELLS

A. Natriuretic Peptides Activate Shark Rectal Gland Cl^- Secretion

The discovery of the potent diuretic, natriuretic, and vasodilatory activities of ANP in the early 1980s suggested that this hormone may have an important role in elasmobranch osmoregulation. Indeed, it was found that bolus injections of ANP or shark heart extracts into the arterial perfusate of isolated rectal glands stimulate Cl^- secretion (Solomon *et al.*, 1985). However, ANP was reported to have no detectable effect on oxygen consumption by collagenase-dispersed SRG cells or on transepithelial voltage of isolated microperfused SRG tubules (Silva *et al.*, 1987). ANP, but not adenosine-stimulated Cl^- secretion, is associated with the appearance of immunoreactive VIP in the venous effluent of the perfused gland. These data, together with observations that ANP-stimulated Cl^- secretion is blocked by inhibitors of neurosecretion such as procaine and Ca^{2+} channel antagonists, led to the hypothesis that ANP activates Cl^- secretion indirectly by stimulating release of VIP from peritubular nerve endings (Fig. 7) (Silva *et al.*, 1987). The presence of scattered nerve fibers containing immunoreactive VIP in the interstitium between secretory tu-

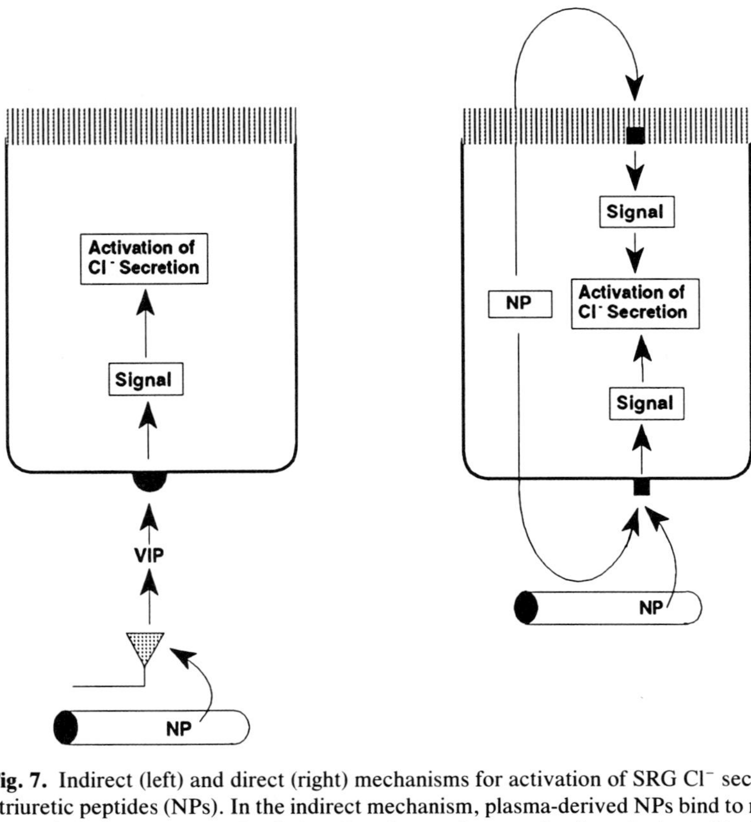

Fig. 7. Indirect (left) and direct (right) mechanisms for activation of SRG Cl^- secretion by natriuretic peptides (NPs). In the indirect mechanism, plasma-derived NPs bind to receptors on peritubular neurons, stimulating the release of vasoactive intestinal peptide (VIP), which subsequently activates Cl^- secretion by binding to epithelial basolateral membrane receptors coupled to the formation of cyclic AMP and inositol phosphates. In the direct mechanism, NPs bind to cognate receptors on either the apical or basolateral membranes of SRG tubular epithelial cells. Cl^- secretion is stimulated by the activation of a complex signaling mechanism involving several second messengers and apical and basolateral transporters (see text for details). NPs may either originate from the plasma or may be synthesized locally by SRG epithelial cells.

bules is cited as support for the indirect activation mechanism (Chipkin *et al.*, 1988). However, the majority of nerve fibers and VIP immunoreactivity in the SRG are located in the fibromuscular capsule and near the central duct and venous sinus, not among the secretory tubules (Chan and Phillips, 1967; Holmgren and Nilsson, 1983; Chipkin *et al.*, 1988). Therefore, VIP measured in venous effluent may never have the opportunity to interact with receptors on secretory epithelial cells. In addition, the physiologic importance of VIP as a secretagogue in the native SRG has been questioned (Shuttleworth and Thorndyke, 1984).

NPs can also act directly on freshly isolated and cultured SRG tubular epithelial cells to stimulate cyclic GMP formation and Cl^- secretion (Fig. 8) (Karnaky *et al.*, 1991; Kennedy, 1991; Lear *et al.*, 1990). The physiologic significance and relative importance of the direct and indirect mechanisms of ANP action are unclear. It is not known, for example, whether identical or different endogenous shark NPs or NP receptors mediate the indirect and direct activation pathways. In some target organs, such as the medullary collecting duct, NPs modulate ion transport by acting directly on

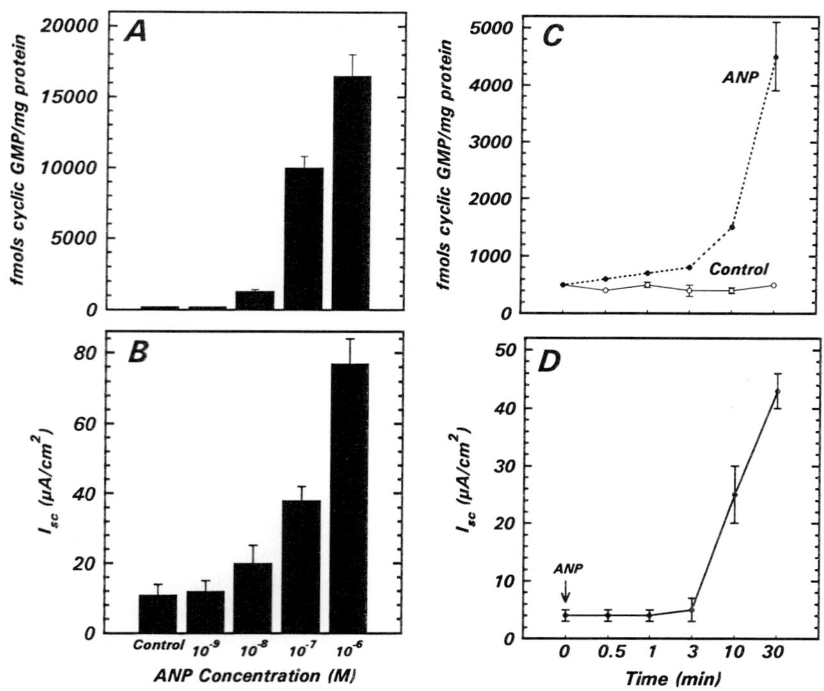

Fig. 8. Comparison of the concentration dependence of ANP-stimulated intracellular cyclic GMP formation (A) and I_{sc} (B) in SRG cultures. Comparison of the time dependence of 10^{-7} *M* ANP-stimulated intracellular cyclic GMP formation (C) and I_{sc} (D) in SRG cultures.

epithelial cells (Rocha and Kudo, 1990). However, in the mammalian intestine, ANP-modulated ion transport does not involve its direct interaction with receptors on epithelial cells (Moriarty *et al.*, 1990).

B. Second Messenger Systems Involved in Natriuretic Peptide Activation of Shark Rectal Gland Cells

NP binding to R1-type receptors activates guanylate cyclase and leads to increased intracellular cGMP in many cell types (Anand-Srivastava and Trachte, 1993). In addition, NP R2-type receptors do not activate guanylate cyclase but are coupled to the mobilization or *de novo* formation of other second messengers such as Ca^{2+} and diacylglycerol or inhibition of cyclic AMP generation (Anand-Srivastava and Trachte, 1993). The latter effect is mediated through pertussis toxin-sensitive G proteins but may also involve activation of cyclic nucleotide phosphodiesterase activity. In cultured SRG cells, NPs stimulate the formation of both cyclic GMP and inositol phosphates (Karnaky *et al.*, 1991; Ecay and Valentich, 1991). In addition, in unpublished studies we have found that the protein kinase C inhibitor staurosporine blocks ANP activation of I_{sc} by 50% without modifying forskolin-stimulated Cl^- secretion. We have also observed that the local anesthetic procaine abolishes the characteristic ANP-induced I_{sc} oscillations and reduces the level of the steady-state I_{sc}. Procaine blocks Ca^{2+} mobilization in several cell types. In addition, procaine also has well-documented inhibitory effects on both Ca^{2+}-dependent and Ca^{2+}-independent K^+ channels. These data suggest that NP activation of SRG cells involves a complex interplay between at least three second messenger systems (cyclic GMP, Ca^{2+}, and diacylglycerol) and two ion channels (Cl^- and K^+).

ANP-activated cyclic GMP formation and I_{sc} stimulation are closely correlated with respect to both ANP concentration and time (Fig. 8). However, the precise role of cyclic GMP in NP-stimulated Cl^- secretion is incompletely defined and may be complex because membrane-permeable analogs such as 8-bromo cyclic GMP elicit only minimal increases in Cl^- secretion when applied at millimolar concentrations. In addition, cyclic GMP analogs fail to elicit the characteristic oscillations in I_{sc} observed following ANP stimulation. Because 1 m*M* 8-bromo cyclic GMP also increases SRG intracellular cyclic AMP levels twofold, cyclic AMP may be responsible for at least part of 8-bromo cyclic GMP-stimulated I_{sc} (Yancey *et al.*, 1991). Therefore, NP-activated cyclic GMP formation alone may be incapable of stimulating SRG Cl^- secretion. Pairing the cyclic GMP signal with those elicited by other messengers may be required for a full secretory response to NPs. The need to coordinate several intracellular signaling events and membrane transporters both temporally and spatially may contribute to the characteristic transient oscillations in I_{sc} following ANP activation of SRG cells.

C. C-Type Natriuretic Peptide Stimulates Shark Rectal Gland Cl^- Secretion

Molecular cloning together with peptide purification and amino acid sequencing show that heart tissue from the dogfish sharks *Squalus acanthias* and *Scyliorhinus canicula* contains a C-type natriuretic peptide (sCNP) (Schofield *et al.*, 1991; Suzuki *et al.*, 1991; Oehlenschlager, 1993). In perfused rectal glands, sCNP is a strong activator of Cl^- secretion (Solomon *et al.*, 1992). A C-type NP has also been purified from killifish brain (Price *et al.*, 1990) and is equipotent with sCNP in stimulating Cl^- secretion in the perfused SRG (Solomon *et al.*, 1992). sCNP and killifish CNP also stimulate I_{sc} in cultured SRG cells with equivalent potency and both are significantly more active than rat ANP (Karnaky *et al.*, 1993). Although evidence for circulating CNP is lacking, the presence of CNP in shark heart tissue and the steep dose–response relation for activation of Cl^- secretion in both the perfused SRG and SRG cultures suggest that sCNP is an important modulator of SRG Cl^- secretion and shark osmoregulation.

In plasma membranes prepared from homogenates of native SRG tissue, CNP binds with high affinity (K_d = 78 pM) to two classes of receptors that have similar affinity for CNP, but different affinities for rat ANP (Gunning *et al.*, 1993). These receptors are thought to be either a novel CNP-specific guanylate cyclase-linked receptor or a modified version of the GC-B receptor. sCNP potently stimulates SRG particulate guanylate cyclase activity, whereas rat ANP and porcine BNP have little or no effect on enzyme activity. The inability of micromolar rat ANP to activate guanylate cyclase activity in isolated SRG plasma membranes contrasts with its rapid and pronounced effect on intracellular cyclic GMP formation in cultured SRG cells where 0.01 μM rANP elicits a significant increase in cGMP (Karnaky *et al.*, 1991). It is possible that in the intact cell, ANP activates the formation of additional second messengers, such as Ca^{2+} and diacylglycerol, which potentiate its effect on receptor-coupled guanylate cyclase activity.

D. Effects of Natriuretic Peptides on Apical and Basolateral Membrane Ion Conductances

Employing intracellular microelectrode techniques, we have assessed the effect of both rat ANP and sCNP on the electrophysiologic properties of SRG cells *in vitro* (Fig. 9). Superfusing cells with ANP elicits a significant fall in fR_a and depolarization of V_a; the latter is depolarized further by a 10-fold reduction in Ringer $[Cl^-]$. These responses are consistent with an ANP-stimulated increase in G_a^{Cl}. However, the magnitude of the effects of ANP on fR_a and V_a are considerably smaller than those of forskolin and Ca^{2+} ionophores (Moran and Valentich, 1991, 1993).

The electrical responses of cultured SRG cells to 5×10^{-9} *M* sCNP are very different from those following ANP, forskolin, or Ca^{2+} ionophore stimulation. Whereas the latter secretagogues elicit dramatic depolarization of V_a and fall in fR_a, sCNP induces a small hyperpolarization of V_a and produces a small increase in fR_a (Fig. 9). Low-Cl^- shark Ringer leads to minimal depolarization of V_a after sCNP stimulation, consistent with only a small increase in G_a^{Cl} or no change from an initial state of low Cl^- permeability.

The relative constancy of fR_a in sCNP-stimulated SRG cells suggests that sCNP rapidly activates a conductive pathway in the basolateral membrane. Since the Na^+/K^+ pump appears to behave as a source of constant current in epithelial cells (i.e., the pump possesses an infinitely high resistance), the pump should not contribute significantly to the conductance

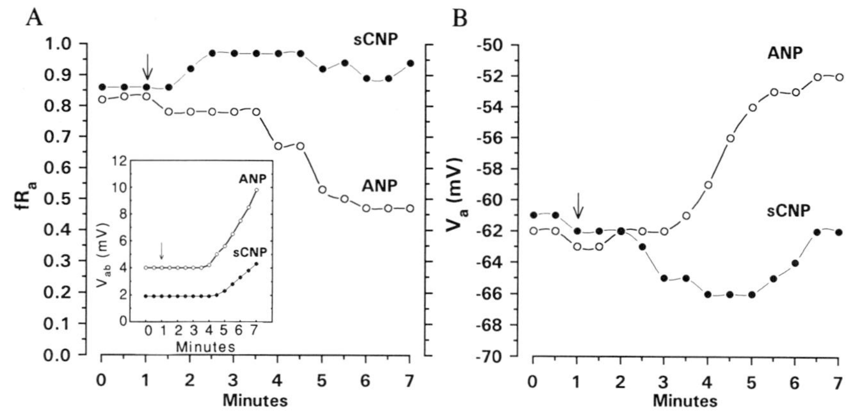

Fig. 9. Effect of 5×10^{-9} *M* shark CNP and 5×10^{-7} *M* rat ANP on membrane electrophysiologic parameters in SRG cultures. Cells were impaled with microelectrodes and the fractional resistance of the apical membrane (fR_a) (A) and apical membrane potential (V_a) (B) were determined as described in Moran and Valentich (1991). Arrows indicate where peptides were added to the superfusate. The inset shows the effect of both peptides on transepithelial voltage (V_{ab}). Although both CNP and ANP stimulate Cl^- secretion as reflected in the change in V_{ab}, their effects on membrane conductive properties differ dramatically. CNP appears to rapidly activate a basolateral membrane K^+ conductance, whereas the initial effect of ANP is to activate an apical membrane Cl^- conductance.

of the basolateral membrane. Therefore, it is likely that sCNP stimulation rapidly increases G_b^K. This is consistent with the observed hyperpolarization of V_b (Fig. 9). Because of the relatively small change in fR_a, it appears that sCNP activates G_a^{Cl} and G_b^K nearly simultaneously. This contrasts with ANP-, cyclic AMP-, and Ca^{2+}-mediated activation where increased G_a^{Cl} is the only initial membrane permeability response and is then followed by an increase in G_b^K (Moran and Valentich, 1991, 1993).

E. Natriuretic Peptides Activate Both Basolateral and Apical Receptors in Shark Rectal Gland Cells

Neuroendocrine inputs to epithelial cells are usually envisioned to interact with cognate receptors located on basolateral plasma membranes. In SRG cells, functional vasoactive intestinal peptide and adenosine receptors are found exclusively on basolateral membranes (Fig. 5). Therefore, the finding that ANP and CNP activate I_{sc} when added to either the apical or basolateral sides of SRG monolayers is unexpected (Karnaky *et al.*, 1991, 1993). Geary *et al.* (1993) reported that in airway epithelium CNP, but not ANP or BNP, receptors coupled to guanylate cyclase are present on both apical and basolateral plasma membranes. In addition, mammalian inner medullary collecting duct cells express both apical and basolateral NP receptors (Paul *et al.*, 1993). In teleost and mammalian intestine (O'Grady *et al.*, 1985; Moriarty *et al.*, 1990) and *Fundulus* operculum (Scheide and Zadunaisky, 1988), NPs modulate ion transport only when added to the basolateral sides of these epithelia. Guanylin and *E. coli* heat-stable enterotoxin receptors, which are structurally and functionally homologous to NP receptors, are found on the apical plasma membranes of mammalian intestinal epithelia (Forte *et al.*, 1993; Kachur *et al.*, 1992). The full physiologic significance of apical receptors in epithelia requires clarification. The expanding list of ligands that activate epithelia from the apical side includes adenosine (Barrett *et al.*, 1989), bradykinin (Denning and Welsh, 1991), vasoactive intestinal peptide (Elgavish *et al.*, 1989), vasopressin (Ando and Asano, 1993), parathyroid hormone (Reshkin *et al.*, 1991), and angiotensin II (Schelling and Linas, 1994).

The presence of receptors for NPs on the apical membranes of SRG cells raises the question of how peptides gain access to these receptors in the native tubule. Because SRG tubules are blind ended, there is no potential "upstream" source of NPs, as in the mammalian nephron. Possible modes of NP delivery to apical receptors include synthesis and secretion by SRG tubular epithelial cells or basolateral to apical transcytosis of peptide delivered to the basolateral membrane from the plasma or by local synthesis and secretion from interstitial neurons or fibroblasts.

Although neither of these mechanisms has been excluded, our preliminary data suggest that SRG epithelial cells are a source of NPs that may be delivered to apical receptors by secretion into the tubular lumen (see the following).

F. Natriuretic Peptide Activity in Shark Rectal Gland Tissue

C18 Sep-Pak chromatography or acetone precipitation of boiled, acetic acid extracts of native SRG tissue yields a partially purified material that (1) stimulates I_{sc} when added to the apical side of SRG monolayers, (2) increases intracellular cyclic GMP levels in SRG cells, and (3) competes with authentic rat ^{125}I-labeled ANP for binding to NP receptors on cultured bovine pulmonary artery endothelial cells (Karnaky *et al.*, 1992). These physiologic and biochemical characteristics strongly suggest that NP-like activity is present in SRG tissue. As shown in Fig. 10A, addition to the apical bath of material eluted by 40% acetonitrile from a Sep-Pak C18 cartridge (SP40) produces a significant increase in I_{sc}, whereas vasoactive intestinal peptide and 2-chloroadenosine are completely ineffective. However, the latter secretagogues elicit marked activation of I_{sc} when added basolaterally. These results show that VIP or adenosine in the SP40 extract could not be responsible for its activation of Cl^- secretion. The effects of SRG extracts prepared by different methods on intracellular cyclic GMP levels are shown in Fig. 10B. The SP40 extract and acid-soluble, acetone-precipitable material, preparations most likely to contain small peptides such as NPs (Oehlenschlager, 1993), stimulated cyclic GMP formation significantly. In contrast, the 10% acetonitrile eluate from the Sep-Pak cartridge (SP10), which contains small molecules such as ATP and adenosine, fails to modify intracellular cyclic GMP levels. Finally, the data in Fig. 10C demonstrate that the SP40 extract elicits a significant reduction in specific ^{125}I-labeled ANP binding to cultured endothelial cells. In contrast, the SP10 fraction has minimal displacement activity.

In sum, these data show that SRG tissue contains NP-like molecules capable of activating Cl^- secretion through autocrine or paracrine mechanisms. ANP and CNP synthesis and secretion by rat kidney epithelial cells *in vitro* have been demonstrated (Ritter *et al.*, 1991; Suzuki *et al.*, 1993; Greenwald *et al.*, 1991). CNP and ANP mRNA have also been detected in mammalian renal tissue by PCR and Northern analysis (Greenwald *et al.*, 1992; Suzuki *et al.*, 1993). These results strongly suggest that NPs are synthesized by epithelial cells and may act as autocrine/paracrine mediators in the nephron.

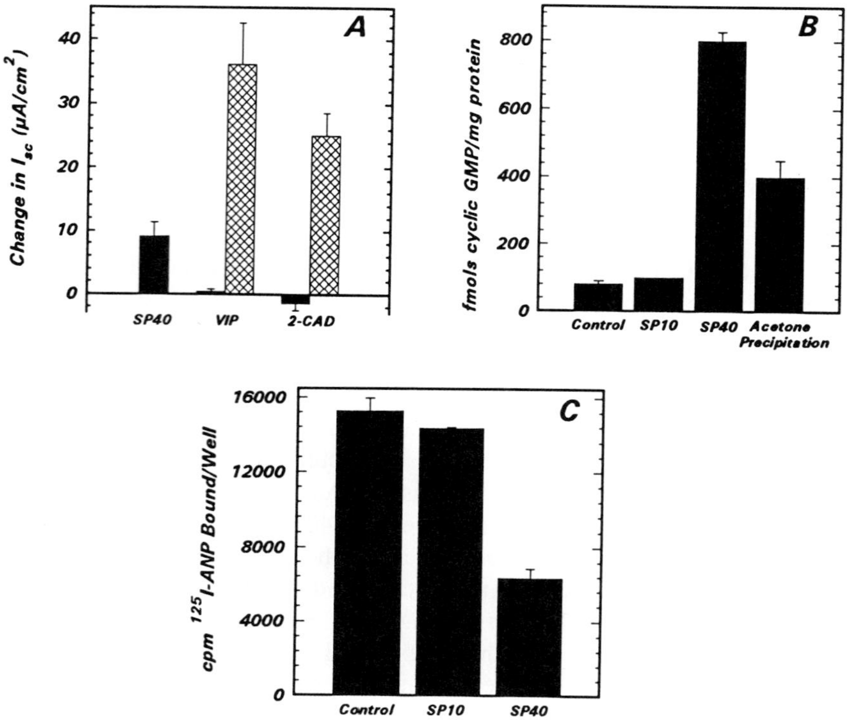

Fig. 10. (A) Change in I_{sc} elicited by SRG SP40 extracts, 10^{-7} *M* vasoactive intestinal peptide (VIP), and 10^{-5} *M* 2-chloroadenosine (2-CAD). Only the SP40 fraction stimulates I_{sc} when added to the apical bath (solid bars), whereas the latter secretagogues are completely ineffective. However, cultures respond to VIP and 2-CAD when they are added to the basolateral bath (hatched bars). Since NPs are the only known peptide secretagogues capable of activating SRG I_{sc} from the apical side, these data suggest that the SP40 fraction contains NP-like peptides. (B) Effects of different SRG fractions on intracellular cyclic GMP formation in SRG cultures. Both the 40% acetonitrile eluate from the Sep-Pak cartridge (SP40) and acid-soluble, acetone-precipitable material stimulate cyclic GMP formation significantly. These fractions should contain small peptides such as NPs. However, the 10% acetonitrile eluate (SP10), which contains small intracellular constituents such as ATP and adenosine, has no effect on cyclic GMP levels. (C) Effects of SRG fractions on ^{125}I-labeled ANP binding to bovine pulmonary artery endothelial cells. Whereas the SP10 fraction has no effect on radioligand binding, the SP40 fraction competes significantly with ^{125}I-labeled ANP for binding to endothelial NP receptors. These data suggest that the SP40 fraction, but not the SP10 fraction, contains biologically active NP-like material capable of binding to NP receptors.

VII. FUTURE DIRECTIONS

Squalus acanthias rectal gland epithelial cells express higher densities of Na^+/K^+ pumps (Hokin, 1978) and $Na^+/K^+/2Cl^-$ cotransport moieties (Haas, 1989) and contain greater quantities of Cl^- channel mRNA (Sullivan *et al.*, 1991) than any other epithelial cell type. In both native tissue (Greger *et al.*, 1984) and cultured cells (Valentich and Forrest, 1991; Moran and Valentich, 1991), secretagogue stimulation increases Cl^- secretory rates and apical membrane Cl^- conductance to a greater degree than in any other epithelium. It is likely that the transmembrane and intracellular signaling networks responsible for activating SRG cells are similarly amplified. The messenger molecules, receptors, and protein kinases comprising these networks in Cl^--secreting epithelial cells are only beginning to be identified. Presently the nature of their integrative interactions are only vague conceptions of the true underlying complexity. The unique specialization of SRG epithelial cells together with the capability of analyzing their transport functions and biochemistry *in vitro* make them a powerful system for unraveling the molecular events that modulate epithelial Cl^- secretion. The functions of ABC transporters in normal epithelial physiology and their responses to xenobiotic exposure, the mechanisms coordinating activation of apical and basolateral transport moieties, and the molecular basis of natriuretic peptide biosynthesis, secretion, and action in epithelia are all problems amenable to analysis using cultured SRG cells.

ACKNOWLEDGMENTS

The research in the authors' laboratories reported in this chapter was supported by grants from the Cystic Fibrosis Foundation (JDV and KJK), American Heart Association (JDV, KJK, and WMM), Markey Foundation (JDV and KJK), National Institutes of Health (JDV), National Institute of Environmental Health Sciences (JDV), the University Research Council of the Medical University of South Carolina (KJK), the USDA (KJK), and a University of Central Arkansas Faculty Research Grant (WMM). SCIOS, Inc. (Mountain View, California) generously supplied the synthetic shark CNP used in these studies. The authors acknowledge the assistance and ideas of numerous colleagues who contributed to the evolution of the work described in this chapter. JDV would like to express his gratitude to John Forrest for conveying his vision and enthusiasm about the utility of a tissue culture model for the dogfish shark rectal gland.

REFERENCES

Anand-Srivastava, M. B., and Trachte, G. J. (1993). Atrial natriuretic factor receptors and signal transduction mechanisms. *Pharmacol. Rev.* **45,** 455–497.

Ando, Y., and Asano, Y. (1993). Functional evidence for an apical V_1 receptor in rabbit cortical collecting duct. *Am. J. Physiol.* **264,** F467–F471.

Barrett, K. E., Huott, P. A., Shah, S.-S., and Dharmsathaphorn, K. (1989). Differing effects of apical and basolateral adenosine on colonic epithelial cell line T84. *Am. J. Physiol.* **256,** C197–C203.

Chan, D. K. O., and Phillips, J. G. (1967). The anatomy, histology and histochemistry of the rectal gland in the lip-shark *Hemiscyllium plagiosum* (Bennet). *J. Anat.* **101,** 137–157.

Chan, K. M., Davies, P. L., Childs, S., Veinot, L., and Ling., V. (1992). P-Glycoprotein genes in the winter flounder, *Pleuronectes americanus:* Isolation of two types of genomic clones carrying 3′ terminal exons. *Biochim. Biophys. Acta* **1171,** 65–72.

Chipkin, S. R., Stoff, J. S., and Aronin, N. (1988). Immunohistochemical evidence for neural mediation of VIP activity in the dogfish rectal gland. *Peptides* (*N.Y.*) **9,** 119–124.

Denning, G. M., and Welsh, M. J. (1991). Polarized distribution of bradykinin receptors on airway epithelial cells and independent coupling to second messenger pathways. *J. Biol. Chem.* **266,** 12932–12938.

Devor, D., Suggs, W. K., Forrest, J. N., Jr., and Frizzell, R. A. (1995). cAMP-activated activation of Cl^- channels in primary cultures of spiny dogfish (*Squalus acanthias*) rectal gland. *Am. J. Physiol.* **268,** C70–C79.

Dubinsky, W., and Monti, L. B. (1986). Resolution of apical from basolateral membrane of shark rectal gland. *Am. J. Physiol.* **251,** C721–C726.

Ecay, T., and Valentich, J. D. (1991). Chloride secretagogues stimulate inositol phosphate formation in shark rectal gland tubules cultured in suspension. *J. Cell. Physiol.* **146,** 407–416.

Elgavish, A., Pillion, D. J., and Meezan, E. (1989). Evidence for vasoactive intestinal peptide receptors in apical membranes from tracheal epithelium. *Life Sci.* **44,** 1037–1042.

Epstein, F. H., Spokes, K., Silva, P., Solomon, R., Hornung, S., Stidham, J. D., McCraw, A. S., Nelson, D. S., and Karnaky, K. J., Jr. (1992). Neuropeptide inhibition of active transport in cultured rectal gland cells of *Squalus acanthias. Bull. Mt. Desert Island Biol. Lab.* **31,** 117.

Ernst, S. A., Hootman, S. R., Schreiber, J. H., and Riddle, C. V. (1981). Freeze-fracture and morphometric analysis of occluding junctions in rectal glands of elasmobranch fish. *J. Membr. Biol.* **58,** 101–114.

Evans, D. (1990). An emerging role for a cardiac peptide hormone in fish osmoregulation. *Annu. Rev. Physiol.* **52,** 43–60.

Eveloff, J., Karnaky, K. J., Jr., Silva, P., Epstein, F. H., and Kinter, W. B. (1979). Elasmobranch rectal gland cell. Autoradiographic localization of [^{3}H]ouabain-sensitive Na,K-ATPase in rectal gland of dogfish, *Squalus acanthias. J. Cell. Biol.* **83,** 16–32.

Feero, W. G., and Valentich, J. D. (1991). Protein kinase C activates chloride secretion in cultured shark (*Squalus acanthias*) rectal gland epithelial cells. *Bull. Mt. Desert Island Biol. Lab.* **30,** 63–64.

Forrest, J. N., Boyer, J. L., Ardito, T. A., Murdaugh, H. V., and Wade, J. B. (1982). Structure of tight junctions during Cl secretion in the perfused rectal gland of the dogfish shark. *Am. J. Physiol.* **242,** C388–C392.

Forte, L., Eber, S. L., Turner, J. T., Freeman, R. H., Fok, K. F., and Currie, M. G. (1993). Guanylin stimulation of Cl^- secretion in human intestinal T_{84} cells via cyclic guanosine monophosphate. *J. Clin. Invest.* **91,** 2423–2428.

Foskett, J. K., Gunter-Smith, P. J., Melvin, J. E., and Turner, R. J. (1989). Physiological localization of an agonist-sensitive pool of Ca^{2+} in parotid acinar cells. *Proc. Natl. Acad. Sci. U.S.A.* **86,** 167–171.

Frizzell, R. A., Field, M., and Schultz, S. G. (1979). Sodium-coupled chloride transport by epithelial tissues. *Am. J. Physiol.* **236,** F1–F8.

Fuller, C. M., and Benos, D. J. (1992). CFTR. *Am. J. Physiol.* **263,** C267–C286.

Geary, C. A., Goy, M. F., and Boucher, R. C. (1993). Synthesis and vectorial export of cGMP in airway epithelium: Expression of soluble and CNP-specific guanylate cyclases. *Am. J. Physiol.* **265,** L598–L605.

Gerbes, A. L., Nathrath, W., Cantin, M., and Denecke, H. (1991). Presence of atrial natriuretic factor prohormone in enterochromaffin cells of the human large intestine. *Gastroenterology* **101,** 424–429.

Gogelein, H., Greger, R., and Schlatter, E. (1987a). Potassium channels in the basolateral membrane of the rectal gland of *Squalus acanthias*. Regulation and inhibitors. *Pfluegers Arch.* **409,** 107–113.

Gogelein, H., Schlatter, E., and Greger, R. (1987b). The "small" conductance chloride channel in the luminal membrane of the rectal gland of the dogfish (*Squalus acanthias*). *Pfluegers Arch.* **409,** 122–125.

Gottesman, M. M., and Pastan, I. (1993). Biochemistry of multidrug resistance mediated by the multidrug transporter. *Annu. Rev. Biochem.* **62,** 385–427.

Greenwald, J. E., Needleman, P., Wilkins, M. R., and Schreiner, G. F. (1991). Renal synthesis of atriopeptin-like protein in physiology and pathophysiology. *Am. J. Physiol.* **260,** F602–F607.

Greenwald, J. E., Ritter, D., Tetens, E., and Rotwein, P. S. (1992). Renal expression of the gene for atrial natriuretic factor. *Am. J. Physiol.* **263,** F974–F978.

Greger, R., and Schlatter, E. (1984). Mechanism of NaCl secretion in the rectal gland of spiny dogfish (*Squalus acanthias*). I. Experiments in isolated *in vitro* perfused rectal gland tubules. *Pfluegers Arch.* **402,** 63–75.

Greger, R., Schlatter, E., Wong, F., and Forrest, J. N., Jr. (1984). Mechanism of NaCl secretion in rectal gland tubules of spiny dogfish (*Squalus acanthias*). III. Effects of stimulation of secretion by cyclic AMP. *Pfluegers Arch.* **402,** 376–384.

Greger, R., Schlatter, E., and Gogelein, H. (1985). Cl^--channels in the apical cell membrane of the rectal gland "induced" by cAMP. *Pfluegers Arch.* **403,** 446–448.

Greger, R., Gogelein, H., and Schlatter, E. (1987a). Potassium channels in the basolateral membrane of the rectal gland of the dogfish (*Squalus acanthias*). *Pfluegers Arch.* **409,** 100–106.

Greger, R., Schlatter, E., and Gogelein, H. (1987b). Chloride channels in the luminal membrane of the rectal gland of the dogfish (*Squalus acanthias*). Properties of the "larger" conductance channel. *Pfluegers Arch.* **409,** 114–121.

Greger, R., Gogelein, H., and Schlatter, E. (1988). Stimulation of NaCl secretion in the rectal gland of the dogfish *Squalus acanthias*. *Comp. Biochem. Physiol., A.* **90,** 733–737.

Grube, D. (1986). The endocrine cells of the digestive system: Amines, peptides, and modes of action. *Anat. Embryol.* **175,** 151–162.

Grzelczak, Z., Dubel, S., Alon, N., Fahim, S., Tsui, L.-C., and Riordan, J. R. (1990). A highly conserved cystic fibrosis gene homologue is expressed in the salt-secreting shark rectal gland. *J. Cell Biol.* **111,** 310a.

Gunning, M., Cuero, C., Solomon, R., and Silva, P. (1993). C-type natriuretic peptide receptors and signaling in rectal gland of *Squalus acanthias*. *Am. J. Physiol.* **264,** F300–F305.

Gutkowska, J., Tremblay, J., Meyer, R., Marcinkiewicz, M., and Nemer, M. (1991). Evidence for atrial natriuretic peptide (ANP) synthesis and the presence of ANP-transducing receptors in the rat olfactory bulb. *J. Neurochem.* **57,** 1855–1861.

Haas, M. (1989). Properties and diversity of (Na-K-Cl) cotransporters. *Annu. Rev. Physiol.* **51,** 443–457.

Hanrahan, J. W., Duguay, F., Sansom, S., Alon, N., Jensen, T., Riordan, J. R., and Grzelczak, Z. (1993). Low-conductance chloride channel activated by cAMP in the rectal gland of the shark *Squalus acanthias* and in cells heterologously expressing shark CFTR. *Bull. Mt. Desert Island Biol. Lab.* **32,** 48–49.

Higgins, C. F. (1992). ABC transporters: From microorganisms to man. *Annu. Rev. Cell Biol. 8,* 67–113.

Hokin, L. E. (1978). Reconstitution of the $Na^+ + K^+$ pump from the purified ($Na^+ + K^+$)-activated adenosine triphosphatase from the rectal gland of *Squalus acanthias*. *In* "Membrane Transport Processes" (D. C. Tosteson, Y. A. Ovchinnikov, and R. Latorre, eds.), pp. 399–416. Raven Press, New York.

Holmgren, S., and Nilsson, S. (1983). Bombesin-, gastrin/CCK-, 5-hydroxytryptamine-, neurotensin-, somatostatin-, and VIP-like immunoreactivity and catecholamine fluorescence in the gut of the elasmobranch, *Squalus acanthias. Cell Tissue Res.* **234,** 595–618.

Hughes, C. E., and Sellitti, D. F. (1991). Thyroidal ANF: A possible mediator of autocrine regulation in the porcine thyroid gland. *Peptides* (*N.Y.*) **12,** 705–714.

Jat, P. S., Noble, M. D., Ataliotis, P., Tanaka, Y., Yannoutsos, N., Larsen, L., and Kioussis, D. (1991). Direct derivation of conditionally immortal cell lines from an H-$2K^b$-tsA58 transgenic mouse. *Proc. Natl. Acad. Sci. U.S.A.* **88,** 5096–5100.

Jefferson, D. M., Valentich, J. D., Marini, F. C., Grubman, S. A., Iannuzzi, M. C., Dorkin, H. L., Li, M., Klinger, K. W., and Welsh, M. J. (1990). Expression of normal and cystic fibrosis phenotypes by continuous airway epithelial cell lines. *Am. J. Physiol.* **259,** L496–L505.

Kachur, J. F., Won-Kim, S., Moummi, C., Fok, K., and Currie, M. G. (1992). Guanylin, a newly discovered activator of intestinal guanylate cyclase, stimulates ion secretion across rat colon. *Gastroenterology* **102,** A738.

Karnaky, K. J., Jr., Valentich, J. D., Currie, M. G., Oehlenschlager, W. F., and Kennedy, M. P. (1991). Atriopeptin stimulates chloride secretion in cultured shark rectal gland cells. *Am. J. Physiol.* **260,** C1125–C1130.

Karnaky, K. J., Jr., Oehlenschlager, W. F., Stidham, J. D., Nelson, D. S., McCraw, A. S., Valentich, J. D., Kennedy, M. P., and Currie, M. G. (1992). An atrial natriuretic peptide-like factor isolated from the dogfish shark (*Squalus acanthias*) rectal gland: Initial characterization. *Bull. Mt. Desert Island Biol. Lab.* **31,** 126–128.

Karnaky, K. J., Jr., Gazley, J. L., Kelmenson, C., French, S., Suggs, W. K., and Forrest, J. N., Jr. (1993). Shark heart C-type natriuretic peptide is a potent chloride secretagogue in monolayers of cultured shark (*Squalus acanthias*) rectal gland cells. *Bull. Mt. Desert Island Biol. Lab.* **32,** 67–68.

Kasai, H. (1990). Cytosolic Ca^{2+} gradients triggering unidirectional fluid secretion from exocrine pancreas. *Nature* (*London*) **348,** 735–738.

Kelley, G. G., Poeschia, E. M., Barron, H. V., and Forrest, J. N., Jr. (1990). A_1 adenosine receptors inhibit chloride transport in the shark rectal gland. Dissociation of inhibition and cyclic AMP. *J. Clin. Invest.* **85,** 1629–1636.

Kelley, G. G., Aassar, O. S., and Forrest, J. N., Jr. (1991). Endogenous adenosine is an autacoid feedback inhibitor of chloride transport in the shark rectal gland. *J. Clin. Invest.* **88,** 1933–1939.

Kennedy, M. P. (1991). Regulation of chloride secretion in cultured dogfish shark rectal gland epithelium: Direct effects of atrial natriuretic peptide. M. S. thesis, College of Charleston, Charleston, SC.

Kurelec, B. (1992). The multixenobiotic resistance mechanism in aquatic organisms. *CRC Crit. Rev. Toxicol.* **22,** 23–43.

Kushman, N., Valentich, J. D., and Dawson, D. C. (1991). Impedance analysis reveals changes in apical and basolateral membrane resistance in shark (*Squalus acanthias*) rectal gland cultures. *Bull. Mt. Desert Island Biol. Lab.* **30,** 60–62.

La, B.-Q., Carosi, S. L., Valentich, J., Shenolikar, S., and Sansom, S. C. (1991). Regulation of epithelial chloride channels by protein phosphatase. *Am. J. Physiol.* **260,** C1217–C1223.

Lear, S., Spokes, K., Taylor, M., Silva, P., and Epstein, F. H. (1990). The effect of ANP on isolated and cultured tubules of the rectal gland of *Squalus acanthias. Bull. Mt. Desert Island Biol. Lab.* **29,** 92–93.

Lear, S., Cohen, B. J., Silva, P., Lechene, C., and Epstein, F. H. (1992). cAMP activates the sodium pump in cultured cells of the elasmobranch rectal gland. *J. Am. Soc. Nephrol.* **2,** 1523–1528.

Marshall, J., Martin, K. A., Picciotti, M., Hockfield, S., Nairn, A. C., and Kaczmarek, L. K. (1991). Identification and localization of a dogfish homolog cystic fibrosis transmembrane conductance regulator. *J. Biol. Chem.* **266,** 22749–22754.

McRoberts, J. A., and Barrett, K. E. (1989). Hormone-regulated ion transport in T_{84} colonic cells. *In* "Functional Epithelial Cells in Culture" (K. S. Matlin and J. D. Valentich, eds.), pp. 235–265. Alan R. Liss, New York.

Moran, W. M., and Valentich, J. D. (1991). Cl^- secretion by cultured shark rectal gland cells. II. Effects of forskolin on cellular electrophysiology. *Am. J. Physiol.* **260,** C824–C831.

Moran, W. M., and Valentich, J. D. (1993). Cl^- secretion by cultured shark rectal gland cells. III. Ca^{2+} regulation of apical membrane Cl^- conductance. *Am. J. Physiol.* **265,** C641–C649.

Morel, G., Chabot, J.-G., Belles-Isles, M., and Heisler, S. (1988a). Synthesis and internalization of atrial natriuretic factor in anterior pituitary cells. *Mol. Cell. Endocrinol.* **55,** 219–231.

Morel, G., Chabot, J.-G., Garcia-Caballero, T., Gossard, F., Dihl, F., Belles-Isles, M., and Heisler, S. (1988b). Synthesis, internalization, and localization of atrial natriuretic peptide in rat adrenal medulla. *Endocrinology* (*Baltimore*) **123,** 149–158.

Morel, G., Chabot, J.-G., Gossard, F., and Heisler, S. (1989). Is atrial natriuretic peptide synthesized and internalized by gonadotrophs? *Endocrinology* (*Baltimore*) **124,** 1703–1710.

Moriarty, K. J., Higgs, N. B., Lees, M., Tonge, A., Wardle, T. D., and Warhurst, G. (1990). Influence of atrial natriuretic peptide on mammalian large intestine. *Gastroenterology* **98,** 647–653.

Morris, A. P., Kirk, K. L., and Frizzell, R. A. (1990). Simultaneous analysis of cell Ca^{2+} and Ca^{2+}-stimulated chloride conductance in colonic epithelial cells (HT-29). *Cell Regul.* **1,** 951–963.

O'Donnell, M. E. (1989). Regulation of Na-K-Cl cotransport in endothelial cells by atrial natriuretic factor. *Am. J. Physiol.* **257,** C36–C44.

O'Donnell, M. E., and Owen, N. E. (1986). Atrial natriuretic factor stimulates Na/K/Cl cotransport in vascular smooth muscle cells. *Proc. Natl. Acad. Sci. U.S.A.* **83,** 6132–6136.

Oehlenschlager, W. F. (1993). Analysis of atrial natriuretic peptide-like activity in diverse tissues and isolation of this activity from shark heart. Ph. D. thesis, Medical University of South Carolina, Charleston.

O'Grady, S. M. (1989). Cyclic nucleotide-mediated effects of ANF and VIP on flounder intestinal ion transport. *Am. J. Physiol.* **256,** C142–C146.

O'Grady, S. M., Field, M., Nash, N. T., and Rao, M. C. (1985). Atrial natriuretic factor inhibits Na-K-Cl cotransport in teleost intestine. *Am. J. Physiol.* **249,** C531–C534.

Palfrey, H. C., Silva, P., and Epstein, F. H. (1984). Sensitivity of cAMP-stimulated salt secretion in shark rectal gland to "loop" diuretics. *Am. J. Physiol.* **246,** C242–C246.

Paul, R. V., Karnaky, K. J., Jr., Budisavljevic, M., Wackym, P. S., and Johnson, R. H.

(1993). ANP-responsive guanylate cyclase in the apical membrane of inner medullary collecting duct (IMCD) cell cultures. *J. Am. Soc. Nephrol.* **4,** 446.

Price, D. A., Doble, K. E., Lee, T. D., Galli, S. M., Dunn, B. M., Parten, B., and Evans, D. H. (1990). The sequencing, synthesis, and biological actions of an ANP-like peptide isolated from the brain of the killifish *Fundulus heteroclitus. Biol. Bull.* (*Woods Hole, Mass.*) **178,** 279–285.

Reshkin, S. J., Forgo, J., and Murer, H. (1991). Apical and basolateral effects of PTH in OK cells—Transport inhibition, messenger production, effects of pertussis toxin, and interaction with a PTH analog. *J. Membr. Biol.* **124,** 227–237.

Ritter, D., Needleman, P., and Greenwald, J. E. (1991). Synthesis and secretion of an atriopeptin-like protein in rat kidney cell culture. *J. Clin. Invest.* **87,** 208–212.

Rocha, A. S., and Kudo, L. H. (1990). Atrial peptide and cGMP effects on NaCl transport in inner medullary collecting duct. *Am. J. Physiol.* **259,** F258–F268.

Scheide, J. I., and Zadunaisky, J. A. (1988). Effect of atriopeptin II on isolated opercular epithelium of *Fundus heteroclitus. Am. J. Physiol.* **254,** R27–R32.

Schelling, J. R., and Linas, S. L. (1994). Angiotensin II-dependent proximal tubule sodium transport requires receptor-mediated endocytosis. *Am. J. Physiol.* **266,** C669–C675.

Schofield, J. P., Jones, D. S., and Forrest, J. N., Jr. (1991). Identification of C-type natriuretic peptide in heart of spiny dogfish shark (*Squalus acanthias.*) *Am. J. Physiol.* **261,** F734–F739.

Schwiebert, E. M., Flotte, T., Cutting, G. R., and Guggino, W. B. (1994). Both CFTR and outward rectifying chloride channels contribute to cAMP-stimulated whole cell chloride currents. *Am. J. Physiol.* **266,** C1464–C1477.

Shuttleworth, T. J. (1988). Salt and water balance—Extrarenal mechanisms. *In* "Physiology of Elasmobranch Fishes" (T. J. Shuttleworth, ed.), pp. 170–199. Springer-Verlag, Berlin.

Shuttleworth, T. J., and Thompson, J. L. (1980). The mechanism of cyclic AMP stimulation of secretion in the dogfish rectal gland. *J. Comp. Physiol.* **140,** 209–216.

Shuttleworth, T. J., and Thorndyke, M. C. (1984). An endogenous peptide stimulates secretory activity in the elasmobranch rectal gland. *Science* **225,** 319–321.

Silva, P., Stoff, J., Field, M., Fine, L., Forrest, J. N., and Epstein, F. H. (1977). Mechanism of active chloride secretion by shark rectal gland: Role of Na-K-ATPase in chloride transport. *Am. J. Physiol.* **233,** F298–F306.

Silva, P., Epstein, J. A., Stevens, A., Spokes, K., and Epstein, F. H. (1983). Ouabain binding in rectal gland of *Squalus acanthias. J. Membr. Biol.* **75,** 105–114.

Silva, P., Stoff, J. S., Leone, D. R., and Epstein, F. H. (1985). Mode of action of somatostatin to inhibit secretion by shark rectal gland. *Am. J. Physiol.* **249,** R329–R334.

Silva, P., Stoff, J. S., Solomon, R. J., Lear, S., Kniaz, D., Greger, R., and Epstein, F. H. (1987). Atrial natriuretic peptide stimulates salt secretion by shark rectal gland by releasing VIP. *Am. J. Physiol.* **252,** F99–F103.

Silva, P., Solomon, R. J., and Epstein, F. H. (1990). Shark rectal gland. *In* "Methods in Enzymology" (S. Fleischer and B. Fleischer, eds.), vol. 192, pp. 754–766. Academic Press, San Diego, CA.

Silva, P., Epstein, F. H., Karnaky, K. J., Jr., Reichlin, S., and Forrest, J. N., Jr. (1993). Neuropeptide Y inhibits chloride secretion in the shark rectal gland. *Am. J. Physiol.* **265,** R439–R446.

Solomon, R., Taylor, M., Dorsey, D., Silva, P., and Epstein, F. H. (1985). Atriopeptin stimulation of rectal gland function in *Squalus acanthias. Am. J. Physiol.* **249,** R348–R354.

Solomon, R., Protter, A., McEnroe, G., Porter, J. G., and Silva, P. (1992). C-type natriuretic peptides stimulate chloride secretion in the rectal gland of *Squalus acanthias*. *Am. J. Physiol.* **262,** R707–R711.

Sullivan, S. K., Swamy, K., and Field, M. (1991). cAMP-activated Cl conductance is expressed in *Xenopus* oocytes by injection of shark rectal gland mRNA. *Am. J. Physiol.* **260,** C664–C669.

Suzuki, E., Hirata, Y., Hayakawa, H., Omata, M., Kojima, M., Kangawa, K., Minamino, N., and Matsuo, H. (1993). Evidence for C-type natriuretic peptide production in the rat kidney. *Biochem. Biophys. Res. Commun.* **192,** 532–538.

Suzuki, R., Takahashi, R., Hazon, N., and Takei, Y. (1991). Isolation of high-molecular-weight C-type natriuretic peptide from the heart of a cartilaginous fish (European dogfish, *Scyliorhinus canicula*). *FEBS Lett.* **282,** 321–325.

Valentich, J. D. (1991a). Primary cultures of shark rectal gland epithelial cells: A model for hormone-sensitive chloride transport. *J. Tissue Cult. Methods* **13,** 149–162.

Valentich, J. D. (1991b). Evidence for a P-glycoprotein-like transporter in the apical plasma membranes of shark (*Squalus acanthias*) rectal gland epithelial cells. *Bull. Mt. Desert Island Biol. Lab.* **30,** 67–68.

Valentich, D. J., and Forrest, J. N., Jr. (1991). Cl^- secretion by cultured shark rectal gland cells. I. Transepithelial transport. *Am. J. Physiol.* **260,** C813–C823.

Valentich, J. D., and Garretson, L. T. (1987). Matrigel promotes morphologic differentiation of cultured shark rectal gland tubular epithelial cells. *J. Cell Biol.* **105,** 140a.

Valentin, J.-P., and Humphreys, M. H. (1993). Urodilatin: A paracrine renal natriuretic peptide. *Semin. Nephrol.* **13,** 61–70.

Vollmar, A., Colbatzky, F., Hermanns, W., and Schulz, R. (1991). Origin and characterization of atrial natriuretic peptide in the rat parotid gland. *Anat. Embryol.* **184,** 331–335.

Welsh, M. J. (1985). Ion transport by primary cultures of canine tracheal epithelium: Methodology, morphology, and electrophysiology. *J. Membr. Biol.* **88,** 149–163.

Yancey, P. H., Ruble, J., and Valentich, J. D. (1991). Effect of chloride secretagogues on cyclic AMP formation in cultured shark (*Squalus acanthias*) rectal gland epithelial cells. *Bull. Mt. Desert Island Biol. Lab.* **30,** 65–66.

8

PRIMARY CULTURES OF TELEOST BRANCHIAL EPITHELIAL CELLS

PETER PÄRT AND ELISABETH BERGSTRÖM

I. INTRODUCTION

The fish gill is a multifunctional organ involved in gas exchange, acid/base regulation, ion regulation, and the excretion of nitrogenous waste. The branchial epithelium is more or less actively involved in all of these processes. This epithelium consists of several cell types; mitochondria-rich cells (chloride cells), mucus cells, pillar cells, and respiratory epithelial cells (sometimes called pavement cells) (Laurent, 1984). These cells are very plastic (Goss *et al.*, 1992). They react rapidly to different kinds of physiological challenges, such as various natural disturbances (hypoxia, acid/base) or in response to environmental pollutants. How these compensatory mechanisms are carried out at the cellular level is unknown and generally we have very little information on the physiology of the branchial cells.

One reason for this is that the methods available to date, mainly *in vivo* experiments or perfused gill preparations, do not have the resolution to elucidate processes at the cellular level. Biochemical preparations (isolated enzymes, membrane or organelle fractions) can do this to some extent, but the results are still more qualitative than quantitative as the cell integrity is disrupted. A successful approach in other areas of physiology has been to use isolated cells and this approach has begun to be introduced into gill physiology. Freshly isolated branchial cells have been used to localize ion-transporting enzymes (Kamiya, 1972; Sargent *et al.*, 1975; Hootman and Philpott, 1978), in studies of metabolism and oxygen consumption (Perry and Walsh, 1989), in studies of ion transport (Battram *et al.*, 1989), in investigations of hormone receptors (Guibbolini *et al.*, 1988), and in studies of xenobiotic metabolism (Kennedy and Walsh, 1994). However, one should be aware that epithelial cells in suspension lose their polarity (Taub, 1985). Polarity—the different properties of the apical and the basolateral membranes—is a key feature of epithelial cells and the loss of the polarity severely limits the conclusions that can be drawn from branchial, or other epithelial, cells in suspension. In addition, the digestion procedures involved in the preparation of cells affect surface proteins—proteins that are likely to be involved in membrane transport processes. Therefore it is clearly advantageous to culture the cells and grow them as monolayers where they may reestablish their polarity, repair their surface proteins, and hence restore their normal functions.

A development in this direction is beginning in branchial physiology. Cell lines isolated from gill tissue and possessing epithelial morphology have been established and are now available from the walking catfish (Noga and Hartman, 1981) and color carp (Ku and Chen, 1992). We have described a method to obtain primary cultures of branchial respiratory epithelial cells from rainbow trout (*Oncorhynchus mykiss*) gills (Pärt *et al.*, 1993) starting with a tryptic digest of the tissue. Avella *et al.* (1994) used another approach, the explant technique, to obtain primary cultures of epithelial cells from sea bass (*Dicentrarchus labrax*) gills. In our view it is a clear advantage to use primary cultures when possible because they generally express the differentiated characteristics of the native tissue whereas cell lines may change their properties over time. In this review we summarize some of our work with primary cultures of branchial cells and describe the techniques and methods we have developed.

II. ESTABLISHMENT OF PRIMARY CULTURES

A. Preparation of Cells

A detailed description of the preparation of branchial cells and the establishment of primary cultures from rainbow trout (*Oncorhynchus mykiss*) is given in Pärt *et al.* (1993). Briefly, prior to dissection, the fish is kept for 15 min in well-aerated, bacteria-free water to flush loosely bound bacteria from the gills. The fish is decapitated and the gill arches are dissected with sterile instruments. Gill filaments are excised from the arches and rinsed twice for 15 min in 10-ml changes of solution I (Table I). The filaments are transferred to 50-ml conical tubes containing 2 ml of trypsinizing solution (solution II, Table I) and incubated on a shaker at 250 revs min^{-1} for 15 min. The cell suspension is aspirated from the tubes and filtered through 80-μm nylon cloth into a stopping solution (solution III, Table I). New trypsin solution is added to the remaining filaments and the procedure is repeated two or three times. The pooled cell suspension from the digestions is centrifuged at 200*g* for 10 min, and the cell pellet is washed twice with 5 ml of solution IV (Table I) by resuspension and centrifugation. After the second wash, the cells are suspended in 10–20 ml of culture medium (solution V, Table I). Some basic information about the tryptic digest is given in Table II. The preparation of cells is carried out at room temperature, 20°C. At higher room temperatures the quality of the cells is severely affected. It is therefore recommended that all solutions except the trypsin be kept on ice. The cells do not seem to suffer from being stored on ice.

B. Culture Conditions

Cells are counted in a hemocytometer and plated in culture dishes at a density of 4–5 $\times$ 10^5 cells cm^{-2} in an appropriate volume of medium. The medium is changed after 24 h and replaced with fresh (solution V, Table I). Nonattached cells are washed away by this procedure. After another 3–4 days the medium is replaced with a medium without antibiotics (solution VI, Table I). The original tryptic digest contains up to 10% red blood cells. They do not attach and they are washed away with the first change of medium. We have not found it necessary to perfuse the gills with saline prior to the preparation of cells in order to clear red cells from the tissue.

The gill cells are routinely grown in room atmosphere, but preferably a more physiological CO_2-enriched atmosphere should be used. In a study

Table I
Solutions for Preparation and Culture of Branchial Epithelial Cells

I	Washing solution for gill filaments	
	PBS[a]	
	PEST[b]	200 $\mu g\ ml^{-1}$
	Gentamicin	400 $\mu g\ ml^{-1}$
	Fungizon[c]	2.5 $\mu g\ ml^{-1}$
II	Trypsin solution	
	PBS	
	Trypsin	0.05%
	EDTA	0.55 m*M*
III	Stop solution	
	PBS	
	Fetal bovine serum (FBS)	10%
IV	Washing solution	
	PBS	
	FBS	2%
V	Culture medium	
	Leibowitz L15	
	Glutamine	2 m*M*
	FBS	5%
	PEST	100 $\mu g\ ml^{-1}$
	Gentamicin	200 $\mu g\ ml^{-1}$
VI	Culture medium without antibiotics	
	Leibowitz L-15	
	Glutamine	2 m*M*
	FBS	5%

[a] PBS: phosphate-buffered saline without Ca^{2+} and Mg^{2+}.
[b] PEST: penicillin/streptomycin.
[c] Fungizon: amphotericin B and Na^+-deoxycholate. Concentration refers to amphotericin B.

of the Na^+/H^+-antiporter we cultured the cells in 0.4% CO_2 in air (P_{CO_2} = 3 mm Hg) with the L15 medium supplemented with 4 mmol liter^{-1} $NaHCO_3$ (Pärt and Wood, 1995). Judging from attachment efficiency, growth, and the general appearance of the cultures, no differences could be observed between cultures grown in room atmosphere or in the CO_2 atmosphere. The optimal temperature for rainbow trout cells is 18°C. At higher temperatures a gradual decline in the quality of the cultures is observed, as shown by low attachment and poor growth.

The attached cells grow and spread in the dishes (Fig. 1). The period of growth is most intense during Days 6–10, an observation that was also confirmed by [^{3}H]thymidine incorporation experiments (Pärt *et al.*, 1993).

Table II
General Information about the Tryptic Digest

Average yield of epithelial cells from 1.5-g wet weight filaments	178×10^6 cells (n = 86)
Viability (eosin exclusion)	86%
Proportion red blood cells of total cell number	5–13%
DASPEI positive (mitochondria rich)	10% (n = 6)
Attachment efficiency (24 h)	35%
DASPEI positive after attachment	0%

A confluent layer is formed that almost covers the entire bottom of the culture dish at Days 10–12 of culture. During this time the cells have gone through two or three cell cycles. The cells stay at this confluent density for about 7 days. Around Day 20, a progressive decline in number starts but there are still cells present after 40 days of culture. The cells are at their peak around Day 12 and the time available for experiments is approximately one week. Results presented here were obtained with 10- to 16-day-old cells.

The cultured cells clearly show characteristics of respiratory (pavement) epithelial cells (Pärt *et al.,* 1993). At confluence they have the general appearance of a "pavement," covering the bottom of the dish (Fig. 2). Furthermore, scanning electron microscopy reveals a typical

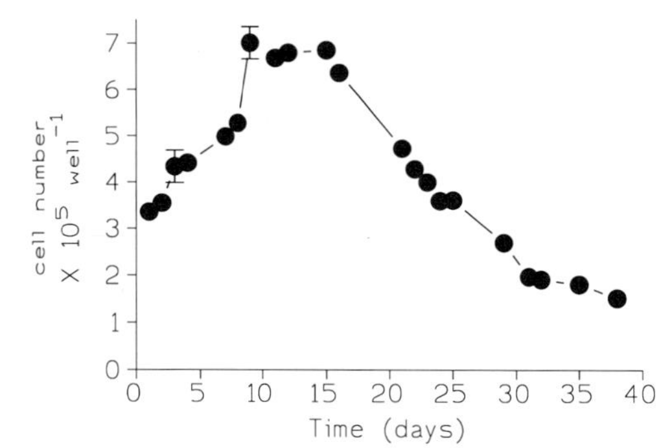

Fig. 1. Growth curve of primary cultures of branchial epithelial cells. Values are mean ± s.e.m. n = 7, Days 1–20; n = 3, Days 21–38. n = number of batches.

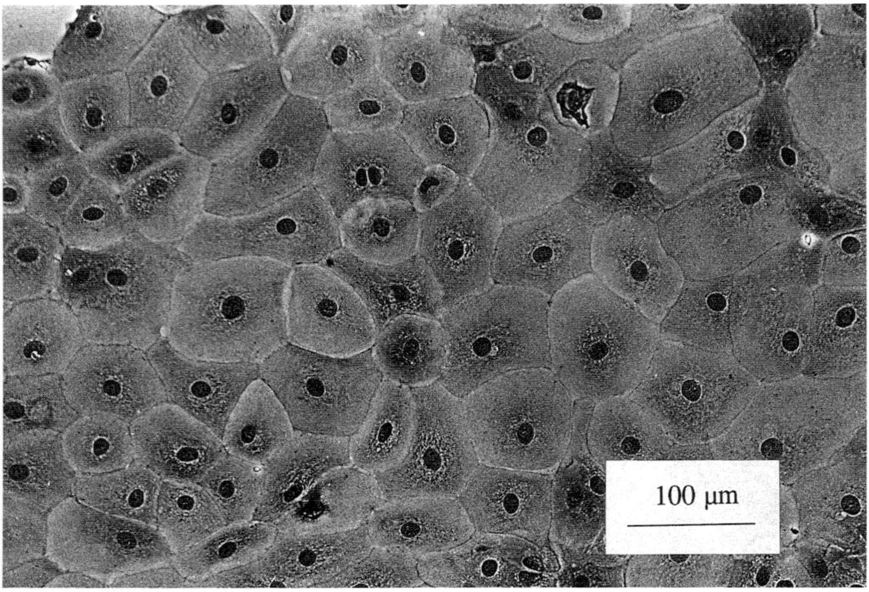

Fig. 2. Photomicrograph of confluent cell layer (Day 12 after seeding).

pattern of microridges on the epithelial surface (Pärt *et al.,* 1993) similar to the pattern reported on the epithelial surface of the gill *in vivo* (Kendall and Dale, 1979; Dunel-Erb and Laurent, 1980).

The tryptic digest contains 6–10% DASPEI-positive cells. Since DASPEI is a vital stain for mitochondria (Bereiter-Hahn, 1976), we assume these mitochondria-rich cells to be chloride cells (Marshall and Nishioka, 1980). This number of mitochondria-rich cells (chloride cells) is in the range reported by others: 5% in freshwater zebrafish (*Brachydanio rero*) (Karlsson, 1983), 8% in freshwater rainbow trout (Salman and Eddy, 1987), and 1% in freshwater tilapia (*Oreochromis mossambicus*) to 13% in seawater toadfish (*Opsanus tau*) (Perry and Walsh, 1989). However, the DASPEI-positive cells do not survive culture conditions. After 24 h, no DASPEI-positive cells are present and the situation is the same in 10-12-day-old confluent cultures. We conclude that chloride cells do not survive the culture conditions probably because the medium is lacking essential ingredients for their survival. Cortisol may be one such ingredient, as indicated by McCormick (1990).

C. Culture Medium

We use Leibowitz L15 medium with 2 mmol liter^{-1} glutamine as culture medium. Several other media have been used with fish cells, such as

M199+Ham's F12 (Dickman and Renfro, 1990), E-MEM (Tocher and Dick, 1990), MEM + Hanks salts (Wohschlag *et al.*, 1989), and RPMI-1640, but L15 works best in our hands. The L15 medium is designed for use in CO_2-free systems. It contains no nominal HCO_3^- and is buffered by its complement of salts (phosphates) and free base amino acids. The pH of the complete medium, including fetal bovine serum (FBS) and antibiotics, is 7.5–7.7.

Addition of FBS is essential. Both attachment and growth are dependent on the presence of FBS. The optimal serum concentration is 5% (Pärt *et al.*, 1993). The beneficial effects of serum in mammalian cell culture are well established and assumed to be related to its content of hormones, growth factors, micronutrients, and so on. These growth-promoting factors are apparently common for both mammalian cells and fish cells because most fish cells or cell lines are stimulated by FBS (Bols and Lee, 1991). Few, if any, homologous growth-promoting additives are known for fish. One example is reported by Collodi and Barnes (1990), who found that extracts from 21-day-old fish embryos stimulated growth in fish cell lines. The natural first choice, fish serum, has been found to be beneficial for some cells (hepatocytes; Kocal *et al.*, 1988) although toxic in others (Fryer *et al.*, 1965; Collodi and Barnes, 1990). We found freshly prepared fish serum to stimulate growth in gill cells, whereas serum that had been stored frozen (−20°C) was toxic (Pärt *et al.*, 1993). That fish serum could be toxic to cells has also been reported by others (Fryer *et al.*, 1965). Since FBS is easily accessible in quantities, and since it works with fish (gill) cells, it will continue to be the first choice until defined media are developed. However, one should be aware that FBS could change the properties of the cells. Tocher *et al.* (1988) showed that growing fish cell lines in medium supplemented with FBS caused the cells to acquire a polyunsaturated fatty acid (PUFA) composition reflecting almost exactly the composition of the serum and different from normal fish tissues. Clearly, if the membrane lipid composition is affected, this could have a profound impact on the behavior of transporting enzymes, ion channels, and also the passive permeability properties of the cells. Supplementation of the medium with fish oil concentrates allowed the fish cell lines to generate PUFA compositions more similar to those of the species from which they arose (Tocher and Dick, 1990).

Antibiotics like streptomycin, gentamicin, and penicillin are added to most cell culture media to control bacterial infections (Bols and Lee, 1991). Many antibiotics are ionophores and, although they are designed to specifically affect bacteria, effects on eukaryote systems cannot be excluded. Indeed, Dickman and Renfro (1990) showed that streptomycin was toxic to primary cultures of renal tubular cells from the winter flounder (*Pseudopleuronectes americanus*). Since the aim with our gill cell work is to study cellular ion and acid/base regulation, we prefer to grow the cells in antibiotic-free conditions. One complicating factor is that the gills are an external tissue and contamination with bacteria during the preparation of the cells is unavoidable. For this reason we culture the cells for the first 4 days in the presence of antibiotics (100 μg liter^{-1} penicillin/streptomycin and 200 μg liter^{-1} gentamicin). Subsequently antibiotics are excluded and the cells are grown in antibiotic-free conditions.

In an effort to increase the efficiency in our culture system, we screened a series of different factors known to promote growth and attachment of epithelial cells. Factors tested were epidermal growth factor (EGF), insulin, transferrin, and cortisol. This screening was done in two ways. The additives were either tested alone or in mixtures or were added in the presence of FBS. The two assays used were [^{3}H]thymidine incorporation, which is a measure of DNA synthesis, and attachment efficiency. Neither transferrin, (5 μg ml^{-1}), insulin (5 μg ml^{-1}), nor EGF (10 ng ml^{-1}) proved to be stimulatory for the cells and none showed the same stimulatory effect as 5% FBS (Pärt *et al.,* 1993). If anything, cortisol and insulin inhibited [^{3}H]thymidine incorporation whereas insulin had a weak positive effect on attachment. The conclusion is that no single substance or growth factor (or a mixture of thereof) has been found that is as efficient as FBS in promoting growth and attachment.

D. Attachment

We have tried to find conditions to optimize attachment. A high attachment efficiency is important for two reasons. First, optimal attachment of cells gives a good yield from the tryptic digest, meaning that one fish can yield several cultures. Second, an efficient attachment and outgrowth is a prerequisite to grow cells on permeable supports in order to obtain tight artificial epithelia. The gill cells attach and grow well on tissue culture quality plastic (Nunclon, Nunc, Denmark, or Falcon, Becton-Dickinson, U.S.A.) and on borosilicate glass. The only permeable support we have found that the cells attach to are Falcon Cyclopore cell culture inserts (Becton-Dickinson, U.S.A.). We have not been able to improve attachment by treating the culture surfaces with proteins found in the basal lamina—a recipe commonly used in mammalian epithelial cell culture. Tests were performed by treating culture dishes with collagen type I and type IV, laminin, Matrigel, and trout skin extract. None of the treatments improved attachment compared to the untreated plastic (Pärt *et al.,* 1993). Matrigel is a mixture of basal laminar proteins containing collagen IV, laminin, entactin, and heparin sulfate proteoglycan (Kiser *et al.* 1990). Trout skin extract was tested because it had been shown to promote attachment of fish hepatocytes (Blair *et al.*, 1990). The observation that collagen failed to improve attachment is puzzling because collagen treatment is almost standard in epithelial cell culture.

From an experimental point of view it is of great advantage that under the conditions described, the cells grow as monolayers and are firmly attached to both glass and plastic supports. These confluent cells have the appearance of a differentiated epithelium and tight junctions are established. Processes studied are almost certainly, therefore, apical processes since access to the basolateral surface is limited. Compared to cells in suspension, attached cells are easy to incubate and easy to wash. Cells grown in 6-, 12-, 24-, or 96-well plates make experiments easy and reproducible. In the following sections we describe experiments made with gill cell cultures where full advantage has been taken of the cells being adherent.

III. INTRACELLULAR MEASUREMENTS

A. Incubation Media

For intracellular measurements, two incubation media have been used: Cortland salmonid saline (Wolf, 1963) in equilibrium with 0.4% CO_2 in air and titrated with $NaHCO_3$ to pH = 7.70 (P_{CO_2} = 3 mm Hg)(18–20°C), and HEPES-Cortland, which is Cortland salmonid saline with 6 m*M* HEPES (*N*′-[2-hydroxyethyl]piperazine-N′-[2-ethanesulfonic acid]) instead of $NaHCO_3$. It is equilibrated with air and pH is adjusted to 7.70 (18–20°C) with NaOH. This medium is nominally bicarbonate free. In the Na^+-free counterparts, Na^+ is replaced by choline, and in the chloride-free media, Cl^- is replaced by gluconate and pH is adjusted with choline-HCO_3 or KOH (KCl reduced accordingly to keep the K^+ concentration constant).

B. Cell Volume

An estimate of cell volume is essential for calculations of intracellular concentrations of ions and metabolites and also for measurement of intracellular pH by the distribution of weak acids (DMO). Cell volume, or more exactly intracellular water space, has been estimated from the distribution of [^{14}C]urea between the extra- and intracellular fluid spaces using the method of Mendoza and Rozengurt (1987). Confluent cultures, 10–15 days old, in 12-well plates were used. The cultures were rinsed three times with HEPES-Cortland and 0.5 ml well^{-1} of incubation medium (Cortland saline or HEPES-Cortland) was added together with 0.3 μCi [^{14}C]urea plus cold urea to give a final urea concentration of 1 m*M*. Incubation time was 60 min. The medium was then aspirated and sampled for radioactivity and the wells rapidly (total time <60 s) rinsed by dipping in five consecutive baths of ice-cold 200 m*M* $MgCl_2$. The cultures were then air-dried and the cells solubilized with 0.5 ml well^{-1} 1% SDS in 300 m*M* NaOH followed by liquid scintillation counting for ^{14}C radioactivity and assay of protein content by the Lowry method.

The foregoing method should be reliable provided urea is not actively accumulated or excluded by the cells. This was checked by comparing the distribution volume of [^{14}C]urea in the cells with that of ^3H_2O. Cells were incubated with a mixture of ^3H_2O and [^{14}C]urea. The medium was aspirated but the cells were not rinsed. The ratio ^{3}H/^{14}C in the medium was compared to that in the cell lysate. The relative distribution between the two labels is shown in Fig. 3. There was a tendency indicating urea accumulation in the cells at urea concentrations lower than 0.5 m*M*, but at 0.5 m*M* and higher no such tendency was found, therefore justifying the use of 1 m*M* [^{14}C]urea+urea in the cell volume measurements. It was also found that no significant urea was lost from the cells provided the rinsing period was kept shorter than 60 s and that the $MgCl_2$ was ice cold. In this respect urea was more useful than ^3H_2O. Double-label experiments that included rinsing showed that the cell volumes calculated from ^3H_2O accumulation were 50% lower than those obtained with [^{14}C]urea, probably

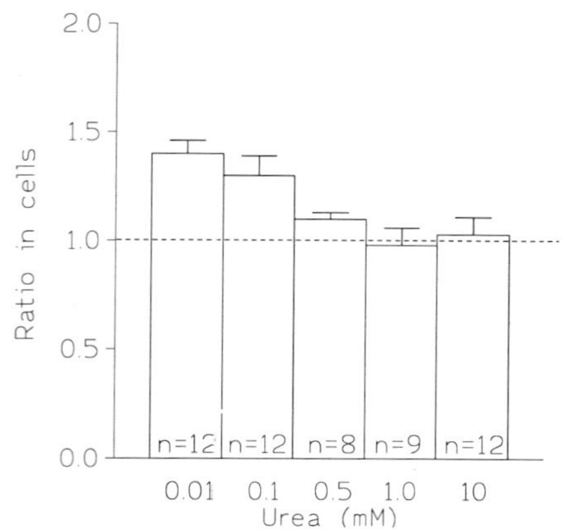

Fig. 3. Relative distribution space of [^{14}C]urea and ^3H_2O. The ratio [^{14}C]urea/^3H_2O = 1 in the incubation medium. A ratio higher than 1 in the cells shows that the intracellular distribution space of [^{14}C]urea is larger than that of ^3H_2O, which indicates that the cells accumulate urea.

because of a significant loss of ^3H label during rinsing. The cell volumes did not differ between cells incubated (30 min) in HEPES-Cortland (pH = 7.70) or in CO_2/HCO_3^--Cortland (P_{CO_2} = 3 mm Hg, pH = 7.70). The volume was 6.3 ± 0.3 μl mg^{-1} cell protein (n = 110).

C. Intracellular Ions

Many of our experiments concerning intracellular pH regulation include substitution of extracellular ions. It is therefore of interest to know the starting intracellular ion concentrations in the cultured cells and how well they correspond to the values measured in the intact branchial tissue. Second, to interpret ion substitution experiments, it is important to know how the intracellular ion concentration changes when the extracellular concentration is modified.

Intracellular Na^+ has been measured by the method of Montrose (1991). The confluent cell layer in a 25-cm^2 culture flask is rapidly rinsed (five washes, <60 s) with Na^+-free HEPES-Cortland. Perchloric acid (10%, 3 ml) is added to the cells. This treatment permeabilizes the cells but they do not detach. The Na^+ content is measured in the perchloric acid on a flame photometer against standards made up in perchloric acid. The remaining cell bodies are solubilized with 1% SDS + 300 m*M* NaOH for measurements of cell protein. The results are expressed as mmol Na^+ mg^{-1} cell protein or converted to mmol Na^+ $liter^{-1}$ cell water by using the protein as a volume correction factor (6.3 μl mg^{-1} cell protein). Intracellular Na^+ was measured in cells cultured in L15, and in cells incubated for 30 min in HEPES-Cortland, for 30 min in HEPES-Cortland + 1 mmol $liter^{-1}$ ouabain, and for 30 min in Na^+-free HEPES-Cortland (Fig. 4a). For intracellular Cl^-, basically the same procedure has been used except that a Cl^--free washing solution (Cl^--free HEPES-Cortland) is used. For measurements of Cl^- in the perchloric acid extract we use the colorimetric method of Zall *et al.* (1956). Intracellular chloride concentrations have also been calculated from radioactive equilibrium experiments with $^{36}Cl^-$ of known specific activity. Chloride concentrations were obtained from cells grown in L15 medium, cells incubated for 30 min in HEPES-Cortland, and cells incubated for 30 min in Cl^--free HEPES-Cortland (Fig. 4b). One should be aware of the risk of external contamination in these kinds of experiments. Solutions have to be freshly made and stored in plastic bottles. Measurements showed that the actual Na^+ concentration in nominally Na^+-free HEPES-Cortland was 40–60 μmol $liter^{-1}$. Experiments with ^3H-labeled PEG (MW = 4000) showed that the volume of trapped incubation medium was 1.2 ± 0.2 μl mg^{-1} cell protein (n = 9) and the intracellular measurements were compensated accordingly. The intracel-

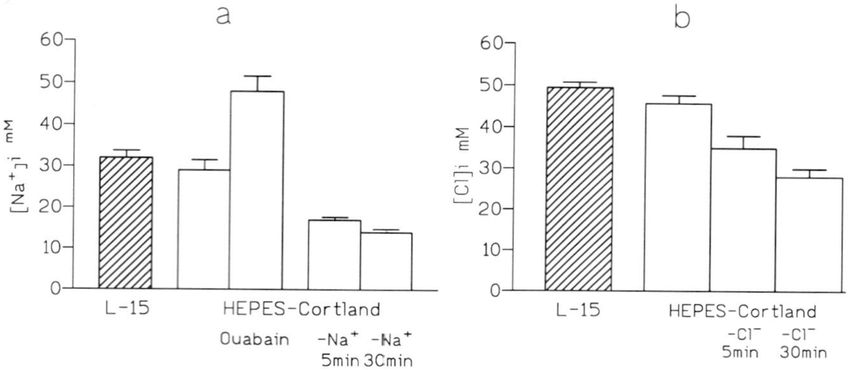

Fig. 4. Intracellular ions in cultured branchial cells. (a) Intracellular Na^+ : L15 = Leibowitz L15 cell culture medium with 5% fetal bovine serum (n = 12). Hepes-Cortland = HEPES-buffered Cortland saline, pH = 7.70 (n = 12). $-Na^+$ = sodium-free HEPES-Cortland (n = 10). Concentration of ouabain = 1 mM (n = 12). (b) Intracellular Cl^- : L15 = Leibowitz L15 cell culture medium with 5% fetal bovine serum (n = 7). Hepes-Cortland = HEPES-buffered Cortland saline, pH = 7.70 (n = 5). $-Cl^-$ = Chloride-free HEPES-Cortland (n = 12).

lular Na^+ concentration was found to be 32 ± 1.8 mM (n = 15) while intracellular Cl^- was higher, 49 ± 1.3 mM (n = 10) in L15 medium. These values are very close to those measured in freshwater rainbow trout gills *in vivo*, Na^+ = 29 mM and Cl^- = 44 mM (Munger *et al.*, 1991), but lower than those reported by Wood and LeMoigne (1991), Na^+ = 55 mM and Cl^- = 75 mM. Maintenance of intracellular Na^+ is dependent on active Na^+/K^+-ATPase. Intracellular Na increased in the presence of ouabain (Fig. 4a). When cells were incubated in Na^+-free media, intracellular Na^+ was reduced by 50% during a 5-min incubation. The loss of Cl^- appears to be slower, as an approximately 50% reduction was observed only after 30 min in Cl^--free conditions (Fig. 4b).

D. Intracellular pH (pHi)

We have used two independent methods based on different principles to measure intracellular pH. In one method, the distribution of a weak acid, ^{14}C-labeled DMO, (5,5-dimethyl-2,4-oxazolidinedione), between the cell and the incubation medium is used to calculate pH, whereas the other method uses the pH-dependent fluorescence of an intracellularly trapped fluorochrome, 2,7-bis(2-carboxyethyl)-5(6)-carboxy-fluorescein (BCECF), to estimate pHi. Both methods yielded almost identical pHi estimates—7.40 (DMO) and 7.43 (BCECF). This value again seems to correspond very well with the pHi of 7.43 measured in intact gill tissue *in vivo* with the DMO technique (Wood and LeMoigne, 1991).

1. DMO Method

The DMO method is based on one critical assumption that the biological membrane is only permeable to the protonated uncharged form of an acid while being impermeable to the ionized form (Roos and Boron, 1981). The substance DMO appears to fulfill this requirement (Waddell and Butler, 1959). The method used in our experiments was essentially that of Mendoza and Rozengurt (1987).

Gill cells in 12-well culture plates were washed three times with the incubation medium (Cortland or HEPES-Cortland), and then 0.5 ml medium + 0.3 μCi ^{14}C-labeled DMO was added to 6 of the 12 wells. The remaining 6 wells were used for measurement of intracellular water by the [^{14}C]urea technique described earlier. The plates were incubated in the appropriate gas atmosphere at 20°C for 30 min (0.3% CO_2 in air for Cortland, air for HEPES-Cortland). At termination, the incubation medium was aspirated and sampled for radioactivity, and the wells were rinsed by five consecutive washes in ice-cold 200 mM $MgCl_2$. The plates were turned upside down and left for 15 min. The medium that collected around the rim of the well was wiped with tissue paper and the wells were left to air-dry for 60 min. Then 500 μl of 1% SDS + 300 mM NaOH was added to solubilize the cells; 200 μl was taken for measurement of protein with the Lowry method. The remaining 300 μl of the solubilized cells was used for scintillation counting. With DMO the rinsing time is even more critical than with urea and care was taken to rinse the plates (wells) within <60 s (<45 s after some training). Initial checks showed that provided the rinsing time was <60 s and the $MgCl_2$ ice cold, less than 7% of the DMO content of the cells was lost. The DMO radioactivity was expressed as dpm mg^{-1} protein. The 6 additional wells on the same plate (same treatment) were incubated with [^{14}C]urea and yielded the intracellular water as μl mg^{-1} protein. Hence the DMO concentration, dpm μl^{-1}, could be calculated. Intracellular pH was calculated using the formula given in Mendoza and Rozengurt (1987) by using the pK_a of 6.29 for DMO (Boutilier *et al.*, 1984).

2. BCECF Method

The method for measuring pHi in gills cells with the fluorescent probe BCECF has recently been reported (Pärt and Wood, 1995). In principle, cells are incubated with an ester form of the fluorochrome BCECF (BCECF/AM)(Rink *et al.*, 1982). This form is membrane-permeable and is easily taken up by the cells. Intracellular esterases cleave the ester, leaving the negatively charged, membrane-impermeable form BCECF trapped in the cell. BCECF has a pH-dependent fluorescence at an emission wavelength of 530 nm and an excitation wavelength of 490–505 nm. A measure of the changes in the intracellular dye concentration is obtained by following the fluorescence at the isobestic excitation wavelength of 440 nm. It is important to have this continuous recording of the dye concentration as leakage and bleaching result in a continuous decline throughout the experiment. To correct for this problem when using BCECF, the fluorimeter is programmed to switch between the two excitation wavelengths and the emission ratio at 530 nm is recorded. Hence the ratio is independent of intracellular dye concentration and directly dependent on intracellular pH.

In our measurements we used a laboratory spectrofluorimeter (Perkin–Elmer LS-50) with a stirred, thermostated cuvette holder for 1-cm cuvettes. The cells were grown on coverslips cut to a size to fit into a perfused 1-cm disposable cuvette. The coverslip was mounted in the cuvette at a 45° angle toward the excitation light in such a fashion that the excitation light is reflected away from the emission collecting system. Fluorescence emission from the cells was collected through the coverslip. The perfusion system allowed very rapid replacements of incubation media. For intracellular calibration, the high-K^+/nigericin method (Thomas *et al.*, 1979) was used following a procedure described by Boyarsky *et al.* (1988). The cells were titrated to different pH's and the fluorescence ratio was recorded.

3. Maintenance and Regulation of pHi

Maintenance of resting pH in Cortland saline and HEPES-Cortland has been studied with the DMO method. The results are summarized in Table III. It appears that resting pHi is dependent on the presence of Na^+ and not dependent on the presence of extracellular Cl^-. Addition of the Na^+/H^+ exchange inhibitor amiloride resulted in intracellular acidification in HEPES-Cortland but not in ordinary HCO_3^--buffered Cortland saline. Addition of the anion exchange inhibitor DIDS had no effect on resting pHi. These results suggest that resting pHi in HEPES-Cortland (HCO_3^--free) is maintained by a classical Na^+/H^+ exchanger—a system active in intracellular pH regulation in almost every animal cell tested to date (for reviews, see Ilundain, 1992; Krapf and Alpern, 1993). In HCO_3^--containing Cortland saline, the situation is not clear and a HCO_3^--dependent mechanism cannot be excluded at present.

To further characterize the Na^+/H^+ exchanger we measured the recovery of pHi from an acid load in HEPES-Cortland. In these experiments the BCECF technique was used because it has a much higher time resolu-

Table III
Intracellular pH in Branchial Epithelial Cells

Treatment	Cortland[a] pHi[b]	n	HEPES-Cortland[c] pHi	n
Control	7.43 ± 0.02	57	7.40 ± 0.03	53
-Na^+	7.20 ± 0.06*	24	7.23 ± 0.03*	20
-Cl^-	7.38 ± 0.03	34	7.41 ± 0.03	18
Amiloride, 0.2 m*M*	7.42 ± 0.04	18	7.17 ± 0.05*	12
DIDS, 0.1 m*M*	7.45 ± 0.02	24	7.33 ± 0.04	12

[a] Cortland salmonid saline in equilibrium with 0.4% CO_2 in air. pCO_2 = 3 mm Hg, pH = 7.70.
[b] Values are means ± standard error of the mean (SEM).
* $P < 0.05$, compared to control, Students *t*-test.
[c] HEPES-buffered Cortland (nominally HCO_3^- free) in equilibrium with air, pH = 7.70.

tion than the DMO technique. The ammonia prepulse method developed by Boron and de Weer (1976) was used to produce intracellular acidosis. Theoretically, exposure of cells to an ammonia-rich solution should cause an abrupt increase in pHi due to rapid entry of NH_3 (which binds intracellular H^+ ions to form NH_4^+), followed by a slow decay due to a gradual influx of NH_4^+ and whatever pHi regulatory mechanisms the cell may mobilize against the alkalosis. Subsequent removal of the external ammonia solution should result in a rapid washout of intracellular NH_3 and dissociation of NH_4^+, leaving behind an excess of accumulated H^+ ions, thereby inducing an abrupt intracellular acidosis. In practice, cells were exposed to 30 mmol liter^{-1} total ammonia as NH_4Cl in the HEPES-Cortland saline at pH = 7.70 for 6–9 min, following which the ammonia solution was rapidly replaced with the test medium (Fig. 5).

Using this protocol, the cells were acidified to pHi around 6.5 to 6.7 but pHi recovered very rapidly, within 3–5 min, to attain a pHi almost identical to that seen before application of the NH_3 prepulse. Our results in HEPES-Cortland (Pärt and Wood, 1995) showed that the pHi recovery from an acid load was Na^+ dependent (Fig. 5a) and amiloride sensitive (Fig. 5b). The recovery was not Cl^- dependent and we conclude that a Na^+/H^+ exchanger is the dominating pHi regulatory mechanism in those cells. By measuring recovery rates from acidification in different Na^+ concentrations, a K_m for Na^+ of 8.3 m*M* was determined. The maximal recovery rate was estimated to be 0.42 pH units min^{-1}, which corresponds to a H^+ extrusion rate of 5.6 mmol liter^{-1} cell water min^{-1} (Pärt and Wood, 1995).

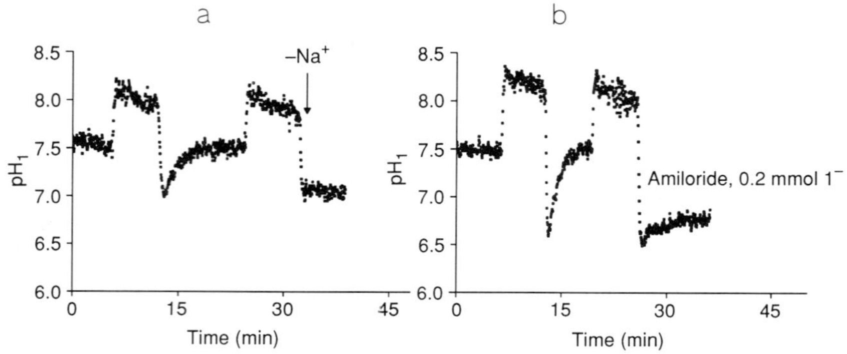

Fig. 5. Recovery from intracellular acidification in branchial cells. Cells were incubated in HEPES-Cortland (pH = 7.70). Intracellular acidification was achieved by the ammonia prepulse technique (the cells were exposed for 6 min to 30 m*M* NH_4Cl). Recovery (a) in the presence and absence of Na^+ and (b) in the absence and presence of 0.2 m*M* amiloride.

The K_m for Na^+ in these cultured cells is clearly higher than the K_m for Na^+ uptake (0.1–0.3 m*M*) obtained in rainbow trout gills *in vivo* (Goss and Wood, 1990). Therefore it is not likely that this mechanism is responsible for Na^+ uptake from fresh water *in vivo*. We speculate that although the Na^+/H^+ exchanger in the cultured cells is present on the apical membrane, *in vivo* in the freshwater trout the exchanger is most likely localized to the basolateral cell membrane, where the Na^+ gradient is favorable for exchange, that is, the exchanger has migrated during isotonic culture on a solid substrate. The Na^+/H^+ mechanism is probably responsible for the regulation of intracellular pH, whereas other mechanisms are likely responsible for Na^+ uptake from the water (see Lin and Randall, Chapter 9, this volume).

E. Buffer Capacity

By using the pHi changes during the ammonia prepulse (Fig. 5) in HEPES-Cortland, the buffer capacity of the cells can be calculated (Roos and Boron, 1981). Two sets of data were used and both gave essentially the same value. In the first set we used the degree of alkalinization induced at the onset of ammonia exposure, and in the other set we used the degree of acidification during washout of the ammonia load. The average nonbicarbonate buffer capacity (β) was 13.4 slykes, which corresponds well with the value of 13.7 obtained for gill homogenates from live trout (Wood and LeMoigne, 1991). This buffer capacity is low compared to that of other fish tissues (brain and muscle) (Wood and LeMoigne, 1991), and it is also low compared to that of most mammalian cells, primary cultures, or cell lines (Pärt and Wood, 1995).

IV. TRANSEPITHELIAL STUDIES

We have managed to grow cells on permeable supports (Wood and Pärt, 1995), whereas earlier we had reported a lack of success in this endeavor (Pärt *et al.*, 1993). Cells are first grown in culture bottles (25 or 80 cm^2) as described earlier; 8- to 10-day-old cells are trypsinated from the bottles and plated on filters (Falcon Cyclopore, 0.45-μm pore size, high or low density, Becton and Dickson, U.S.A.). The optimal plating density was found to be 5×10^5 cells cm^{-2}. The cells attach, grow, and spread to entirely cover the filter after 4 to 6 days. Chop-stick electrodes (EVOM, World Precision Instruments, U.S.A.) were used to measure transepithelial resistance (*R*), which is a measure of the tightness of the epithelial sheet, particularly of the tight junctions.

A gradual increase in resistance is apparent from Day 2 after plating on filters and the resistance is fully developed by Day 6 (Fig. 6), giving values of 3000 to 5000 Ω cm^2. The resistance increases up to 10-fold when the apical solution is changed to fresh water (Wood and Pärt, 1995) but decreases gradually over the following 48 h to approach the value in the saline–saline situation. These resistance values are among the highest ever recorded for cultured cells or intact tissues. They are matched in fish by the *R* value of 10,400 Ω cm^2 found for the cleithrum skin epithelium of freshwater rainbow trout with fresh water on the apical side (Marshall *et al.*, 1992). For comparison, the opercular membrane of seawater-adapted

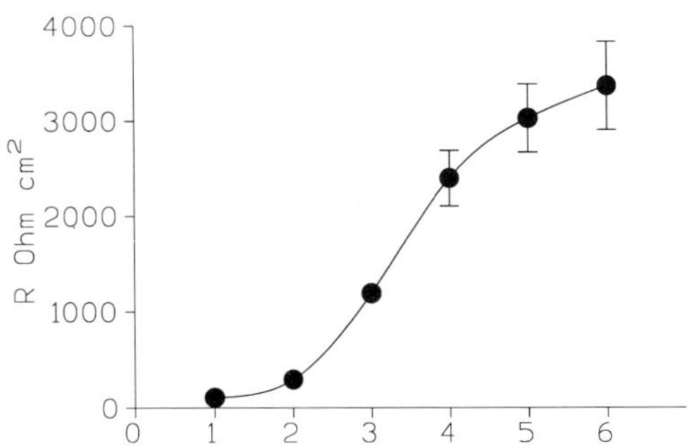

Fig. 6. Transepithelial resistance in cultured branchial epithelium. Branchial cells were seeded on filters and the resistance (*R*) was followed over time.

Fundulus heteroclitus has resistances in the range 130–170 Ω cm^2 (Karnaky *et al.*, 1977), but this is considered to be a "leaky" epithelium. Our findings confirm that the pavement cell epithelium is an extremely tight epithelium as stated by Isaia (1984). The fact that the cultured epithelium still maintains a high resistance after 48 h with fresh water on the apical side attests to the durability of the preparation.

V. CULTURED GILL CELLS IN TOXICOLOGY

The fish gill is the primary target and uptake site of many pollutants in the aquatic environment (Evans, 1987), making the use of cultured branchial cells an interesting prospect in aquatic toxicology. Though they could be used in elucidating effects of toxicants on branchial function, cultured cells could also serve as a permeability model for the barrier function of the gill. These ideas are certainly in line with the current trend in toxicology toward cellular methods (Isomaa *et al.*, 1994). Kennedy and Walsh (1994) used freshly isolated branchial cells from the gulf toadfish, *Opsanus beta,* to investigate benzo[*a*]pyrene uptake at different temperatures, effects on membrane fluidity, and the metabolism of benzo[*a*]pyrene in the gills. They found evidence for both phase I (arylhydrocarbon hydroxylase) and phase II enzymes (glutathione-*S*-transferase). We have used cultured confluent layers of gill cells to investigate cadmium uptake and the effect of Cl^- complexation, Ca^{2+} concentration, and organic complexants (ethylxanthate) (Block and Pärt, 1992). These studies were partially hampered by the difficulty in controlling the exposure situation specifically in terms of cadmium speciation in the medium—a problem common in *in vitro* studies of toxicological effects. Culture media are very complex with respect to their composition and many of the organic and inorganic component ingredients may affect the speciation of the relevant chemical. Nevertheless, gill cells offer an excellent opportunity in this context. When grown on filters they can withstand exposure to very dilute media on the apical side, thereby reducing the uncertainties with respect to speciation. It seems likely that such artificial branchial epithelia will become a useful system for testing of the bioavailability and membrane permeability of chemicals. Together with the relevantly modest requirements of such preparations with respect to culture conditions (atmosphere and temperature), cultured branchial cells will likely have a great potential in *in vitro* toxicology.

ACKNOWLEDGMENTS

The volume editors, C. M. Wood and T. J. Shuttleworth, are gratefully acknowledged for critical comments and language correction of the manuscript. Financial support from the Swedish Natural Science Research Council, the Swedish Environmental Protection Agency, the Magnus Bergwall Foundation, and the C. F. Lundström Foundation is acknowledged.

REFERENCES

Avella, M., Berhaut, J., and Payan, P. (1994). Primary culture of gill epithelial cells from the sea bass *Dicentrarchus labrax. In Vitro Cell Dev. Biol.* **30A,** 41–49.

Battram, J. C., Eddy, F. B., Chang, Y. J., and Fidler, J. (1989). Chloride transport by isolated gill cells of the fresh water adapted rainbow trout (*Salmo gairdneri*). *Comp. Biochem. Physiol., A* **94,** 439–445.

Bereiter-Hahn, J. (1976). Dimethylaminostyrylmethyl-pyridiniumiodine (DASPMI) as a fluorescent probe for mitochondria *in situ. Biochim. Biophys. Acta* **423,** 1-14.

Blair, J. B., Miller, M. R., Pack, D., Barnes, R., Teh, S., and Hinton, D. E. (1990). Isolated trout liver cells: Establishing short-term primary cultures exhibiting cell-to-cell interactions. *In Vitro Cell Dev. Biol.* **26,** 237–249.

Block, M., and Pärt, P. (1992). Uptake of ^{109}Cd by cultured gill epithelial cells from rainbow trout (*Oncorhynchus mykiss*). *Aquat. Toxicol.* **23,** 137–151.

Bols, N. C., and Lee, L.E. J. (1991). Technology and uses of cell cultures from the tissues and organs of bony fish. *Cytotechnology* **6,** 163–187.

Boron, W. F., and de Weer, P. (1976). Intracellular pH transients in squid giant axon caused by CO_2, NH_3 and metabolic inhibitors. *J. Gen. Physiol.* **67,** 91–112.

Boutilier, R. G., Heming, T. A., and Iwama, G. K. (1984). Appendix: Physicochemical parameters for use in fish respiratory physiology. *In* " Fish Physiology" (W. S. Hoar and D. J. Randall, eds.), Vol. 10A, pp 403–430. Academic Press, New York.

Boyarsky, G., Ganz, M. B., Sterzel, R. B., and Boron, W. F. (1988). pH regulation in single glomerular mesangial cells. I. Acid extrusion in the presence and absence of HCO_3^-. *Am. J. Physiol.* **255,** C857–C869.

Collodi, P., and Barnes, D. W. (1990). Mitogenic activity from trout embryos. *Proc. Natl. Acad. Sci. U.S.A.* **87,** 3498–3502.

Dickman, K. G., and Renfro, J. L. (1990). Streptomycin toxicity in primary cultures of flounder renal proximal tubule cells. *In Vitro Cell Dev. Biol.* **26,** 571–578.

Dunel-Erb, S., and Laurent, P. (1980). Ultrastructure of marine teleost gill epithelia: SEM and TEM study of the chloride cell apical membrane. *J. Morphol.* **165,** 175–186.

Evans, D. H. (1987). The fish gill: Site of action and model for toxic effect of environmental pollutants. *Environ. Health Perspect.* **71,** 54–58.

Fryer, J. L., Yusha, A., and Pilcher, K. S. (1965). The *in vitro* cultivation of tissue and cells of Pacific salmon and steelhead trout. *Ann. N.Y Acad. Sci.* **126,** 566–586.

Goss, G. G., and Wood, C. M. (1990). Na^+ and Cl^- uptake kinetics, diffusive effluxes and acid equivalent fluxes across the gills of rainbow trout. I. Responses to environmental hyperoxia. *J. Exp. Biol.* **152,** 521–547.

Goss, G. G., Laurent, P., and Perry, S. F. (1992). Evidence for a morphological component in acid–base regulation during environmental hypercapnia in the brown bullhead (*Ictalurus nebulosus*). *Cell Tissue Res.* **268,** 539–552.

Guibbolini, M. E., Henderson, I. W., Mosley, W., and Lahlou, B. (1988). Arginine vasotocin binding to isolated branchial cells of the eel: Effect of salinity. *J. Mol. Endocrinol.* **1,** 125–130.

Hootman, S. R., and Philpott, C. W. (1978). Rapid isolation of chloride cells from pinfish gill. *Anat. Rec.* **190,** 687–702.

Ilundain, A. (1992). Intracellular pH regulation in intestinal and renal epithelial cells. *Comp. Biochem. Physiol., A* **101,** 413–424.

Isaia, J. (1984). Water and nonelectrolyte permeation. *In* "Fish Physiology" (W. S. Hoar and D. J. Randall, eds.), Vol. 10B, pp 1–38. Academic Press, New York.

Isomaa, B., Lilius, H., and Råberg, C. (1994). Aquatic toxicology *in vitro:* A brief review. *ATLA* **22,** 243–253.

Kamiya, M. (1972). Sodium-potassium-activated adenosinetriphosphatase in isolated chloride cells from eel gills. *Comp. Biochem. Physiol., B* **43,** 611–617.

Karlsson, L. (1983). Gill morphology in the zebrafish, *Brachydanio rerio* (Hamilton-Buchanan). *J. Fish Biol.* **23,** 511–524.

Karnaky, K. J., Jr., Degnan, K. J., and Zadunaisky, J. A. (1977). Chloride transport across isolated opercular epithelium of killifish: A membrane rich in chloride cells. *Science* **195,** 203–205.

Kendall, M. W., and Dale, J. E. (1979). Scanning and transmission electron microscopic observations of rainbow trout (*Salmo gairdneri*) gill. *J. Fish. Res. Board Can.* **36,** 1072–1079.

Kennedy, C. J., and Walsh, P. J. (1994). The effect of temperature on the uptake and metabolism of benzo[*a*]pyrene in isolated gill cells of the gulf toadfish, *Opsanus tau. Fish Physiol. Biochem.* **13,** 93–103.

Kiser, C. S., Rahemtulla, F. and Månsson-Rahemtulla, B. (1990). Monolayer culture of rat parotid acinar cells without basement membrane substrates. *In Vitro Cell Dev. Biol.* **26,** 878–888.

Kocal, T., Quinn, B. A., Smith, I. R., and Ferguson, H. W. (1988). Use of trout serum to prepare primary attached monolayer cultures of hepatocytes from rainbow trout (*Salmo gairdneri*). *In vitro Cell Dev. Biol.* **24,** 304–308.

Krapf, R., and Alpern, R. J. (1993). Cell pH and transepithelial H^+/HCO_3^- transport in the renal proximal tubule. J. Membrane Biol. **131,** 1–10.

Ku, C. C., and Chen, S. N. (1992). Characterization of three cell lines derived from color carp *Cyprinus carpio. J. Tissue Cult. Meth.* **14,** 63–72.

Laurent, P. (1984). Gill internal morphology. *In* "Fish Physiology" (W. S. Hoar and D. J. Randall, eds.), Vol 10A, pp 73–183. Academic Press, New York.

Marshall, W. S., and Nishioka, R. S. (1980). Relation of mitochondria-rich chloride cells to active chloride transport in the skin of a marine teleost. *J. Exp. Zool.* **214,** 147–156.

Marshall, W. S., Bryson, S. E., and Wood, C. M. (1992). Calcium transport by the isolated skin of rainbow trout. *J. Exp. Biol.* **166,** 297–316.

McCormick, S. D. (1990). Cortisol directly stimulates differentiation of chloride cells in tilapia opercular membranes. *Am. J. Physiol.* **259,** R857–R863.

Mendoza, S. A., and Rozengurt, E. (1987). Measurement of ion flux and concentration in fibroblast cells. *In* "Methods in Enzymology" (D. Barnes and D. A. Sirbasku, eds.), Vol. 146, pp 384–399. Academic Press, Orlando, FL.

Montrose, M. H. (1991). Measurement of intracellular sodium and potassium in cultured epithelial cells. *J. Tissue Cult. Methods* 13, 211–216.

Munger, R. S., Reid, S. D., and Wood, C. M. (1991). Extracellular fluid volume measurements in tissues of the rainbow trout (*Oncorhynchus mykiss*) *in vivo* and their effects on intracellular pH and ion calculations. *Fish Physiol. Biochem.* **9,** 313–323.

Noga, E. J., and Hartman, J. X. (1981). Establishment of walking catfish (*Clarius batrachus*) cell lines and development of a channel catfish (*Ictalurus punctatus*) virus vaccine. *Can. J. Fish. Aquat. Sci.* **38,** 925–930.

Pärt, P., and Wood, C. M. (1995). Na^+/H^+ exchange in cultured epithelial cells from fish gills. *J. Comp Physiol.* (submitted).

Pärt, P., Norrgren, L., Bergström, E., and Sjöberg, P. (1993). Primary cultures of epithelial cells from rainbow trout gills. *J. Exp. Biol.* **175,** 219–232.

Perry, S. F., and Walsh, P. J. (1989). Metabolism of isolated fish cells: Contribution of epithelial chloride cells. *J. Exp. Biol.* **144,** 507–520.

Rink, T. J., Tsien, R. Y., and Pozzan, T. (1982). Cytoplasmic pH and free Mg^{2+} in lymphocytes. *J. Cell Biol.* **95,** 189–196.

Roos, A., and Boron, W. F. (1981). Intracellular pH. *Physiol. Rev.* **61,** 296–434.

Salman, N. A., and Eddy, F. B. (1987). Response of chloride cell numbers and gill Na^+/K^+ ATPase activity of freshwater rainbow trout (*Salmo gairdneri* Richardson) to salt feeding. *Aquaculture* **61,** 41–48.

Sargent, J. R., Thomson, A. J., and Bornancin, M. (1975). Activities and localization of succinic dehydrogenase and Na^+/K^+-activated adenosine triphosphatase in the gills of fresh water and sea water eels (*Anguilla anguilla*). *Comp. Biochem. Physiol., B* **51,** 75–79.

Taub, M. (1985). "Tissue Culture of Epithelial Cells." Plenum, New York and London.

Thomas, J. A., Bushbaum, R. N., Zimniak, A., and Racker, E. (1979). Intracellular pH measurements in Erlich ascites tumor cells utilizing spectroscopic probes generated *in situ*. *Biochemistry* **18,** 2210–2218.

Tocher, D. R., and Dick, J. R. (1990). Incorporation and metabolism of (*n*-3) and (*n*-6) polyunsaturated fatty acids in phospholipid classes in cultured Atlantic salmon (*Salmo salar*) cells. *Comp. Biochem. Physiol., B* **96,** 73–80.

Tocher, D. R., Sargent, J. R., and Frerichs, G. N. (1988). Fatty acid compositions of established cell lines after long-term culture in mammalian sera. *Fish Physiol. Biochem.* **5,** 219–227.

Waddell, W. J., and Butler, T. C. (1959). Calculation of intracellular pH from the distribution of 5,5-dimethyl-2,4-oxazolidinedione (DMO). Application to skeletal muscle of the dog. *J. Clin. Invest.* **38,** 720–729.

Wohlschlag, N., de los Santos, L., and Dehn, P. F. (1989). Preparation of primary cultures of adult and juvenile catfish (*Ictalurus punctatus*) hepatocytes. *J. Tissue Cult. Methods* **12,** 53–55.

Wolf, K. (1963). Physiological salines for fresh-water teleosts. *Prog. Fish-Cult.* **25,** 135–140.

Wood, C. M., and LeMoigne, J. (1991). Intracellular acid–base responses to environmental hyperoxia and normoxic recovery in rainbow trout. *Respir. Physiol.* **86,** 91–113.

Wood, C. M., and Pärt, P. (1995). In preparation.

Zall, D. M., Fisher, D., and Garner, M. Q. (1956). Photometric determination of chlorides in water. *Anal. Chem.* **28,** 1665–1678.

9

PROTON PUMPS IN FISH GILLS

HONG LIN AND DAVID RANDALL

I. INTRODUCTION: GENERAL MODELS OF OSMOREGULATION AND ACID–BASE REGULATION IN FISH GILLS

The gills of the fish are the primary site of gas exchange, acid–base regulation, and osmoregulation (Maetz and Garcia-Romeu, 1964; McDonald *et al.*, 1989; Randall, 1990). The gill lamellar epithelium is permeable to O_2, CO_2, and NH_3, the transfer of which depends largely on passive diffusion (Randall, 1990). Ion transfer across the gill epithelium, on the

CELLULAR AND MOLECULAR APPROACHES
TO FISH IONIC REGULATION

other hand, is usually mediated by active or passive ion transport processes. In freshwater teleosts, branchial absorption of Na^+ and Cl^- from the hypoosmotic environment compensates for the constant loss of ions from the body by diffusion. The mechanisms for Na^+ and Cl^- uptake are independent (Maetz and Garcia-Romeu, 1964) and the counterions for Na^+ and Cl^-, presumably H^+ (NH_4^+) and HCO_3^-, respectively, are extruded to the water simultaneously (Perry and Randall, 1981; McDonald *et al.*, 1989). These electroneutral ion-exchange pathways have been suggested to be involved in acid–base regulation and a portion of carbon dioxide and ammonia excretion.

The gill lamellar epithelium separating the blood from the external water consists mainly of three cell types—mucous (also called goblet), epithelial (also called pavement), and chloride cells (Laurent and Dunel, 1980). Mucous cells secrete mucus that forms a thin layer on the gill surface. Epithelial cells are permeable to respiratory gases such as O_2, CO_2, and NH_3 and play a prominent role in gas exchange. Chloride cells, also known as mitochondria-rich cells, house most of the energy-consuming ion transport pathways such as Na^+-K^+-ATPase and are involved in osmoregulation by euryhaline teleosts (Pisam *et al.*, 1987). In freshwater fish, the epithelial and chloride cells are joined by tight junctions. These act as a minimal barrier for diffusion of gases, but have a high resistance to the transfer of ions and water. Thus gill epithelia in freshwater fish are considered to be "tight" epithelia (Sardet, 1980).

A. Na^+/H^+ Exchanger Model in Freshwater Fish

Na^+/H^+ (NH_4^+) exchange in the apical membrane was concluded to be the major pathway for sodium uptake and proton excretion and an optional excretory pathway for ammonia (Wright and Wood, 1985). Amiloride is a very potent and specific inhibitor of sodium transport, including both the sodium conductive channels and Na^+/H^+ (NH_4^+) exchange, in a wide variety of cellular and epithelial transport systems, by competing for the Na^+ transport site (Benos, 1982). Addition of amiloride to water reduced sodium uptake in the gill epithelium of trout by 84% (Perry and Randall, 1981) and 94% (Wright and Wood, 1985). The *in vivo* studies of Wright and Wood (1985) demonstrated that sodium influx and ammonia efflux across trout gills are coupled with approximately 1 : 1 stoichiometry in the external water pH range of 4–9 and concluded that a flexible combination of NH_3 diffusion and Na^+/NH_4^+ exchange was the major mechanism of ammonia excretion. The inhibition of Na^+ uptake by 0.1 m*M* amiloride in the external media was associated with only a 23% reduction in ammonia excretion, with no change in ammonia or ammonium ion gradients across the gill epithelium. Wilson *et al.* (1994) could find no evidence for an apical Na^+/NH_4^+ exchange and concluded that any coupling of ion transport was secondary to pH changes in the gill water microenvironment. The degree of coupling between sodium absorption and ammonia excretion was also found to be rather loose in goldfish, carp, and trout *in vivo* (Payan, 1978). Sodium uptake from a dilute medium such as fresh water will usually require an active transport process. Wood and LeMoigne (1991) reported intracellular sodium concentrations of 52 m*M* for the trout epithelium. The values reported by Eddy and Chang (1993) for the Atlantic salmon were somewhat higher. These values are much higher than that recorded for frog skin of 6.2 m*M* using double-barreled ion-sensitive microelectrodes (Harvey and Ehrenfeld, 1986). The sodium levels in the gill epithelium are much higher than that in fresh water (usually <1 m*M*) and the sodium electrochemical gradient across the apical membrane could not possibly drive the Na^+/H^+ exchange (Avella and Bornancin, 1989). Wright (1991) has calculated that protons could drive the Na^+/H^+ exchange as long as intracellular pH was 0.3 units below external pH. Fish can excrete acid into water of pH 6 (Lin and Randall, 1990) and yet the gill epithelial pH has been determined to be around pH 7.4 (Wood, 1991). Thus it seems unlikely that protons drive Na^+/H^+ exchange in fish gills.

The trout gill NH_3 permeability coefficient of 6×10^{-3} cm s^{-1} (Avella and Bornancin, 1989) is intermediate between values reported for the toad bladder and mammalian kidney tubule. It has been suggested that ammonia excretion, although dominated by NH_3 diffusion (Hillaby and Randall, 1979; Cameron and Heisler, 1983), is also mediated by Na^+/NH_4^+ exchange on the apical surface (Payan, 1978; Wright and Wood, 1985). Indeed, the NH_4^+ concentration gradient might provide the driving force for Na^+/H^+ (NH_4^+) exchange. Avella and Bornancin (1989), however, reexamined the mechanism of ammonia excretion and sodium absorption using an isolated-perfused head preparation and found that ammonia excretion was basically dependent on passive NH_3 diffusion. The stimulating effect of NH_4^+ on sodium flux could be explained in terms of a pH effect of the ammonia addition (Cameron and Kormanik, 1982), and so Avella and Bornancin (1989) and Heisler (1990) concluded that the balance of evidence was against the presence of Na^+/NH_4^+ exchange across the apical surface of trout gills. Avella and Bornancin (1989) considered the trout gill to be similar to other tight epithelia, such as frog skin and toad bladder, in that passive sodium uptake from water is indirectly coupled to an active electrogenic proton transport system.

B. Proton Pump and Na^+ Channel Model in Freshwater Fish

Na^+/H^+ exchange is only one of the two fundamentally different mechanisms that have been proposed to account for Na^+ and H^+ transport in opposite directions in ion-transporting epithelia. Na^+/H^+ countertransport, a passive and electroneutral process, exists in certain isolated cells such as sea urchin eggs and red blood cells, and in "leaky" epithelia such as rabbit gallbladder, small intestine, and renal proximal tubule. The other mechanism is that of an active proton transport, which has been described in "tight" epithelia such as the turtle and toad urinary bladder (Steinmetz, 1986; Al-Awqati, 1978), frog skin (Ehrenfeld *et al.*, 1985), and mammalian renal collecting tubule (Ait-Mohamed *et al.*, 1986). This proton transport is initiated by an electrogenic proton-translocating-ATPase that pumps hydrogen ions to one side of the membrane and generates a negative potential in the other side of the membrane that, in some cases, drives sodium flux via a sodium conductive channel (Ehrenfeld *et al.*, 1985). Potts (1994) has reviewed these mechanisms and pointed out that a proton pump coupled to a passive influx of sodium can theoretically show saturation kinetics and mimic carrier-mediated competitive inhibition.

The gill epithelium of freshwater fish is considered to be a tight epithelium (Sardet, 1980; Avella and Bornancin, 1989). It resembles the freshwater frog skin and turtle urinary bladder epithelia in many features. Bartels (1989) suggested that, in lamprey, the morphological characteristics of the epithelial cell were identical to those of the frog skin granular cell. Functionally they are both capable of acid–base regulation and electrolyte transport. Both freshwater frogs and freshwater fish are hyperosmotic to their aqueous surroundings and face the problem of continual loss of body salt to the environment. The lost salts have to be replaced through an active transport system across the epithelia in the skin of frogs or the gills of fish. Morphologically they contain analogous cell types. The epithelium of turtle urinary bladder consists of basal cells, granular cells, and carbonic anhydrase (CA)-rich cells. CA-rich cells, which contain numerous mitochondria and tubulovesicular membrane structures, are responsible for H^+ secretion in turtle bladder (Madsen and Tisher, 1985). The CA-rich cell could be equivalent to the chloride cell in the gill epithelium, although carbonic anhydrase is distributed in both chloride and pavement cells in fish gills (Kultz and Jurss, 1993).

The apical membrane of frog skin comprises a *stratum granulosum* firmly interconnected by tight junctions that form a barrier separating the apical bathing solution from the basolateral solution (Nielsen, 1982), similar to those in gill lamellae. The outermost living cell layer of the frog skin epithelium is composed of cuboidal granular (GR) cells and mitochondria-rich (MR) cells (Ehrenfeld *et al.*, 1989). The latter cells can be identified by (1) their long, flask-like shape and narrow apical pole beneath the *stratum corneum,* (2) the exclusive localization of carbonic anhydrase, and (3) the rod-shaped intraplasma membrane particles. The intercalated cells in mammalian renal collecting tubule also share similar characteristics described here for the MR cell of frog epithelium and are responsible for proton transport across the kidney tubular epithelium (Brown *et al.*, 1988). Avella *et al.* (1987) observed a correlation between chloride cell number and sodium influx across trout gill. Thus, it is logical to suggest that the same ion transport mechanism, namely, H^+-ATPase indirectly coupled with a sodium channel, might exist in gill chloride cells to account for the proton and sodium transport. Carbonic anhydrase, however, is also found in pavement cells in fish gills and Goss (1993) was unable to establish a clear relationship between fractional area of chloride cells and sodium flux in fish. Thus the H^+-ATPase indirectly coupled with a sodium channel might be more generally distributed in fish gills compared with frog skin and turtle bladder.

Differences do exist between the gill epithelium and other tight epithelia. Carbonic anhydrase, for example, is generally distributed in the chloride cells, epithelial cells, and mucous cells in fish gills (Rahim *et al.*, 1988), but restricted to MR cells of frog skin (Rosen and Friedley, 1973) and CA-rich cells of turtle bladder (Madsen and Tisher, 1985). Pisam *et al.* (1993) have described two types of chloride cells in freshwater teleosts, the α-chloride cell and the β-chloride cell, the former in contact with arterial blood and the latter facing the central venous sinus. The β-cells degenerate during transfer to seawater, whereas the α-cells enlarge and become the typical seawater chloride cell. The percentage of chloride cells in gill epithelium of freshwater fish is only 1% (Perry and Walsh, 1989), whereas the ratio of MR to GR cells in frog skin is much higher and, since frog skin plays a minor role in gas transfer, MR cells can represent between 13 and 60% of the exposed apical area depending on salt adaptation (Ehrenfeld *et al.*, 1989). The location and distribution of ion transport pathways in different cell types in fish gill epithelium, therefore, might not be exactly the same as that in frog skin epithelium.

II. PROTON PUMPS IN GENERAL

A. Classification of Proton Pump

Proton-translocating-ATPases are integral membrane proteins that vectorially translocate H^+ from one surface to the other (Pedersen and

Carafoli, 1987). They can be classified into three categories: mitochondrial H^+-ATPase (F-type), vacuolar H^+-ATPase (V-type), and plasma membrane H^+-ATPase (P-type). Mitochondrial H^+-ATPase utilizes the proton gradient generated by the cytochrome chain in the inner mitochondrial membrane for ATP synthesis and provides the energy source for other ATPases in the cell. Vacuolar H^+-ATPase and mitochondrial H^+-ATPase share a number of important structural properties, including complexity of subunit composition, and probably are derived from a common evolutionary ancestor (Forgac, 1989). Vacuolar H^+-ATPase consumes ATP and actively pumps protons against an electrochemical gradient into the vacuoles. The function of plasma membrane H^+-ATPase is similar to that of vacuolar H^+-ATPase but it has a lower molecular weight and a simpler subunit structure, with only 2 subunits instead of 18 as in mitochondrial H^+-ATPase and 16 as in vacuolar H^+-ATPase (Forgac, 1989).

The H^+-ATPases in the plasma membrane of eukaryotic cells are classified as phosphorylated ion motive enzymes because they form a covalent phosphorylated intermediate as part of the reaction cycle (Pedersen and Carafoli, 1987). Na^+ and K^+-ATPase and Ca^{2+}-ATPase are also phosphorylated ion motive ATPases. Vanadate, a transition-state analog of phosphate, inhibits P-type ATPase by blocking the formation of phosphorylated intermediates in the ATPase.

Plasma membrane H^+-ATPase and vacuolar H^+-ATPase couple to mitochondrial H^+-ATPase in a master–slave relationship. Mitochondrial H^+-ATPase functions obligatorily in the direction of ATP synthesis and supplies ATP to the plasma membrane H^+-ATPase and vacuolar H^+-ATPase, both of which function obligatorily in the direction of ATP hydrolysis to transport protons actively (Pedersen and Carafoli, 1987).

B. Characteristics of Proton Pumps in Other Tight Epithelia

Many epithelial membranes have the capacity to transport hydrogen ions. In the "tight" epithelial membranes, such as in mammalian renal collecting tubules (Gluck and Al-Awqati, 1984; Ait-Mohamed *et al.*, 1986), turtle urinary bladder (Steinmetz and Andersen, 1982; Steinmetz, 1985), and frog skin (Ehrenfeld *et al.*, 1990), the P-type or V-type proton-translocating-ATPase, also called electrogenic proton pump, are responsible for the transport of protons.

The proton pump or proton-ATPase, either vacuolar or plasma membrane type, has several important properties: the proton pump is electrogenic (Steinmetz and Andersen, 1982; Ehrenfeld *et al.*, 1985), relying on the energy of ATP hydrolysis. The stoichiometry of the pump is estimated to be 3 H^+/ATP (Al-Awqati and Dixon, 1982). Proton pump operation is regulated primarily by the proton electrochemical gradient across the membrane (Steinmetz, 1986). Cellular acid–base conditions are the second major determinant of the rate of proton transport (Cannon *et al.*, 1985; Steinmetz, 1985).

H^+-ATPase is bound to membranes or packaged in cytoplasmic vesicles of carbonic anhydrase-rich cells in turtle bladder (Arruda *et al.*, 1990) and intercalated cells in rat kidney (Brown *et al.*, 1988). Proton pumps can be inserted in the membrane by exocytosis of vesicles containing the ion motive enzyme or removed from the membrane by endocytosis of segments of the membrane in which H^+-ATPases are concentrated, and the process can be induced by environmental stimuli (Brown, 1989; Schwartz and Al-Awqati, 1985; Stetson, 1989).

A number of inhibitors have been used to study proton-ATPases both *in vitro* and *in vivo*. Vanadate is an inhibitor of plasma membrane ATPase. It inhibits not only H^+-ATPase in plasma membrane, but many other transport ATPases, including Na^+-K^+-ATPase and Ca^{2+}-ATPase. Vanadate ion, VO_4^{3-}, acting as a phosphate transition analog, blocks the formation of phosphorylated intermediates in these ATPases (Pedersen and Carafoli, 1987). Vanadate in the serosal solution of toad bladder (Beauwens *et al.*, 1981) and turtle bladder decreases proton secretion markedly under anaerobic and aerobic conditions (Arruda *et al.*, 1981). Proton excretion through frog skin is completely abolished when 1 mM vanadate is applied to the serosal site (Ehrenfeld *et al.*, 1985). However, vanadate exerts no inhibitory effect on either the proton-ATPase activity or the proton transport in mammalian renal tubules (Gluck and Caldwell, 1987; Turrini *et al.*, 1989) because the H^+-ATPase in mammalian kidney belongs to the vacuolar type.

N-Ethylmaleimide (NEM) is another metabolic inhibitor that affects V-type and P-type H^+-ATPase, with much more potent inhibition on vacuolar H^+-ATPase (1–2 μM) than plasma membrane H^+-ATPase (0.1–1 mM). Mitochondrial ATPase is virtually resistant to NEM (Forgac, 1989). NEM is an alkylating agent that is relatively selective for sulfhydryl groups (SH-) and inhibits H^+-ATPase in an ATP-protectable manner. So-called NEM-sensitive ATPase is found in all segments of mammalian kidney (Ait-Mohamed *et al.*, 1986; Gluck and Al-Awqati, 1984; Gluck and Caldwell, 1987). Proton excretion across frog skin is also inhibited by 1 mM NEM (Ehrenfeld *et al.*, 1990). The drug *p*-chloromercuri-benzenesulfonate (PCMBS) is also a SH-group reagent and affects proton-ATPase in rat and bovine kidney (Turrini *et al.*, 1989; Gluck and Al-Awqati, 1984).

Dicyclohexylcarbodiimide (DCCD) can bind to a subunit (the DCCD binding protein) of the hydrophobic channel portion and inhibit H^+-

ATPase in mitochondria, vacuoles, and plasma membranes (Pedersen and Carafoli, 1987). The sensitivity was highest in F-type, followed by V-type and then P-type (Forgac, 1989). H^+-ATPase in mammalian kidney (Ait-Mohamed *et al.*, 1986), turtle bladder (Steinmetz and Andersen, 1982), and frog skin (Ehrenfeld *et al.*, 1985) is sensitive to DCCD. Diethylstilbestrol (DES) has a very similar effect on H^+-ATPase as DCCD (Pedersen and Carafoli, 1987) and 0.1 m*M* DES was found to inhibit proton excretion in frog skin (Ehrenfeld *et al.*, 1990).

7-Chloro-4-nitrobenz-2-oxa-1,3-diazole (NBD-Cl) is also an alkylating agent that potently inhibits H^+-ATPases of all types (Forgac, 1989). Proton-ATPase in mammalian kidney was reported to be sensitive to 10–20 μM NBD-Cl (Gluck and Caldwell, 1987; Turrini *et al.*, 1989).

Bafilomycin is a macrolide antibiotic that appears to be a very specific and potent inhibitor to vacuolar H^+-ATPase (Bowman *et al.*, 1988). Mitochondrial H^+-ATPase is resistant to it and plasma membrane H^+-ATPase is moderately sensitive to it. It can be used as a valuable tool for distinguishing between the three different types of H^+-ATPases. Another reagent demonstrating unique specificity to V-type H^+-ATPase is potassium nitrate (KNO_3) (Bowman, 1983). Vacuolar H^+-ATPase from *Neurospora* was inhibited by KNO_3 with a half-maximal inhibition seen at 50 m*M*, whereas plasma membrane and mitochondrial H^+-ATPase from the same organism were completely resistant to KNO_3 up to 100 m*M* (Bowman, 1983).

III. PROTON PUMPS IN FISH GILLS

A. The Evidence for Branchial Proton Excretion

Freshwater rainbow trout were shown to excrete a substantial amount of protons across the gill under certain experimental conditions (Lin and Randall, 1991). In these studies, total CO_2, total ammonia, pH, and buffering capacity of the inspired and expired water and water flow rate in rainbow trout were measured and branchial proton excretion rate was calculated from these data. When water pH was approximately neutral, [HCO_3^-] in water entering the gills was always greater than [HCO_3^-] in water leaving the gills, indicating that there was no CO_2 converted to HCO_3^- as water passed over the gills. Instead, bicarbonate levels were reduced owing to HCO_3^- dehydration induced by proton excretion or bicarbonate absorption, perhaps via Cl^-/HCO_3^- exchange. SITS (4-acetamido-4′-isothiocyanatostilbene-2-2′-disulfonic acid), an inhibitor of erythrocytic Cl^-/HCO_3^- exchange, however, had no effect on acid excretion across fish gills (Lin and Randall, 1991), indicating that either there was no bicarbonate absorption or the apical membrane anion exchanger was SITS insensitive. This indicated that branchial proton excretion probably not only caused expired water acidification but also dehydrated HCO_3^- as water passed over the gills. Therefore, the expired water acidification reported by Wright *et al.* (1986) and Lin and Randall (1990) may not have been due to the hydration of excreted CO_2 but could have been due to acid excretion across the gills. The excreted acid was partially consumed by HCO_3^- dehydration and ammonia protonation, and the rest contributed to expired water acidification. The experiments of Lin and Randall (1991) reported very high levels of proton excretion, much higher than those reported by McDonald and Wood (1981) using a different method over a longer time interval of measurement. Comparisons using both methods on the same fish gave similar but very variable results (Xoan Fuentes, unpublished results). The high rates of proton excretion observed by Lin and Randall (1991) were probably due to stimulation of proton excretion resulting from an experimental elevation of water sodium and calcium levels, required for stable pH measurements, just before the experiments began.

B. The Role of the Proton Pump in Branchial Proton Excretion

Increasing amiloride concentration in the external medium induced a reduction in proton excretion, but more than 50% of the net proton excretion was still sustained even in 1 m*M* amiloride. Similar phenomena have been observed in frog skin. Proton excretion in frog skin was inhibited by 0.5 m*M* amiloride by 35% but was not affected by 0.05 m*M* amiloride, whereas sodium uptake was completely abolished. Ammonium ions can replace potassium on the Na^+-K^+-ATPase and thus enter the cell and form NH_3 and protons (Evans *et al.*, 1989). The deprotonation of NH_4^+ could supply the proton pump and NH_3 could diffuse passively across the apical membrane into the water. Although much less sensitive than the Na^+ channel, Na^+-K^+-ATPase in the basolateral membrane can be inhibited by amiloride that has entered the cell when applied in high concentrations to the apical side (Knauf *et al.*, 1976; Kleyman and Cragoe, 1988). Thus, the reduced proton and ammonia excretion in 0.5 and 1 m*M* amiloride treatments could be accounted for by the inhibitory effect of amiloride on Na^+-K^+ (NH_4^+)-ATPase in the basolateral membrane (Lin and Randall, 1991). In support of this contention, Evans *et al.* (1989) showed that amiloride did not affect ammonia excretion if the perfused head of the toadfish was pretreated with ouabain, which blocks Na^+-K^+(NH_4^+)-ATPase.

Lin and Randall (1991) found that more than 50% of the net proton excretion across the gill epithelium of trout was inhibited by 0.1 m*M* vanadate applied to the apical membrane. De Sousa and Grosso (1981) showed that applying 1 m*M* vanadate to the outer surface did not affect the Na^+-K^+-ATPase in the basolateral membrane of frog skin. Arruda *et al.* (1981) showed that vanadate had no effect on the backleak of proton or bicarbonate secretion but had a direct effect on H^+-transporting-ATPase in turtle bladder. Thus, Lin and Randall (1991) concluded that the reduction in proton excretion was induced by the inhibitory effect of vanadate on an H^+-translocating-ATPase in the apical membrane. The reason that the proton excretion was not completely abolished was, presumably, because of the difficulty of vanadate reaching the action site from the apical side (Arruda *et al.*, 1981) even though the vanadate was placed in the water.

Klungsoyr (1987) isolated vesicular membrane preparations from trout gills that accumulated protons and contained a magnesium-dependent ATPase. Justesen *et al.* (1993) and Lin and Randall (1993) demonstrated the existence of an *N*-ethylmaleimide (NEM)-sensitive ATPase in crude homogenates of trout gill tissue. Since the assay of Lin and Randall (1993) was carried out in the presence of EGTA, a Ca^{2+} chelator that should abolish Ca^{2+}-ATPase activity, azide, a mitochondrial H^+-ATPase inhibitor, and ouabain, a Na^+-K^+-ATPase inhibitor, the contribution of unrelated ATPase activity was considered to be minimal. In addition, the NEM-sensitive ATPase activity was also inhibited by other proton-ATPase blockers such as DCCD, DES, PCMBS, and bafilomycin. It was concluded that the NEM-sensitive ATPase activity was generated by a proton-translocating-ATPase.

H^+-ATPase has been noted to be either bound to membranes or packaged in cytoplasmic vesicles (Arruda *et al.*, 1990). The crude homogenates prepared by Lin and Randall (1993) contained mainly the membrane fraction of gill cells, and they concluded that the ATPase activity in the crude homogenate of gill tissue that was sensitive to 1 m*M* of NEM probably originated from plasma membrane H^+-ATPase. To determine whether the H^+-ATPase in gill homogenates is a P-type or a V-type, specific inhibitors were tested by Lin and Randall (1993). Bafilomycin, a very specific and potent inhibitor of vacuolar H^+-ATPase (Bowman *et al.*, 1988), significantly inhibited ATPase activity of fish gills only at concentrations above 25 μM, whereas vacuolar H^+-ATPase was completely blocked at concentrations as low as 0.1 μM (Bowman *et al.*, 1988). Potassium nitrate, another inhibitor used to distinguish vacuolar H^+-ATPase from plasma membrane H^+-ATPase (Bowman, 1983), caused less than a 30% reduction in fish gill ATPase activity at a concentration of 100 m*M*, a dosage that is sufficient to inhibit 80% of the vacuolar H^+-ATPase. The gill ATPase sensitive to NEM however, was sensitive to vanadate. Vanadate (0.1 m*M*) suppressed 60% of the ATPase activity in gill tissue. The resistance of fish gill ATPase to bafilomycin and nitrate and the sensitivity to vanadate indicate that the H^+-ATPase in fish gills is a plasma membrane type and not a vacuolar type.

The localization of proton pumps (H^+-ATPase) in gill epithelia of rainbow trout (*Oncorhynchus mykiss*) was elucidated by immunofluorescence microscopy using rabbit polyclonal antibodies against the 70-kDa subunit of clathrin-coated vesicle H^+-ATPase (vacuolar type) from bovine brain (Lin *et al.*, 1994). In the gill epithelium of freshwater trout, the immunostaining was uniformly distributed along the lamellae and generally concentrated in apical regions (Fig. 1). It was concluded, therefore, that H^+-

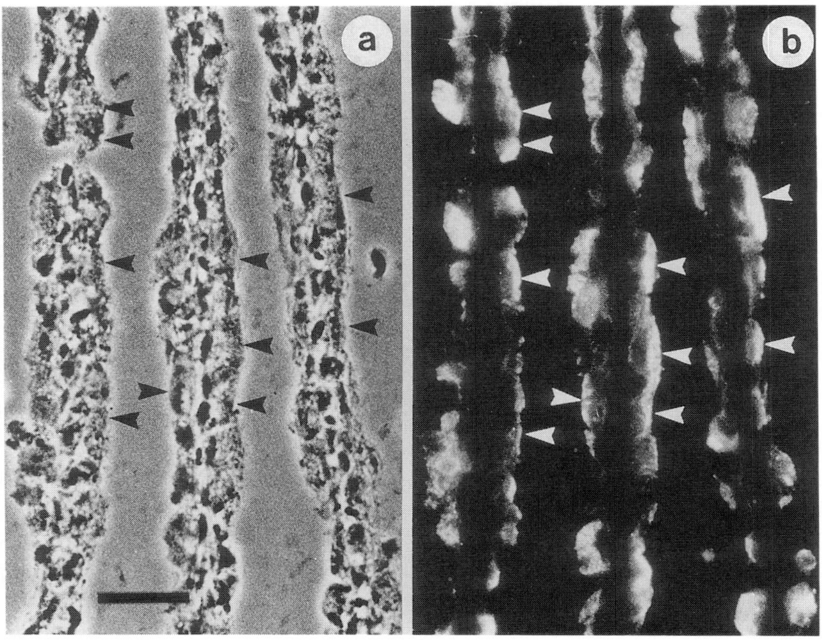

Fig. 1. Paired phase contrast (a) and fluorescence (b) micrographs of fixed frozen sections of gill from a trout adapted to fresh water. The section was treated with a fluorescent immunological probe for the 70-kDa subunit of the mammalian proton-ATPase. Fluorescence is distributed along the apical membrane of the gill epithelium. Bar = 20 μm. (From Lin *et al.*, 1994).

ATPase is located in the apex of both chloride cells and epithelial cells of freshwater fish. Hypercapnic treatment resulted in a nonpolarized and restrictive distribution of H^+-ATPase in the chloride cell. No fluorescent staining was observed in the gill epithelium of seawater-adapted rainbow trout except in some unidentified anucleate surface material. The presence of the 70-kDa subunit in fish gill epithelia was confirmed by Western blot. These results supported the proposed role of proton pump in sodium uptake in freshwater fish, and demonstrated that the H^+-ATPase in fish gills is a vacuolar type antigenically similar to the H^+-ATPase in mammalian brain and kidney.

The findings of this study did not support the earlier suggestion that the H^+-ATPase in fish gill epithelium is a plasma membrane (P-type) H^+-ATPase (Lin and Randall, 1993), which was based on the inhibitor sensitivity of gill H^+-ATPase. The study of Lin and Randall (1993) was carried out on crude homogenates and the specificity of most of the inhibitors is affected by the degree of purification of the enzyme homogenates (Bowman, 1983; Forgac, 1989). Immunofluorescence studies (Lin *et al.*, 1994) indicate that the fish gill epithelium appears to contain a vacuolar-type H^+-ATPase that is analogous to that from the clathrin-coated vesicle of bovine brain, the intercalated cell of rat and human kidney, and the vacuoles of plants, fungi, and Archaebacteria (Sudhof *et al.*, 1989; Kim *et al.*, 1992).

Acetazolamide, a traditional carbonic anhydrase inhibitor, inhibits proton excretion in fish gills (Lin and Randall, 1991), as well as in frog skin and turtle bladder (Ehrenfeld and Garcia-Romeu, 1977; Steinmetz, 1986). The apical addition of 0.01 m*M* ethoxzolamide, a lipid-soluble CA inhibitor, blocked net H^+ excretion as well as Na^+ absorption in frog skin (Harvey and Ehrenfeld, 1986). Acetazolamide was demonstrated to inhibit luminal acidification in turtle bladder by stimulating the endocytosis of apical membrane (Dixon *et al.*, 1988; Graber *et al.*, 1989). The inhibition appeared to be independent of cell pH, which ruled out the possibility of a secondary effect due to carbonic anhydrase inhibition. The inhibitory effect of acetazolamide on *in vivo* proton excretion in trout (Lin and Randall, 1991) could not be reproduced in an *in vitro* ATPase assay (Lin and Randall, 1993), indicating that acetazolamide has no direct effect on H^+-ATPase itself. This indicates that intracellular CO_2 hydration is important in maintaining the proton supply to the proton-ATPase.

Ammonium ions may be another source of protons for the proton-ATPase in fish gills (Lin and Randall, 1991). Ammonia elimination, however, was not affected by vanadate or acetazolamide, indicating that proton and ammonia efflux from the gill epithelium are through different pathways. Thus, ammonium entry into the gill epithelium may affect proton excretion through the supply of protons, but variations in proton excretion do not appear to affect ammonia excretion. In addition, ammonium cannot be the sole source of protons because proton excretion can be more than twice ammonia excretion in some instances, indicating that carbon dioxide hydration is probably the more important source of protons feeding the proton-ATPase in fish gills.

A possible relationship of the proton pump and the other ion transport pathways in the gill epithelium of freshwater fish is summarized in Fig. 2. The primary electrogenic proton pump in the apical membrane consumes ATP, excludes hydrogen ions from the intracellular compartment to the external water, and generates a negative potential that serves as a driving force for sodium uptake from fresh water via a sodium conductive channel. CO_2 hydration in the intracellular compartment provides one of the sources of protons to the proton pump. HCO_3^- generated by CO_2 hydration can be either excreted to the external water by a Cl^-/HCO_3^- exchanger or recycled to the serosal side via an anion channel. The deprotonation of NH_4^+ that entered the cell from the serosal side through the Na^+-$K^+(NH_4^+)$-ATPase pathway in the basolateral membrane provides another source of protons to the pump. The resulting NH_3 produced will diffuse across the apical membrane and be excreted to the water. Na^+ and perhaps Ca^{2+} enter the cell from the apical side via the conductive channels and act as the counterion for the proton pump. Cl^- transferred

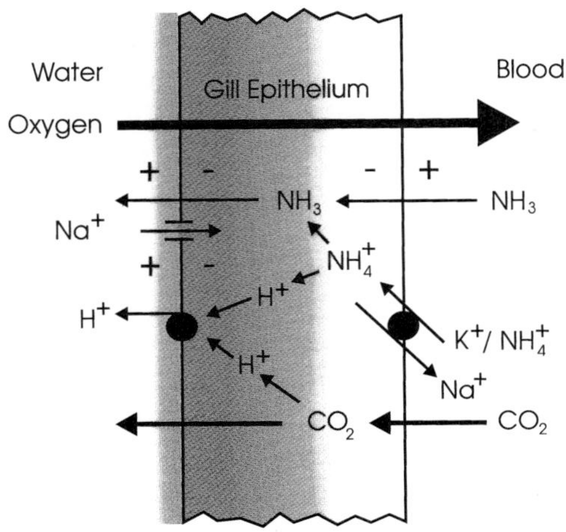

Fig. 2. Model of gas, sodium, and proton transfer across the gill epithelium of a freshwater teleost. ATP-driven pumps are denoted by solid circles and passive diffusion by arrows. The shading indicates the distribution of carbonic anhydrase activity.

across the apical membrane can also affect the electrical potential generated by the pump and therefore affect the proton pump's operation.

0.1 m*M* amiloride had no effect on the putative fish gill proton pump in open-circuit conditions. In frog skin epithelium under open-circuit conditions, however, application of 0.01 m*M* amiloride caused a hyperpolarization of membrane potential, a fall in intracellular sodium ion activity, an inhibition of H^+ excretion, and a decrease in pHi (Harvey and Ehrenfeld, 1986). Accumulation of negative potential in the apical membrane of fish gills was also expected when 0.1 m*M* amiloride was added to the external water and blocked the sodium conductive channels, but this did not result in an inhibition of proton excretion (Lin and Randall, 1991). Amiloride, however, also inhibited chloride influx across fish gills (Perry and Randall, 1981), which might alleviate the buildup of the negative potential against the proton pump. This explanation was supported by the fact that addition of 9-anthroic acid (9-AA) caused a transient recovery of branchial proton excretion under sodium-free conditions. Apical addition of 0.1 m*M* 9-AA inhibited HCO_3^+ secretion initiated by the proton pump in turtle bladder by blocking the recycling of Cl^- via a chloride channel in the apical membrane (Stetson *et al.*, 1985). Application of 9-AA to the apical side of fish gills might inhibit chloride influx and therefore temporarily reduce the electrical gradient against which the proton pump must operate. The effects of amiloride, 9-anthroic acid (a chloride channel blocker), and sodium-free water on branchial proton excretion can be explained by an electrical linkage between Na^+, Cl^-, and H^+ in the gill epithelial compartment. Under open-circuit conditions the negative potential generated by the proton pump will slow down its own operation, unless counterions are available to balance the potential. Na^+ influx across the apical membrane is usually playing this role but reduced Cl^- influx across the apical membrane or increased Cl^- efflux across the basal membrane would have the same effect, and the balance of all of these effects would determine the electric potential imposed on the proton pump. Thus it is possible that there are chloride channels in the apical and/or basal membranes of the gill epithelium.

Chloride transport across the gills has been shown to be SITS sensitive (Perry and Randall, 1981), indicating the presence of a chloride/bicarbonate exchange mechanism in the gill epithelium. If this was located in the basal membrane of the pavement cell, then this cell would be similar to the A-type, acid-secreting cell of the mammalian kidney. Although there are high levels of carbonic anhydrase inside the epithelial cells, none is available to the plasma (Rahim *et al.*, 1988). If there was carbonic anhydrase on both sides of the membrane, as well as an anion channel through the membrane, acid would be rapidly transferred from blood to epithelium.

Thus the absence of carbonic anhydrase on the blood side of the basal membrane is consistent with the presence of a chloride/bicarbonate anion-exchange mechanism on the basal membrane associated with a low proton permeability of the epithelium. The freshwater trout gill chloride cell is thought to possess a chloride/bicarbonate exchange process in the apical membrane (Goss *et al.*, 1992) that may be insensitive to SITS as Lin and Randall (1991) observed no effect of SITS on acid excretion across the gills in short-term experiments. Perry and Randall (1981), however, observed reductions in chloride influx and acid–base effects over a longer time period. One possible explanation of these apparently conflicting results is that the apical anion exchanger in the chloride cell is insensitive to SITS but the basal anion exchanger on the pavement cells is inhibited by SITS. The chloride cell has a somewhat more diffuse localization of the proton-ATPase than the pavement cells (Lin *et al.*, 1994) and, if it has a SITS-insensitive anion exchanger, then it resembles the B-type cell of the mammalian kidney. If gill pavement cells are the functional equivalent of the A-type cell and the chloride cell the equivalent of the B-type cell of the mammalian kidney, then the pavement cell is the acid-excreting cell and the chloride cell is a base-excreting system. However, this is no more than a speculation.

IV. REGULATION OF THE PROTON PUMP

A. pH Gradient

The electrochemical gradient for protons across the membrane is a fundamental regulator of proton transport mediated by the proton pump. This is evidenced by the relationship between net proton excretion across fish gills and the inspired water pH illustrated in Fig. 3. When external water pH was lower than 5.5, the ΔpH across the epithelial membrane is >2.4 pH units (assuming blood pH of fish was 7.9, see Lin and Randall, 1990). The electrochemical gradient for protons was too great for the proton pump to work against, therefore no hydrogen ions were extruded across the gill epithelium. The reversal of gill transepithelial potential associated with low water pH (McWilliams and Potts, 1978; Ye *et al.*, 1991) probably resulted from the shutdown of the proton pump. When water pH is between 5.5 and 8.0, net proton excretion increased linearly with inspired water pH because more protons were transported by the active pump as the proton electrochemical gradient opposing the pump decreased. Above inspired water pH of 8.0, the proton electrochemical

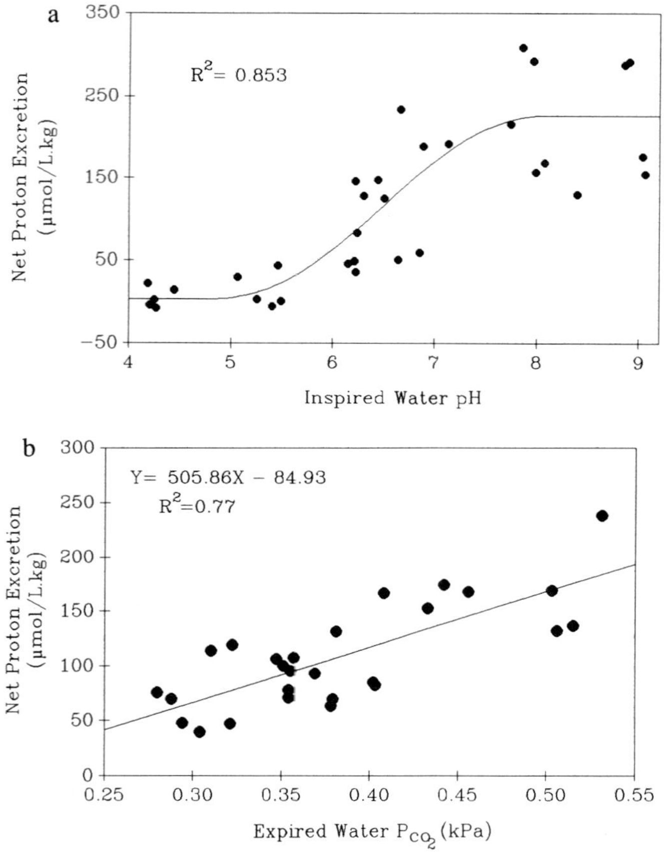

Fig. 3. The relationship between branchial net proton excretion and (a) inspired water pH expressed by a five degree regression curve (R = 0.896) and (b) expired water CO_2 levels. (From Lin and Randall, 1991.)

gradient was favorable to proton excretion, however, branchial proton excretion leveled off, probably because the electrogenic proton pump was now operating at its maximal capacity.

B. CO_2 Availability

Elevated carbon dioxide levels in expired water, which is an indicator of the CO_2 level in venous blood, appear to enhance proton excretion (Fig. 3; Lin and Randall, 1991). This was also observed in toad bladder (Al-Awqati, 1978), turtle bladder (Arruda *et al.*, 1990), and frog skin (Ehrenfeld and Garcia-Romeu, 1977). In all of these epithelia, carbon dioxide provides a source of hydrogen ions for the proton pump when it is hydrated, the reaction catalyzed by carbonic anhydrase in the intracellular compartment. The same theory probably applies to the fish gill epithelium because acetazolamide, a carbonic anhydrase inhibitor, also inhibits proton excretion in fish gills (Lin and Randall, 1991).

Consistent with the observations on proton excretion, hypercapnia treatment in freshwater fish induced a rapid increase in the NEM-sensitive ATPase activity, which stabilized at a level twice that of normocapnia (Fig. 4; Lin and Randall, 1993). Similar hypercapnia treatment in freshwater catfish was reported to cause a marked increase in Na^+ influx, which might be correlated with the increased H^+-ATPase activity (Goss, 1993). The elevated H^+-ATPase activity could be induced by CO_2-mediated proton pump insertion via exocytosis, as demonstrated in turtle bladder epithelium (Cannon *et al.*, 1985; Arruda *et al.*, 1990). High CO_2 levels reduced the intracellular pH of the proton-secreting cell, which increased the intracellular Ca^{2+} concentration and in turn stimulated the fusion of cytoplasmic vesicles containing proton pump into the apical membrane. This would then correct the intracellular acidosis, indicating that the proton pump in freshwater fish gills might play a role in acid–base regulation. Although

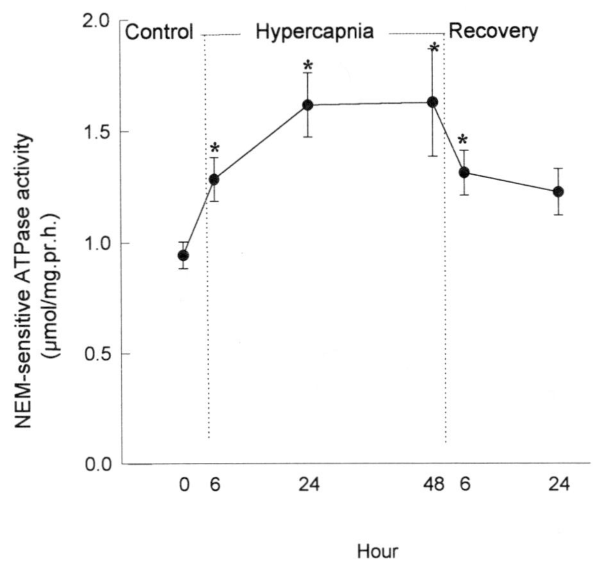

Fig. 4. NEM-sensitive ATPase activity in the gill tissue of freshwater rainbow trout during 48 hr of hypercapnia treatment and 24 hr of recovery. (From Lin and Randall, 1993.)

difficult to quantify, immunofluorescence of the 70-kDa subunit of the proton-ATPase appeared to decrease in the pavement cells in trout exposed to hypercapnia but increased in the chloride cells (Lin *et al.*, 1994). This indicates that modulation of anion exchange and chloride cell function may be more important in acid–base regulation, whereas modulation of pavement cell proton-ATPase has more to do with sodium homeostasis.

C. Sodium Concentrations in Water

Sodium concentration in the external medium is also an important regulator of the operation of the proton pump in fish gills. Acute removal of sodium from the external water resulted in a marked reduction of branchial proton excretion (Fig. 5), probably due to the lack of a counter-ion to diminish the negative potential generated by the proton pump on the inner side of the apical membrane (Lin, 1993). This then increased the electrochemical gradient against which the proton pump must operate and accordingly proton excretion was reduced.

Chronic adaptation of freshwater rainbow trout to different water sodium levels results in different amounts of proton-ATPase recruited onto the gill epithelium, with low sodium levels associated with high H^+-ATPase activity in the gill membrane fractions (Fig. 6; Lin and Randall, 1993).

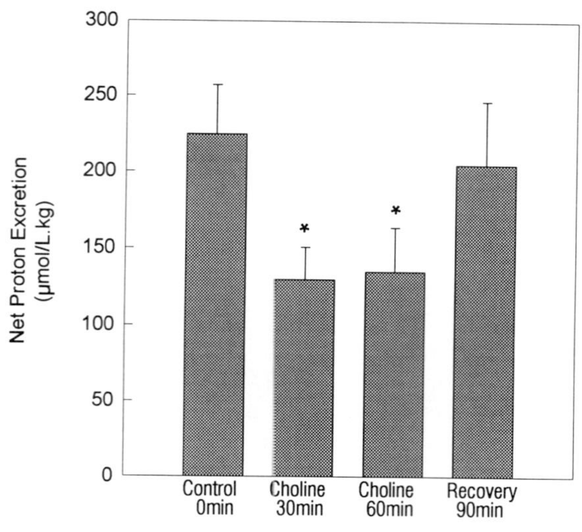

Fig. 5. Branchial proton excretion rate of rainbow trout in control and sodium-free water. Asterisk indicates a significant difference between control and treatment values ($P < 0.05$). Bars show standard errors. (From Lin, 1993.)

NEM-sensitive ATPase activity was significantly lower in fish acclimated to 100 m*M* of Na^+. Seawater-adapted rainbow trout had only one-third of the NEM-sensitive ATPase activity of their freshwater counterparts (Lin and Randall, 1993) and there was marked reduction in immunofluorescence associated with the presence of the 70-kDa subunit of the proton-ATPase (Lin *et al.*, 1994). The fact that fish acclimated to water with low Na^+ level have higher H^+-ATPase activity suggests that the functional significance of the H^+-ATPase in gill pavement cells of freshwater fish is to generate an electrochemical gradient to energize Na^+ uptake from a dilute medium. Despite the fact that the gills of marine fish are organized to deal with the high salt influx, there may still be coupled Na^+/H^+ and HCO_3^-/Cl^- exchange for acid–base regulation (Evans, 1993). There is evidence in hagfish (Evans, 1984) and dogfish (Evans, 1982) that sodium uptake and proton excretion are coupled. There is also some evidence that chloride and bicarbonate transfer are coupled in hagfish (Evans, 1979). It is not known if sodium uptake is coupled to acid excretion via a proton-ATPase in seawater fish or via an amiloride-sensitive Na^+/H^+ exchange process. The sodium gradient is adequate to drive the Na^+/H^+ exchange process in seawater fish. In general, much less is known with respect to acid–base regulation in seawater fish than in freshwater animals.

D. Calcium Concentrations in Water

Variations in Ca^{2+} level in freshwater environments were reported to affect the gill morphology and sodium influx in rainbow trout (Avella *et al.*, 1987). Na^+ influx increased 2.5 times in fish acclimated to FW + 10 m*M* $CaCl_2$ for 15 days and new globular chloride cells appeared and proliferated in the secondary lamellae. High external calcium concentration, however, had a marked stimulating effect on H^+-ATPase activity in freshwater fish gills when the sodium level was low (Fig. 6; Lin and Randall, 1993). When the external medium has a high calcium concentration, Ca^{2+} could enter the cell via a Ca^{2+} channel and reduce the potential gradient generated by the H^+-ATPase and therefore Na^+ influx. The resulting concentration in the cell, however, might stimulate H^+-pump insertion from intracellular vesicles into the apical membrane and so increase Na^+ influx. Exocytosis of proton pump-containing vesicles into the cell membrane is Ca^{2+} dependent in turtle bladder epithelium (Adelsberg and Al-awqati, 1986). The fact that Ca^{2+} level had no effect on proton-ATPase activity in gill epithelium when water Na^+ concentration was high indicates that water sodium level has the predominant regulatory effect on proton pump recruitment.

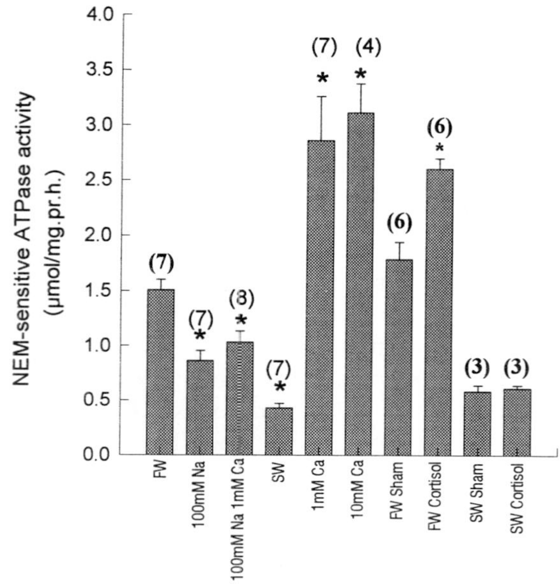

Fig. 6. NEM-sensitive ATPase activity in the gill tissue of rainbow trout acclimated to various Na^+ and Ca^{2+} levels in the external media for 10–14 days, and following elevation of plasma cortisol concentration in both freshwater (FW) and seawater (SW) rainbow trout due to 7 days of chronic cortisol infusion. Asterisk indicates a significant difference between control and treatment values ($P < 0.05$). Bars show standard errors and numbers in parentheses indicate sample size. (From Lin and Randall, 1993.)

E. Cortisol

Mineralocorticoid hormones are involved in the regulation of proton transport via proton pump activity in other tight epithelia. Aldosterone treatments, either long term (7 days) or short term, were reported to stimulate H^+ secretion mediated by H^+-ATPase in the collecting duct of the mammalian kidney (Garg and Narang, 1988; Mujais, 1987). The functionally parallel steroid hormone in fish is cortisol. Chronic cortisol infusion in freshwater rainbow trout caused a 170% increase in plasma cortisol level and a 30% increase in NEM-sensitive ATPase activity in gill tissue (Fig. 6; Lin and Randall, 1993). Seawater-adapted animals, on the other hand, showed no response of NEM-sensitive ATPase activity to a similar cortisol treatment, although their plasma cortisol levels increased fourfold. Perry and Laurent (1989) have shown that plasma cortisol levels rose transiently in fish exposed to deionized water. Daily intramuscular injections of cortisol for 10 days caused an increase in Na^+ uptake (Perry and Laurent, 1989). The 30% increase of H^+-ATPase activity observed in freshwater trout following chronic cortisol treatment is probably responsible for this increased Na^+ uptake. Similar cortisol treatment had no effect on seawater-acclimated rainbow trout, indicating that Na^+ concentration is the predominant regulator of the H^+-ATPase in fish gills.

Ca^{2+} and cortisol stimulate the recruitment of H^+-ATPase to the apical membrane when external Na^+ levels are low, and this is probably associated with the gill morphological changes induced by Ca^{2+} (Avella *et al.*, 1987) and cortisol (Perry and Laurent, 1989) under similar experimental conditions.

V. OTHER ATPases IN FISH GILLS AND THEIR INTERACTIONS WITH PROTON PUMPS

High levels of Na^+-K^+-ATPase (see McCormick, Chapter 11, this volume) have been measured in both freshwater and marine fish. This enzyme is generally considered to be located on the basolateral membrane and functions to remove sodium from the gill epithelium into the blood. In freshwater fish it presumably operates in series with the apical proton pump to energize the uptake of sodium from the water.

Na^+-ATPase was first described by Balm *et al.* (1988) in the branchial membranes of tilapia. It was also called $Na^+/H^+(-NH_4^+)$-ATPase because one of the major characteristics of this ATPase was a sensitivity to NH_4^+. Other properties that distinguished it from Na^+-K^+-ATPase was that it was inhibited by K^+ and amiloride. The authors speculated that Na^+-ATPase was located in the basolateral membrane and that it might play a role in active Na^+ transport across fish gills. Except for the fact that this enzyme is considered to be located on the basal membrane, it is possible that this Na-ATPase is in fact a H^+-ATPase indirectly coupled with Na^+ transport. The Na^+-ATPase and H^+-ATPase have a similar pH optimum of 8.0, and the ratio of Na^+ATPase to Na^+-K^+-ATPase activity altered as tilapia was adapted to either seawater or a low-pH environment (Balm *et al.*, 1988).

Ventrella *et al.* (1992) has studied a Na^+-ATPase in gill membrane fractions of rainbow trout and reported in agreement with Balm *et al.*, (1988), that this ATPase was Na^+ dependent, Mg^{2+}-ATP dependent, and ouabain-insensitive, and had a higher activity in fish adapted to fresh water than to brackish water. They reported, however, that this Na^+-ATPase was enriched in microsomal as well as basolateral fractions. There were two pH optima for Na^+-ATPase in trout, 5.2 and 8.2–8.5.

Flik and Verbost (1993) and Flik *et al.* (Chapter 12, this volume) have reviewed calcium transfer across fish gills. Ca^{2+}-ATPase is located on the basal membrane of chloride cells in freshwater teleosts. Calcium enters the cell through channels in the apical membrane of the chloride cell and is then pumped out of the cell into the blood. There are high levels of proton-ATPase, probably associated with the tubular system of the chloride cell, perhaps involved with acid–base regulation. How this proton-ATPase may or may not interact with the Ca^{2+}-ATPase is not known.

Bornancin *et al.* (1980) reported the presence of a Cl^--HCO_3^--ATPase in the microsomal fraction of the gills of rainbow trout. The enzyme activity was measured in the presence of 0.1 m*M* ouabain and was activated by HCO_3^-, Cl^-, and SO_3^-. SCN^- inhibited the activation by HCO_3^- and Cl^-. In the study of Lin and Randall (1993), 30% of the ATPase activity was NEM-insensitive and was of unknown origin. Unidentified NEM-insensitive ATPase was also detected in mammalian kidney (Ait-Mohamed *et al.,* 1986; Garg and Narang, 1988). The Cl^--HCO_3^--ATPase reported by Bornancin *et al.* (1980) might account for the NEM-insensitive ATPase activity in the gill tissue crude homogenates reported by Lin and Randall (1993). The exact location of this anion-dependent ATPase is unknown except that it is in the plasma membrane. It is possible that Cl^--HCO_3^--ATPase is involved in Cl^-/HCO_3^- exchange and acid–base regulation of freshwater fish and is the anion transporter, insensitive to SITS, in the apical membrane of the chloride cell.

REFERENCES

Adelsberg, J. V., and Al-Awqati, Q. (1986). Regulation of cell pH by Ca^{++}-mediated exocytotic insertion of H^+-ATPases. *J. Cell Biol.* **102,** 1638–1645.

Ait-Mohamed, A. K., Marsy, S., Barlet, C., Khadouri, C., and Doucet, A. (1986). Characterization of *N*-ethylmaleimide-sensitive proton pump in the rat kidney. *J. Biol. Chem.* **261**(27), 12526–12533.

Al-Awqati, Q. (1978). H^+ transport in urinary epithelia. *Am. J. Physiol.* **235,** F77–F88.

Al-Awqati, Q., and Dixon, T. E. (1982). Effect of electrochemical gradients on active H^+ transport in an epithelium. *Curr. Top. Membr. Trans.* **16,** 163–174.

Arruda, J. A. L., Sabatini, S., and Westenfelder, C. (1981). Vanadate inhibits urinary acidification by the turtle bladder. *Kidney Int.* **20,** 772–779.

Arruda, J. A. L., Dytko, G., and Talor, Z. (1990). Stimulation of H^+ secretion by CO_2 in turtle bladder: Role of intracellular pH, exocytosis, and calcium. *Am. J. Physiol.* **258,** R222–R231.

Avella, M., and Bornancin, M. (1989). A new analysis of ammonia and sodium transport through the gills of the freshwater rainbow trout (*Salmo gairdneri*). *J. Exp. Biol.* **142,** 155–175.

Avella, M., Masoni, A., Bornancin, M., and Mayer-Gostan, N. (1987). Gill morphology and sodium influx in the rainbow trout (*Salmo gairdneri*) acclimated to artificial freshwater environments. *J. Exp. Zool.* **241,** 159–169.

Balm, P., Goossen, N., Rijke, S. V. D., and Wendelaar Bonga, S. (1988). Characterization of transport Na-ATPases in gills of freshwater tilapia. *Fish Physiol. Biochem.* **5**(1), 31–38.

Bartels, H. (1989). Freeze-fracture study of the pavement cell in the lamprey gill epithelium. Analogy of membrane structure with the granular cell in the amphibian urinary bladder. *Biol. Cell.* **66,** 165–171.

Beauwens, B., Crabbé, J., and Rentmeesters, M. (1981). Effects of vanadate on the functional properties of the isolated toad bladder. *J. Physiol.* (*London*) **310,** 293–305.

Benos, D. J. (1982). Amiloride: A molecular probe of sodium transport in tissues and cells. *Am. J. Physiol.* **242,** C131–C145.

Bornancin, M., De Renzis, G., and Naon, R. (1980). Cl^--HCO_3-ATPase in gills of the rainbow trout: Evidence for its microsomal localization. *Am. J. Physiol.* **238,** R251–R259.

Bowman, E. J. (1983). Comparison of the vacuolar membrane ATPase of *Neurospora crassa* with the mitochondrial and plasma membrane ATPases. *J. Biol. Chem.* **258**(24), 15238–15244.

Bowman, E. J., Siebers, A., and Altendorf, K. (1988). Bafilomycins: A class of inhibitors of membrane ATPases from microorganisms, animal cells, and plant cells. *Proc. Natl. Acad. Sci. U.S.A.* **85,** 7972–7976.

Brown, D. (1989). Membrane recycling and epithelial cell function. *Am. J. Physiol.* **256,** F1–F12.

Brown, D., Hirsch, S., and Gluck, S. (1988). Localization of a proton-pumping ATPase in rat kidney. *J. Clin. Invest.* **82,** 2114–2126.

Cameron, J. N., and Heisler, N. (1983). Studies of ammonia in the rainbow trout: Physico-chemical parameters, acid–base behaviour, and respiratory clearance. *J. Exp. Biol.* **105,** 107–125.

Cameron, J. N., and Kormanik, G. A. (1982). The acid–base responses of gills and kidneys to infused acid and base loads in the channel catfish, *Ictalurus punctatus*. *J. Exp. Biol.* **99,** 143–160.

Cannon, C., Adelsberg, J. V., Kelly, S., and Al-Awqati, Q. (1985). Carbon-dioxide-induced exocytotic insertion of H^+ pumps in turtle-bladder luminal membrane: Role of cell pH and calcium. *Nature* (*London*) **314**(4), 443–446.

De Sousa, R. C., and Grosso, A. (1981). The mode of action of vasopressin: Membrane microstructure and biological transport. *J. Physiol.* (*Paris*) **77,** 643–669.

Dixon, T. E., Clausen, C., and Coachman, D. (1988). Constitutive and transport-related endocytotic pathways in turtle bladder epithelium. *J. Membr. Biol.* **102,** 49–58.

Eddy, F. B., and Chang, Y. J. (1993). Effects of salinity in relation to migration and development in fish. *In* "The Vertebrate Gas Transport Cascade; Adaptations to Environment and Mode of Life" (J. Eduardo and P. W. Bicudo, eds.). CRC Press, Boca Raton, FL.

Ehrenfeld, J., and Garcia-Romeu, F. (1977). Active hydrogen excretion and sodium absorption through isolated frog skin. *Am. J. Physiol.* **233,** F46–F54.

Ehrenfeld, J., Garcia-Romeu, F., and Harvey, B. J. (1985). Electrogenic active proton pump in *Rana esculenta* skin and its role in sodium ion transport. *J. Physiol.* (*London*) **359,** 331–355.

Ehrenfeld, J., Lacoste, I., and Harvey, B. J. (1989). The key role of the mitochondria-rich cell in Na^+ and H^+ transport across the frog skin epithelium. *Pfluegers Arch.* **414,** 59–67.

Ehrenfeld, J., Lacoste, I., Garcia-Romeu, F., and Harvey, B. (1990). Interdependence of Na^+ and H^+ transport in frog skin. *In* "Animal Nutrition and Transport Processes. 2. Transport, Respiration and Excretion: Comparative and Environmental Aspects" (J. P. Truchot and B. Lahlou, eds.), *Comp. Physiol.,* Vol. 6, pp. 152–170. Karger, Basel.

Evans, D. H. (1979). Fish. *In* "Comparative Physiology of Osmoregulation in Animals" (G. M. O. Maloiy, ed.), p. 305. Academic Press, New York.

Evans, D. H. (1982). Mechanisms of acid extrusion by two marine fishes; the teleost, *Opsanus beta,* and the elasmobranch, *Squalus acanthias. J. Exp. Biol.* **97,** 289.

Evans, D. H. (1984). Gill Na^+/H^+ and Cl^-/HCO_3^- exchange systems evolved before the vertebrates entered fresh water. *J. Exp. Biol.* **113,** 464.

Evans, D. H. (1993). Osmotic and ionic regulation. *In* "The Physiology of Fishes" (D. H. Evans, ed.), CRC Mar. Sci. Ser., pp. 315–341. CRC Press, Boca Raton, FL.

Evans, D. H., More, K. J., and Robbins, S. L. (1989). Modes of ammonia transport across the gill epithelium of the marine teleost fish *Opsanus beta. J. Exp. Biol.* **144,** 339–356.

Flik, G., and Verbost, P. M. (1993). Calcium transport in fish gills and intestine. *J. Exp. Biol.* **184,** 17–29.

Forgac, M. (1989). Structure and function of vacuolar class of ATP-driven proton pumps. *Physiol. Rev.* **69**(3), 765–796.

Garg, L. C., and Narang, N. (1988). Effects of aldosterone on NEM-sensitive ATPase in rabbit nephron segments. *Kidney Int.* **34,** 13–17.

Gluck, S., and Al-Awqati, Q. (1984). An electrogenic proton-translocating adenosine triphosphatase from bovine kidney medulla. *J. Clin. Invest.* **73,** 1704–1710.

Gluck, S., and Caldwell, J. (1987). Immunoaffinity purification and characterization of vacuolar H^+-ATPase from bovine kidney. *J. Biol. Chem.* **262**(32), 15780–15789.

Goss, G. G. (1993). Interrelationships between morphological and physiological mechanisms of acid–base regulation in freshwater fish. Ph.D. Thesis, University of Ottawa.

Goss, G. G., Laurent, P., and Perry, S. F. (1992). Evidence for a morphological component in acid–base regulation during environmental hypercapnia in the brown bullhead (*Ictalurus nebulosus*). *Cell Tissue Res.* **268,** 539–552.

Graber, M., Dixon, T., Coachman, D., and Devine, P. (1989). Acetazolamide inhibits acidification by the turtle bladder independent of cell pH. *Am. J. Physiol.* **256,** F923–F931.

Harvey, B. J., and Ehrenfeld, J. (1986). Regulation of intracellular sodium and pH by the electrogenic H^+ pump in frog skin. *Pfluegers Arch.* **406,** 362–366.

Heisler, N. (1990). Mechanisms of ammonia elimination in fishes. *In* "Animal Nutrition and Transport Processes. 2. Transport, Respiration and Excretion: Comparative and Environmental Aspects" (J. P. Truhot and B. Lahlou, eds.), Comp. Physiol., Vol. 6, pp. 137–151. Karger, Basel.

Hillaby, B. A., and Randall, D. J. (1979). Acute ammonia toxicity and ammonia excretion in rainbow trout (*Salmo gairdneri*). *J. Fish. Res. Board Can.* **36,** 621–629.

Justesen, N. P. B., Dall-Larsen, T., and Klungsoyr, L. (1993). Proton transport and chloride cells in the gill of rainbow trout (*Oncorhynchus mykiss*). *Can. J. Fish. Aquat. Sci.* **50,** 988–995.

Kim, J., Welch, W. J., Cannon, J. K., Tisher, C. C., and Madsen, K. M. (1992). Immunocytochemical response of type A and type B intercalated cells to increased sodium chloride delivery. *Am. J. Physiol.* **262,** F288–F302.

Kleyman, T. R., and Cragoe, E. J., Jr. (1988). Amiloride and its analogs as tools in the study of ion transport. *J. Membr. Biol.* **105,** 1–21.

Klungsoyr, L. (1987). Magnesium ion activated ATPase and proton transport in vesicles from the gill of rainbow trout (*Salmo gairdneri*). *Comp. Biochem. Physiol., B* **88,** 1125–1134.

Knauf, H., Simon, B., and Wais, U. (1976). Non-specific inhibition of membrane-ATPase by amiloride. *Naunyn-Schmiedeberg's Arch. Pharmacol.* **292,** 189–192.

Kultz, D., and Jurss, K. (1993). Biochemical characterization of isolated branchial mitochondria-rich cells of *Oreochromis mossambicus* acclimated to freshwater or hypersaline seawater. *J. Comp. Physiol., B* **163,** 406–412.

Laurent, P., and Dunel, S. (1980). Morphology of gill epithelia in fish. *Am. J. Physiol.* **238,** R147–R159.

Lin, H. (1993). Proton ATPase in fish gills. Ph. D. thesis. University of British Columbia, Vancouver, Canada.

Lin, H., and Randall, D. J. (1990). The effect of varying water pH on the acidification of expired water in rainbow trout. *J. Exp. Biol.* **149,** 149–160.

Lin, H., and Randall, D. J. (1991). Evidence for the presence of an electrogenic proton pump on the trout gill epithelium. *J. Exp. Biol.* **161,** 119–134.

Lin, H., and Randall, D. J. (1993). Proton-ATPase activity in the crude homogenates of fish gill tissue: Inhibitor sensitivity and environmental and hormonal regulation. *J. Exp. Biol.* **180,** 163–174.

Lin, H., Pfeiffer, D. C., Vogl, A. W., Pan, J., and Randall, D. J. (1994). Immunolocalization of proton-ATPase in the gill epithelia of rainbow trout. *J. Exp. Biol.* **195,** 169–183.

Madsen, K. M., and Tisher, C. C. (1985). Structure–function relationship in H^+-secreting epithelia. *Fed. Proc. Fed. Am. Soc. Exp. Biol.* **44,** 2704–2709.

Maetz, J., and Garcia-Romeu, F. (1964). The mechanism of sodium and chloride uptake by the gills of a freshwater fish, *Carassius auratus*. II. Evidence for NH_4^+/Na^+ and HCO_3^-/Cl^- exchanges. *J. Gen. Physiol.* **47,** 1209–1227.

McDonald, D. G., and Wood, C. M. (1981). Branchial and renal acid and ion fluxes in the rainbow trout, *Salmo gairdneri,* at low environmental pH. *J. Exp. Biol.* **93,** 101–118.

McDonald, D. G., Tang, Y., and Boutilier, R. G. (1989). Acid and ion transfer across the gills of fish: Mechanisms and regulation. *Can. J. Zool.* **67,** 3046–3054.

McWilliams, P. G., and Potts, W. T. W. (1978). The effects of pH and calcium concentrations on gill potentials in the brown trout, *Salmo trutta. J. Comp. Physiol.* **126,** 277–286.

Mujais, S. K. (1987). Effects of aldosterone on rat collecting tubule *N*-ethylmaleimide-sensitive adenosine triphosphatase. *J. Lab. Clin. Med.* **109,** 34–39.

Nielsen, R. (1982). A coupled electrogenic Na^+–K^+ pump for mediating transepithelial sodium transport in frog skin. *Curr. Top. Membr. Transp.* **16,** 87–108.

Payan, P. (1978). A study of the Na^+/NH_4^+ exchange across the gill of the perfused head of the trout (*Salmo gairdneri*). *J. Comp. Physiol.* **124,** 181–188.

Pedersen, P. L., and Carafoli, E. (1987). Ion motive ATPases. I. Ubiquity, properties, and significance to cell function. *Trends Biochem.* **12,** 146–150.

Perry, S. F., and Laurent, P. (1989). Adaptational response of rainbow trout to lowered external NaCl concentration: Contribution of the branchial chloride cell. *J. Exp. Biol.* **147,** 147–168.

Perry, S. F., and Randall, D. J. (1981). Effects of amiloride and SITS on branchial ion fluxes in rainbow trout (*Salmo gairdneri*). *J. Exp. Zool.* **215,** 225–228.

Perry, S. F., and Walsh, P. (1989). Metabolism of isolated fish gill cells: Contribution of epithelial chloride cells. *J. Exp. Biol.* **144,** 507–520.

Pisam, M., Caroff, A., and Rambourg, A. (1987). Two types of chloride cells in the gill epithelium of a freshwater-adapted euryhaline fish: *Lebistes reticulatus;* their modifications during adaptation to saltwater. *Am. J. Anat.* **179,** 40–50.

Pisam, M., Auperin, B., Prunet, P., Rentier-Delrue, F., Martial, J., and Rambourg, A. (1993). Effects of prolactin on α and β chloride cells in the gill epithelium of the saltwater adapted tilapia "*Oreochromis niloticus*." *Anat. Rec.* **235,** 275–284.

Potts, W. T. W. (1994). Kinetics of sodium uptake in freshwater animals: A comparison of ion-exchange and proton pump hypotheses. *Am. J. Physiol.* **266,** R315–R327.

Rahim, S., Delaunoy, J. P., and Laurent, P. (1988). Identification and immunocytochemical localization of two different carbonic anhydrase isoenzymes in teleostean fish erythrocyte and gill epithelia. *Histochemistry* **89,** 451–459.

Randall, D. J. (1990). Control and co-ordination of gas exchange in water breathers. *Adv. Comp. Environ. Physiol.* **6,** 253–278.

Rosen, S., and Friedley, N. J. (1973). Carbonic anhydrase activity in *Rana pipiens* skins: Biochemical and histochemical analysis. *Histochemie* **36,** 1–4.

Sardet, C. (1980). Freeze fracture of the gill epithelium of euryhaline teleost fish. *Am. J. Physiol.* **238,** R207–R212.

Schwartz, G. J., and Al-Awqati, Q. (1985). Carbon dioxide causes exocytosis of vesicles containing H^+ pumps in isolated perfused proximal and collecting tubules. *J. Clin. Invest.* **75,** 1638–1644.

Steinmetz, P. R. (1985). Epithelial hydrogen ion transport. *In* "The Kidney: Physiology and Pathophysiology" (D. W. Seldin and G. Giebiach, eds.), pp. 1441–1458. Raven Press, New York.

Steinmetz, P. R. (1986). Cellular organization of urinary acidification. *Am. J. Physiol.* **251,** F173–F187.

Steinmetz, P. R., and Andersen, O. S. (1982). Electrogenic proton transport in epithelial membranes. *J. Membr. Biol.* **65,** 155–174.

Stetson, D. L. (1989). Turtle urinary bladder: Regulation of ion transport by dynamic changes in plasma membrane area. *Am. J. Physiol.* **257,** R973–R981.

Stetson, D. L., Beauwens, R., Palmisano, J., Mitchell, P. P., and Steinmetz, P. R. (1985). A double-membrane model for urinary bicarbonate secretion. *Am. J. Physiol.* **249,** F546–F552.

Sudhof, T. C., Fried, V. A., Stone, D. K., Johnston, P. A., and Xie, X.-S. (1989). Human endomembrane H^+ pump strongly resembles the ATP-synthetase of Archaebacteria. *Proc. Natl. Acad. Sci. U.S.A.* **86,** 6067–6071.

Turrini, F., Sabolic, I., Zimolo, Z., Moewes, B., and Burckhardt, G. (1989). Relation of ATPases in rat renal brush-border membranes to ATP-driven H^+ secretion. *J. Membr. Biol.* **107,** 1–12.

Ventrella, V., Trombetti, F., Pagliarani, A., Trigari, G., Pirini, M., and Borgatti, A. R. (1992). Salinity dependence of the ouabain-insensitive Mg-dependent Na-ATPase in gills of rainbow trout (*Oncorhynchus mykiss* Walbaum) adapted to fresh and brackish water. *Comp. Biochem. Physiol., B* **101,** 1–7.

Wilson, R. W., Wright, P. M., Munger, S., and Wood, C. M. (1994). Ammonia excretion in freshwater rainbow trout (*Oncorhynchus mykiss*) and the importance of gill boundary layer acidification: Lack of evidence for Na^+/NH_4^+ exchange. *J. Exp. Biol.* **191,** 37–58.

Wood, C. M. (1991). Branchial ion and acid–base transfer in freshwater teleost fish: Environmental hyperoxia as a probe. *Physiol. Zool.* **64**(1), 68–102.

Wood, C. M., and LeMoigne, J. (1991). Intracellular acid–base responses to environmental hyperoxia and normoxic recovery in rainbow trout. *Respir. Physiol.* **86,** 91–113.

Wright, P. A., and Wood, C. M. (1985). An analysis of branchial ammonia excretion in the freshwater rainbow trout: Effects of environmental pH change and sodium uptake blockade. *J. Exp. Biol.* **114,** 329–353.

Wright, P. A., Heming, T. A., and Randall, D. J. (1986). Downstream pH changes in water flowing over the gills of rainbow trout. *J. Exp. Biol.* **126,** 499–512.

Wright, S. H. (1991). The interface of animal and aqueous environment: Strategies and constraints on the maintenance of solute balance. *In* "Biochemistry and Molecular Biology of Fishes" (P. W. Hochachka and T. P. Mommsen, eds.), Vol. 1, pp. 165–180. Elsevier, Amsterdam.

Ye, X., Randall, D. J., and He, X. (1991). The effect of acid water on oxygen consumption, circulating catecholamines and blood ionic and acid–base status in rainbow trout (*Salmo gairdneri,* Richardson). *Fish Physiol. Biochem.* **9**(1), 23–30.

10

ULTRASTRUCTURAL AND MORPHOMETRIC STUDIES ON ION AND ACID–BASE TRANSPORT PROCESSES IN FRESHWATER FISH

GREG GOSS, STEVE PERRY, AND PIERRE LAURENT

I. INTRODUCTION

In recent years there has been significant progress in elucidating the mechanisms of ion and acid–base regulation in freshwater fish. It has become apparent that many ion- and acid–base-transporting epithelia with functions analogous to those of the freshwater fish gill are able to alter their morphology in response to variations in their function. In acidic/basic equivalent regulating epithelia, such as the mammalian renal cortical collecting duct (Madsen and Tisher, 1984), the frog skin (Rick, 1992), and the turtle bladder (Kniaz and Arruda, 1990, 1991; Rich *et al.*, 1990), there are discrete changes in the cellular ultrastructure and morphology in response to acid–base disturbances. The freshwater fish gill is likely to undergo similar changes, yet until recently the possibility that the gill remodels during acid–base disturbances had not been investigated. It is well documented that the freshwater fish gill dynamically alters Na^+/acidic equivalent and Cl^-/basic equivalent exchanges in response to acid–base disturbances (Maetz, 1972; Cameron, 1976; Wood *et al.*, 1984), but the transporter(s) involved, their localizations, and the mechanisms controlling them remain uncertain. This review will highlight recent advances in the understanding of how changes in gill morphology, which result from acid–base disturbances, play a key role in acid–base regulation. In addition, we will attempt to provide a framework for future research to further clarify the mechanisms by which fish regulate ion and acid–base balance.

A. Gill Structural and Ultrastructural Characteristics

The gill epithelium contains at least five cell types (Conte, 1969; Dunel-Erb *et al.*, 1982; Laurent, 1984). Mucous cells (MCs) probably are not directly involved in either ion or acid–base regulation, although they may have an indirect role in modulating gill ion transport by providing a microenvironment rich in ions (Handy, 1989). Other cell types found in fish epithelia include serotonergic neuroepithelial cells (Dunel-Erb *et al.*, 1982; Bailly *et al.*, 1989, 1992) and undifferentiated or stem cells for which no definitive roles have been described (Laurent, 1984). This review will focus on the potential roles of the two major cell types present in the fish gill; the pavement or respiratory cell (hereafter abbreviated PC) and the mitochondria-rich chloride cell (hereafter abbreviated CC).

Pavement or respiratory cells make up the largest single fraction, often comprising more than 95% of the gill epithelium. These surface structure and ultrastructural features of these cells have been extensively studied (Hughes, 1979; Olson and Fromm, 1973; Dunel and Laurent, 1980). Their shape is generally squamous, although there exists a small population of roughly columnar pavement cells to which a function has not yet been attributed (Laurent, 1984). Lamellar PCs generally contain few mitochondria, which are considered necessary in actively transporting cells, although they do have other ultrastructural features suggestive of high metabolic activity, including a well-developed Golgi apparatus, an abundance of rough endoplasmic reticulum, and numerous vesicles (Laurent and Dunel, 1980). Their external morphology is characterized by a wide range of morphologies ranging from long elaborate ridges to microvilli-like projections (Perry *et al.*, 1992). The function of these varying morphologies has not been determined, although they are thought to play a role in mucous adhesion (Hughes, 1979; Laurent, 1985).

The primary cell type thought to be responsible for the transepithelial transport of ions in freshwater fish is the chloride cell, also called the ionocyte or mitochondria-rich cell (Laurent, 1984, 1989; Perry and Laurent, 1989). These cells are generally much larger than PCs and maintain their basolateral surface in close proximity to the arteriovenous filamental circulation. The CC has a number of distinct ultrastructural features characteristic of ion-transporting cells. These include large numbers of mitochondria and a dense network of tubules that are believed to be continuous with the basolateral membrane (Doyle and Gorecki, 1961; Philpott, 1966). It was shown *in vitro* that CCs have higher rates of oxidative metabolism in comparison to other gill cell types (Perry and Walsh, 1989), which is likely related to the high levels of Na^+/K^+-ATPase activity in the CCs (Sargent *et al.,* 1975; Naon and Mayer-Gostan, 1983; Perry and Walsh, 1989). A distinguishing feature of the CC is the presence of a vesiculotubular system in the apical portion of the CC, where mitochondria are generally scarce. For a more detailed review of the general characteristics of the fish gill, the reader is referred to Laurent (1984).

B. Structural and Functional Similarities with Other Acid–Base-Secreting Epithelia

The freshwater fish gill shares a number of common functional and morphological features with some well-characterized epithelial surfaces, including the frog skin, toad and turtle urinary bladder, and the cortical collecting duct of the mammalian kidney. These epithelia have an apical (mucosal) surface bathed in a very hypotonic solution (water or urine), whereas the basal (serosal) surfaces are exposed to extracellular fluid. These epithelia function effectively in the dynamic uptake of ions and the excretion of acidic and basic equivalents (Gluck *et al.,* 1982; Madsen and Tisher, 1984, 1986; Kniaz and Arruda, 1990). In addition to its primary role as a gas-exchange organ, the gill serves essentially the same function in freshwater fish as the aforementioned epithelia. The apical surface is bathed by the water whereas the basolateral surface is exposed to extracellular fluid, creating a high concentration gradient for passive flow of ions. Morphologically, the composition of cells is also similar in freshwater fish compared to the aforementioned epithelia. PCs have a slightly different nomenclature depending on the system under study. They have been called the principal cell (cortical collecting duct) or the granular cell (frog skin and toad/turtle bladder). The morphological/physiological features of the principal/granular cells in other species are similar to those listed previously for the PC. The principal/granular cells are the sites of Na^+ uptake through apical membrane Na^+ channels (frog skin: Ehrenfeld and Garcia-Romeu, 1977; Harvey and Ehrenfeld, 1988; Ehrenfeld *et al.,* 1990; turtle bladder: Steinmetz *et al.,* 1967; collecting duct: Benos, 1982; Stokes, 1982; Kristensen and Ussing, 1985), but not the site of either apical Cl^-/HCO_3^- exchange or apical H^+-ATPase activity (Steinmetz,

1987; Weiner and Hamm, 1991). Although the individual components for the Na^+ uptake/H^+ excretion mechanisms are physically separated, they can be indirectly coupled by the apical membrane potential and the requirement of the need to maintain electroneutrality. Therefore, active extrusion of H^+ via an ATPase mechanism is coupled with passive Na^+ entry via a conductive Na^+ channel (Avella and Bornancin, 1989).

In addition to the pavement-type cells, these epithelia possess a mitochondria-rich cell type similar to the CC of the fish gill. These cells have been called intercalated cells (IC) in the kidney or mitochondria-rich (MR) cells in the toad/turtle bladder and frog skin and have been shown to be rich in carbonic anhydrase activity (Rosen, 1970, 1972; Rosen and Freidley, 1973; Dimberg *et al.*, 1981; Rahim *et al.*, 1988). In the MR cells of the turtle bladder and the IC cells of the collecting duct, there is a further subdivision into two distinct cell types (α- and β-subtypes) based on apical membrane characteristics (Stetson and Steinmetz, 1985). Currently, the frog skin is thought to consist of a homogeneous group of α-type MR cells (Ehrenfeld *et al.*, 1990) although subdivision on the basis of intracellular ion concentrations has been suggested (Rick, 1992).

In the cortical collecting duct, the α-type (acid-excreting) cell is characterized by an apical H^+-ATPase, basolateral Cl^-/HCO_3^- exchanges (and Cl^- conductance), and an ability to rapidly endocytose the apical (luminal) membrane in acidic cytoplasmic vesicles (Gluck *et al.*, 1982; Schwartz and Al-Aqwati, 1985; Schwartz *et al.*, 1985; Al-Aqwati, 1986; Brown *et al.*, 1988a,b). Morphologically, they display numerous microplicae on their apical surface and possess numerous rod-shaped intramembrane particles on the apical membrane (Stetson and Steinmetz, 1985). During hypercapnia (elevated plasma P_{CO_2}), these cytoplasmic vesicles can rapidly reinsert themselves into the luminal membrane (Cannon *et al.*, 1985) and they display more extensive microplicae on their apical surface (Fritsche *et al.*, 1991). It has been shown immunocytochemically that these vesicles contain the H^+-ATPase (Brown *et al.*, 1987) and, therefore, this system allows for rapid modulation of proton secretion.

A second type of MR cell is the β-type (base-secreting), which is characterized as having apical Cl^-/HCO_3^- exchanges and basolateral H^+-ATPase (Stetson and Steinmetz, 1985), that is, a reversed polarity to that of the α-type cell. These cells do not endocytose vesicles (Schwartz *et al.*, 1985, 1988) and are characterized by their ability to bind peanut agglutinin to surface receptors (Schuster *et al.*, 1986). These cells do not have apical microplicae (Fritsche *et al.*, 1991) and do not exhibit apical membrane-associated rod-shaped particles (Stetson and Steinmetz, 1985).

C. Morphology and Acid–Base Regulation

Although the responses of the gill epithelial Cl^-/HCO_3^- and Na^+/H^+ (H^+ pump/Na^+ channel) exchanges to changes in internal acid–base status have been extensively studied, the mechanisms by which the responses occur and their regulation remain unresolved (Potts, 1994). Modification of the morphological characteristics of other "tight" epithelia has been shown to occur during acid–base disturbances. In turtle bladder, regulation of acid excretion is accomplished by rapid insertion of H^+ pumps into the luminal membrane, thereby altering the apical surface area of the α mitochondria-rich (MR) cells during acidosis (Gluck *et al.*, 1982). Similar changes occur in α-type intercalated cell in the mammalian kidney (Madsen and Tisher, 1986; Satlin and Schwartz, 1989) and in the frog skin (Ehrenfeld *et al.*, 1990). Cameron and Iwama (1987) first suggested a relationship between gill morphology and acid–base regulation in fish based on results of a study using hypercapnic catfish (*Ictalurus punctatus*). The theory behind morphological alteration of the gill epithelia as a mechanism of acid–base regulation is straightforward. If the transporters responsible for acid–base regulation are present on a particular cell type, then by alteration of the apical exposure of those cell types, the availability and, hence, the activity of the transporters can be modulated.

II. PHYSIOLOGICAL AND MORPHOLOGICAL RESPONSES TO ACID–BASE DISTURBANCES

A. Chloride Cells

1. Changes in CC Morphology during Acid–Base Disturbances

The results presented in this review stem mainly from studies in our lab involving the induction of acid–base disturbances in freshwater fish. The means used to induce these acid–base disturbances and their physiological consequences have been extensively characterized and are only briefly described here. Furthermore, we discuss the results from three different species: the rainbow trout, *Oncorhynchus mykiss,* the brown bullhead catfish, *Ictalurus nebulosus,* and the American eel, *Anguilla rostrata*. Each of these species presents different life strategies and their responses to acid–base disturbances provide insight into the mechanisms responsible for acid–base regulation.

Exposure to hypercapnia (elevated P_{CO_2}; e.g., Cameron, 1976; Perry *et al.*, 1987) or hyperoxia (elevated P_{CO_2}; e.g., Wood and Jackson, 1980; Wood *et al.*, 1984; Goss and Wood, 1990a) are two common means of inducing an increase in arterial P_{CO_2} resulting in a respiratory acidosis. This is compensated for over 48–96 h by elevation in the plasma [HCO_3^-]. When the hyperoxic or hypercapnic stimulus is removed, P_{CO_2} declines rapidly, leaving an excess of [HCO_3^-] in the extracellular compartment, resulting in a metabolic alkalosis. Alternatively, infusion of "fixed" acids [e.g., HCl: Boutilier *et al.*, 1986; Goss and Wood, 1991; $(NH_4)_2SO_4$: McDonald and Prior, 1988] or bases (e.g., $NaHCO_3$: Claiborne and Heisler, 1986; McDonald and Prior, 1988; Goss and Wood, 1990b) induces purely metabolic alterations in the acid–base status of the fish (i.e., P_{CO_2} is virtually constant) and is useful in avoiding the possible effects of changes in the perfusive and convective properties of the gill surface that may occur as a result of hypercapnia or hyperoxia (Wood and Jackson, 1980).

A summary of the effects of various acid–base disturbances on the unidirectional uptake of Na^+ ($J_{in}^{Na^+}$) and Cl^- ($J_{in}^{Cl^-}$) and gill morphology (chloride cell fractional area, CCFA) is shown in Fig. 1 and 2. In Fig. 1, a respiratory acidosis induced by either hypercapnia or hyperoxia, in either catfish or trout, resulted in similar responses. The rate of $J_{in}^{Na^+}$ was increased initially, whereas $J_{in}^{Cl^-}$ and CCFA were decreased simultaneously. The magnitude of the response was greater in catfish than in trout. $J_{in}^{Na^+}$ was initially stimulated (150% increase; Fig. 1A) whereas $J_{in}^{Cl^-}$ was rapidly reduced to 15% of the initial control value (Fig. 1B). By 48 h, the acidosis was compensated and $J_{in}^{Na^+}$ was reduced to a level near that of $J_{in}^{Cl^-}$. These alterations in ion fluxes were accompanied by a large reduction in CCFA, which was reduced to 50% of the initial value after 6 h and only 10% of control after 48 h of hypercapnia (Fig. 1C).

In trout, initial exposure to hypercapnia resulted in no significant changes in either $J_{in}^{Na^+}$ or $J_{in}^{Cl^-}$ (Figs. 1D and 1E) despite a 40% drop in CCFA (Fig. 1F). In the posthypercapnic period, there was a large increase in CCFA on the filament epithelium in concert with an increase in $J_{in}^{Cl^-}$ and a drop in $J_{in}^{Na^+}$. Similar responses were noted during exposure to environmental hyperoxia (Figs. 1G–1I). By 70 h, the acidosis was compensated and $J_{in}^{Na^+}$ was reduced equivalent to $J_{in}^{Cl^-}$. Exposure to hyperoxia also resulted in a 40% decrease in CCFA by 70 h. In the posthyperoxic period, $J_{in}^{Cl^-}$ and CCFA were both increased dramatically whereas $J_{in}^{Na^+}$ was reduced further. During infusion of $NaHCO_3$, CCFA increased by 70%, coinciding with a large increase in $J_{in}^{Cl^-}$ and a decrease in $J_{in}^{Na^+}$ (Fig. 2).

Thus, a clear pattern has emerged with respect to gill CC morphology during acid–base disturbances. During acidosis, the surface area of CCs

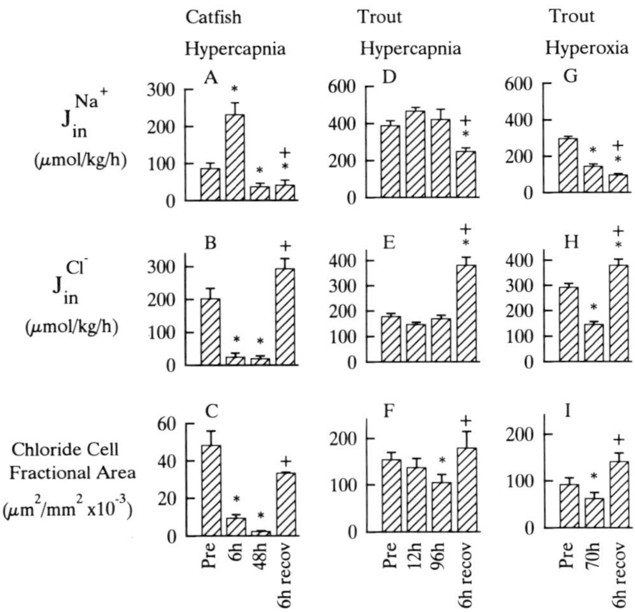

Fig. 1. The effect of hypercapnia and its subsequent removal (A–C: catfish, *Ictalurus nebulosus*, P_{CO_2} = 15 Torr, n = 6; D–F: rainbow trout, *Oncorhynchus mykiss*, P_{CO_2} = 7.5 Torr, n = 6) or hyperoxia and its subsequent removal (G–I: 70 h, $P_{O_2} > 500$ Torr, trout, *Oncorhynchus mykiss*, n = 7) on whole-body influx of chloride ($J_{in}^{Cl^-}$), whole-body influx of sodium ($J_{in}^{Na^+}$), and chloride cell fractional area (CCFA). Asterisk indicates significantly different from control values, plus sign indicates significantly different from previous point (recovery only), $P < 0.05$. Means ± 1 S.E.M. Fluxes for catfish hypercapnia were from original data of Goss *et al.* (1992); fluxes for trout hypercapnia/hyperoxia experiments were calculated from original data using ambient [NaCl] and measured K_m and J_{max} values. (Data from Goss and Perry, 1993; Goss *et al.*, 1994a.)

exposed to the water is reduced, whereas during alkalosis the surface area is increased. Given that the CC is the presumed site of branchial Cl^-/HCO_3^- exchange, such morphological responses can explain, at least in part, the modulation of Cl^-/HCO_3^- exchange during acid–base disturbances. The loss of CC surface area during acidosis serves to reduce the numbers of Cl^-/HCO_3^- exchangers exposed to the water and thus lower the rate of HCO_3^- excretion, which is the predominant response to correct acidosis. The increase in CC surface area during alkalosis serves to increase the numbers of Cl^-/HCO_3^- exchange sites and therefore enhances HCO_3^- excretion.

The eel is unusual among freshwater fish because it lacks appreciable Cl^- uptake, a feature that is clearly evident (Fig. 2H) (Kirsch, 1972; Hyde

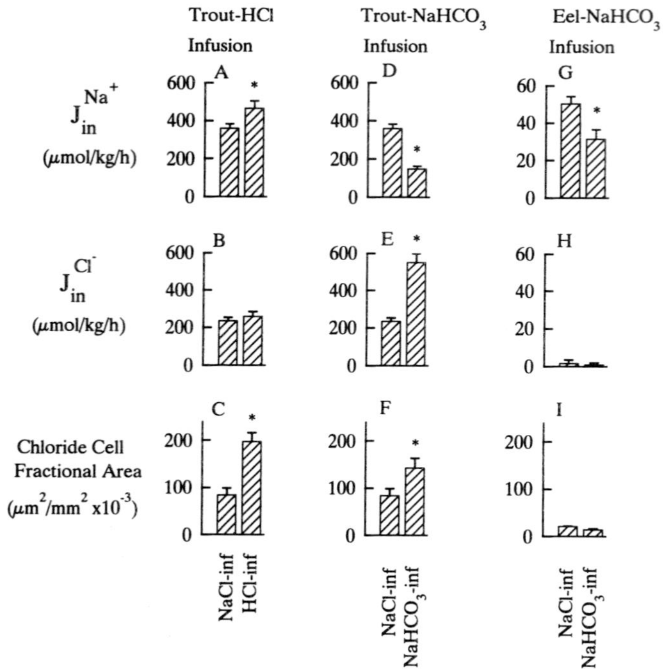

Fig. 2. The effects of infusion of 140 mmol/liter NaCl (sham control, $n = 10$, 393 ± 14 μmol/kg/h), 140 mmol/liter $NaHCO_3$ ($n = 7$, 409 ± 15 μmol/kg/h), and 70 mmol/liter HCl plus 70 mmol/liter NaCl ($n = 7$, 419 ± 18.2 μmol/kg/h) in rainbow trout, *Oncorhynchus mykiss* (A–F), and infusion of $NaHCO_3$ ($n = 7$, 957 ± 121 μmol/kg/h) in American eel, *Anguilla rostrata* (G–I), on whole-body influx of chloride ($J_{in}^{Cl^-}$), whole-body influx of sodium ($J_{in}^{Na^+}$), and chloride cell fractional area. Asterisk indicates significantly different from control values, $P < 0.05$. Means ± 1 S.E.M. Fluxes for infusion experiments were calculated from original data using ambient [NaCl] and measured K_m and J_{max} values. (Data from Goss *et al.*, 1994a; Goss and Perry, 1994.)

and Perry, 1987, 1989). $J_{in}^{Cl^-}$ is less than 1% of the values reported for trout under resting conditions. In addition, the eel is unable to increase branchial Cl^-/HCO_3^- exchange as a mechanism to compensate metabolic alkalosis. The eel relies on other mechanisms to regulate blood pH, including adjustments of the branchial Na^+ uptake mechanism and efflux regulation (Hyde and Perry, 1987, 1989; Goss and Perry, 1994). In addition to the differences in the fluxes, there are also significant differences between trout and eel in the surface area of CCs exposed to the water. Eels display few, small CCs on the gill filament epithelium, and hence CCFA is much lower than in trout (Perry *et al.*, 1992). The combination of very low

CCFA and negligible $J_{in}^{Cl^-}$ reinforces our contention that the CC is the site of $J_{in}^{Cl^-}$ (via Cl^-/HCO_3^- exchange).

The responses of rainbow trout to hypercapnic/hyperoxic acidosis and posthypercapnic/posthyperoxic alkalosis are similar to those of the catfish but attenuated in magnitude. Therefore, we believe that alteration in gill morphology is likely a general response to acid–base disturbances and suggest that the differences in magnitude noted between catfish and trout may be the result of different life histories. The rainbow trout is an active, pelagic species that would rarely encounter environmental hypercapnia, whereas the catfish is a benthic species inhabiting ponds and lakes that are subject to large, rapid fluctuations in P_{CO_2}. Therefore, the catfish is more likely to require a rapid, coordinated response to alteration in environmental P_{CO_2}. When the fish wishes to reduce $J_{in}^{Cl^-}$ as a means of compensating acidosis, removal of the Cl^-/HCO_3^- exchangers from the external surface through covering by adjacent PCs or removal altogether from the membrane is an available mechanism. This will act to reduce the number of transporters available and serve to lower the rate of Cl^-/HCO_3^- exchange. The converse would be true during a metabolic alkalosis.

Figure 3 shows representative scanning electron microscope (SEM) and transmission electron microscope (TEM) photographs of catfish exposed to 48 h of hypercapnia and the 6-h recovery period, respectively, whereas Fig. 4 shows representative SEM and TEM photographs from rainbow trout exposed to hyperoxia. In both conditions, exposure to a respiratory acidosis resulted in a reduction in CCs on the filament epithelium (Figs. 3B, 3E, 4B, and 4E) compared to control conditions (Figs. 3A, 3B, 4A, and 4B), whereas a posthypercapnic/posthyperoxic metabolic alkalosis resulted in an increase in the appearance of CCs on the filament epithelium (Figs. 3C, 3F, 4C, and 4F).

2. Effect of Varying CC Fractional Area on Rates of Acid–Base Compensation

Adaptation to ion-poor fresh water (Hobe *et al.*, 1984; Perry and Laurent, 1989; Avella and Bornancin, 1990), cortisol treatment (Perry and Wood, 1985; Laurent and Perry, 1990; Madsen, 1990a), growth hormone treatment or combined cortisol/growth hormone treatment (Madsen, 1990b), and cortisol/thyroxine treatment (Madsen, 1990c) have all been shown to affect the number of CCs on the gill epithelial surface. Laurent and Perry (1990) demonstrated that elevation of plasma cortisol through daily intramuscular injection resulted in increased rates of uptake for Cl^-. The increases in $J_{in}^{Cl^-}$ were correlated with increases in filament CC

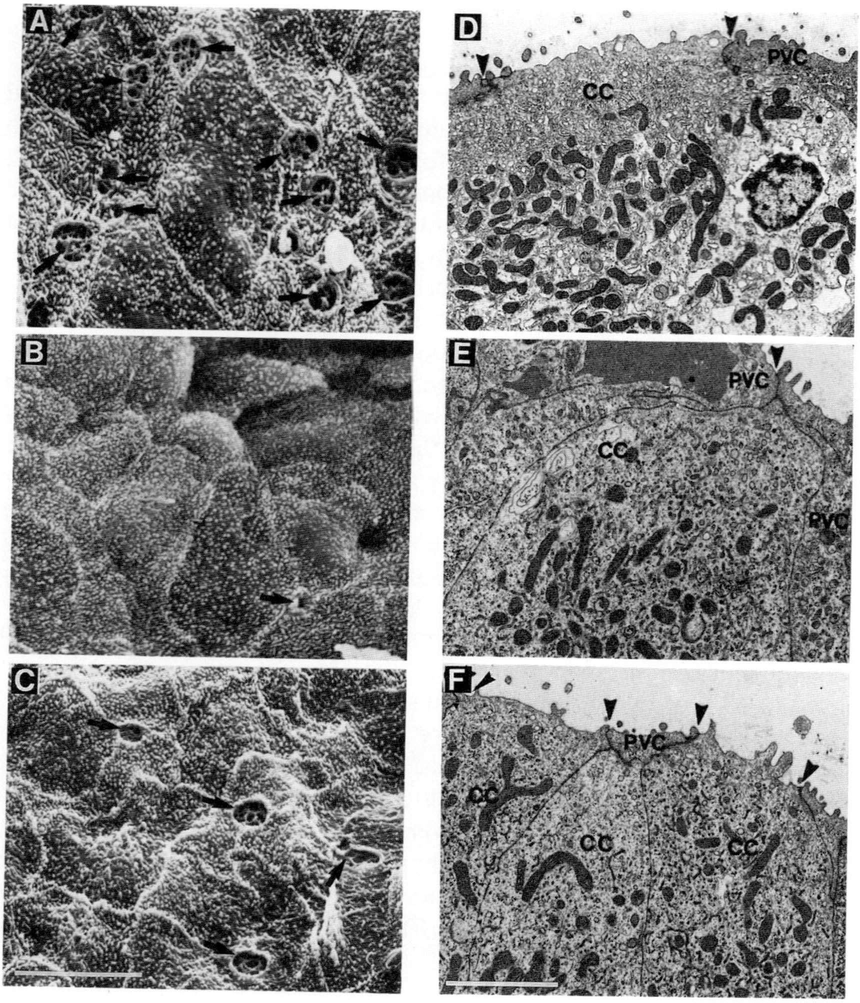

Fig. 3. Representative scanning electron microscope (SEM, A–C) and transmission electron microscope (TEM, D–F) photographs of the filamental epithelium for catfish, *Ictalurus nebulosus,* under control conditions (A,D), after 48 h exposure to hypercapnia (B,E), and 6 h after removal of the hypercapnic stimulus (C,F). Note the dramatic reduction in the CC apical surface area and change in the appearance of the CCs after exposure to hypercapnia, whereas the 6 h posthypercapnia CC fractional area is increased. Note the atypical appearance of the CCs on the catfish epithelium in the posthypercapnic period. In A–C, arrows indicate chloride cells; CC, chloride cell; PVC, pavement cell. (A–C) bar = 10 μm, (D–F) bar = 1 μm. Arrowheads indicate cell–cell junctions (D–F only). See text for further details. [From "Relationships between ion and acid–base regulation in freshwater fish," by Goss, G. G., Wood, C. M., Perry, S. F., and Laurent, P. *J. Exp. Zool.* **263,** 143–159. Copyright © 1992 by Wiley-Liss, Inc. Reprinted by permission of John Wiley & Sons, Inc.]

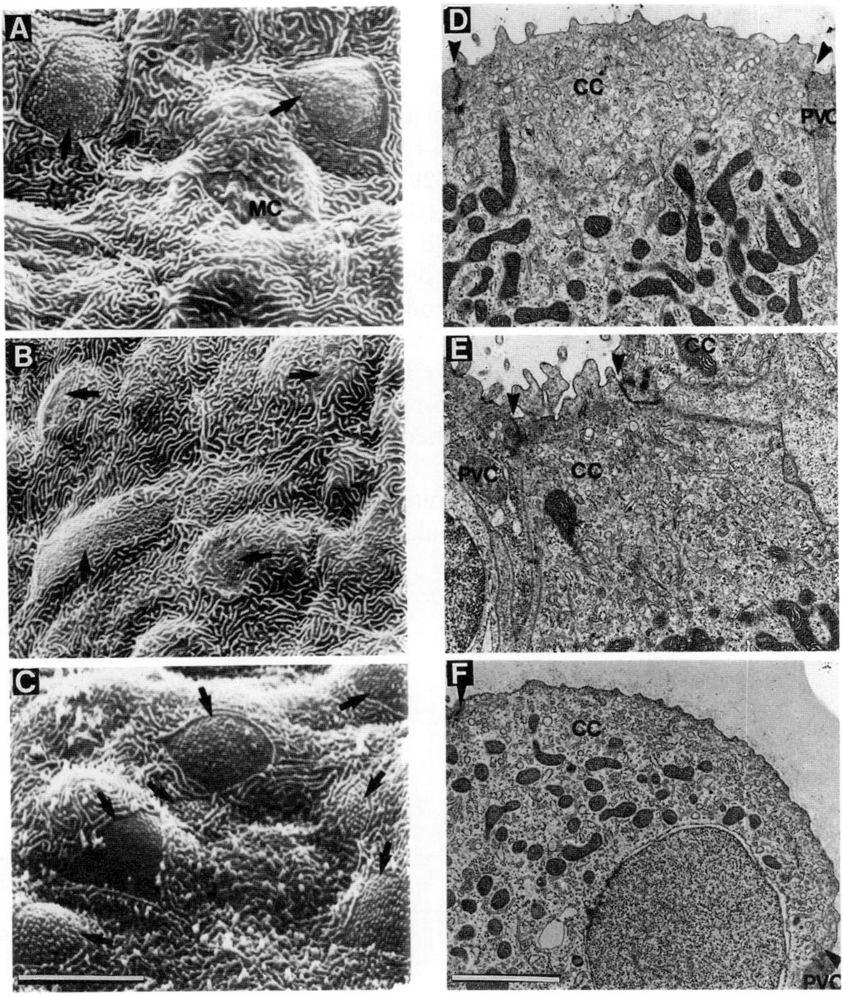

Fig. 4. Representative scanning electron microscope (SEM, A–C) and transmission electron microscope (TEM, D–F) photographs of the filamental epithelium for rainbow trout, *Oncorhynchus mykiss,* under control conditions (A,D), after 70 h exposure to hyperoxia (B,E), and 6 h after removal of the hyperoxic stimulus (C,F). Note the reduction in the CC apical surface area after prolonged exposure to hyperoxia, whereas the 6 h posthyperoxia CC fractional area is increased. In A–C, arrows indicate chloride cells; CC, chloride cell; PVC, pavement cell; MC, mucus cell. (A–C) bar = 10 μm, (D–F) bar = 1 μm. Arrowheads indicate cell–cell junctions (D–F only). See text for further details. [From "Relationships between ion and acid–base regulation in freshwater fish," by Goss, G. G., Wood, C. M., Perry, S. F., and Laurent, P. *J. Exp. Zool.* **263,** 143–159. Copyright © 1992 by Wiley-Liss, Inc. Reprinted by permission of John Wiley & Sons, Inc.]

fractional area. We investigated whether an increase in CCFA might allow an increased rate of compensation from an acid–base disturbance (Perry and Goss, 1994). Fish were injected with a combination of cortisol/ovine growth hormone (cort/oGH) over a period of 10 days according to the protocol of Madsen (1990b). This resulted in a large increase in CCFA from 83,951 ± 9789 $\mu m^2/mm^2$ to 296,512 ± 14,287 $\mu m^2/mm^2$ (Fig. 5A). A severe metabolic alkalosis was induced by intra-arterial infusion of 140 m*M* $NaHCO_3$ (800 μmol/kg/h for 24 h) in both control and cort/oGH-treated fish. However, the degree of alkalosis in the cort/oGH-treated fish was attenuated. In control fish, blood pH rose from 8.05 to 8.53 (ΔpH = 0.48) due to an increase in plasma [HCO_3^-] from 5.9 to 22.2 mmol/liter. In the cort/oGH-treated fish, the alkalosis was less severe; blood pH only rose from a preinfusion value of 7.78 to 8.11 after 24 h of infusion (ΔpH = 0.33), whereas plasma [HCO_3^-] only increased from 5.5 to 13.9 mmol/liter.

The protective effect of an increased CCFA was supported further by the fact that branchial basic equivalent excretion was increased substan-

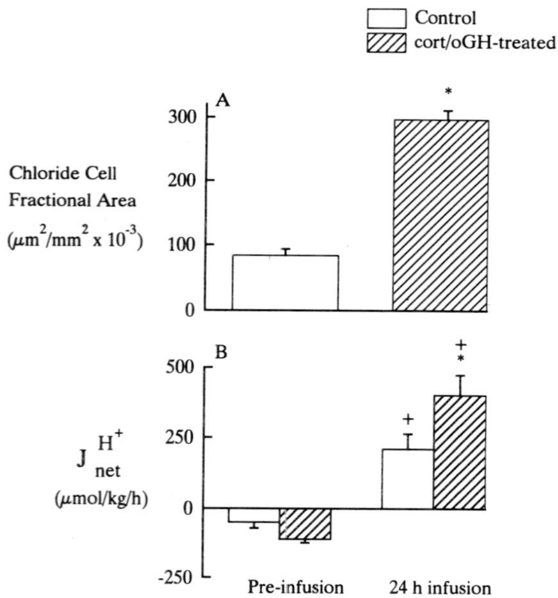

Fig. 5. The effects of 10 days cortisol/ovine growth hormone treatment on (top) the fractional area of chloride cells exposed on the filamental epithelium and (bottom) the rate of acid uptake/base excretion ($J_{net}^{H^+}$) prior to and after 24 h infusion of 140 m*M* $NaHCO_3$. Asterisk indicates significantly different from control values, plus sign indicates significantly different from preinfusion value, $P < 0.05$. (Data from Goss and Perry, 1994.)

tially in the cort/oGH-treated fish (Fig. 5B). Under control preinfusion conditions, there was a net acidic equivalent loss ($-J_{net}^{H^+}$) due to the continual metabolic production of protons. This was slightly higher in the cort/oGH-treated fish, likely due to increased metabolism. After 24 h of infusion, there was an increase in net basic equivalent excretion ($+J_{net}^{H^+}$) in both groups; however, this value was greater in the cort/oGH-treated animals. These data suggest that an increased CCFA preadapts fish to excrete excess base, a predicted response if the CC is the site of HCO_3^- excretion.

The response of the trout gill to intravascular infusion of acid is inconsistent with the morphological model for acid–base regulation. Continuous infusion of HCl produced an unexpected 135% rise in the CCFA in rainbow trout (Fig. 2C). The physiological significance of this response is unclear given our contention for the role of the CC in base (HCO_3^-) excretion via Cl^-/HCO_3^- exchange. Clearly, an increase in the CCFA would act to increase Cl^-/HCO_3^- exchange capacity, which would be detrimental to the regulation of the metabolic acidosis. However, this observation helps to explain the previous observations of Goss and Wood (1991) that HCl infusion was not associated with reduced Cl^- uptake, but instead the acidosis was compensated by large increases in Na^+ uptake and hence "acid" excretion. A possible explanation for the increase in CCFA is that acid infusion increases plasma cortisol levels owing to the stressful nature of the protocol and the well-documented effect of increased stress on cortisol release (Pickering *et al.,* 1991). Increases in plasma cortisol are known to cause increases in CCFA (Perry and Wood, 1985; Laurent and Perry, 1990; Madsen, 1990a). An alternative explanation is that the CCs on the trout gill epithelium are not a homogeneous population of β-type CCs but instead are a mixed population of both β- and α-type MR cells. The increase in CCFA noted during acid infusion may therefore represent a selective increase in α-type CCs (see Section II,A,4).

3. Effect of Environmental pH on Gill Morphology

The importance of variations in gill morphology in response to exposure to variation in environmental pH has received a great deal of attention. Exposure to low or high pH results in changes in the driving gradients for acidic and basic equivalents. During exposure to low environmental pH, there is an increase in the electrochemical gradients favoring net retention/gain of acidic equivalents and a net loss of basic equivalents from the fish. The converse is true during exposure to elevated environmental pH. These conditions create special long-term challenges to the fish to maintain blood acid–base status.

Acute exposure to acidified water (pH 5.0–5.5) and/or ion-poor water has been shown to increase the number and distribution and to change the morphology of the chloride cells on the gills of fathead minnows (*Pimephales promelas;* Leino and McCormick, 1984) and rainbow trout (*Oncorhynchus mykiss;* Perry and Laurent, 1989). At lower environmental pH, pathological changes (peeling or sloughing of cells, cellular necrosis) begin to occur in the gill structure. However, more acid-resistant species such as the yellow perch (*Perca flavacens*) do not display these pathological features until much lower pH (pH 4.0; Leino *et al.,* 1987). Species-specific differences in acid tolerance have been correlated to both the density of chloride cells displayed on the gill and the length of the tight junctions between adjacent gill pavement cells (McDonald *et al.,* 1991).

In addition to the changes in the driving gradients, many parallel effects during exposure to low pH complicate the interpretation of the noted changes in gill morphology. Environmental Ca^{2+} is known to have a protective effect in fish exposed to acid water (Playle *et al.,* 1989) and has been shown to reduce the damage occurring during acid exposure (Leino *et al.,* 1987). A further complication is the common presence of toxic metal cations in acidified water that have been shown to exacerbate the ionoregulatory disturbances (e.g., Al^{3+}: Playle *et al.,* 1989; Audet and Wood, 1993; Cu^{2+}: Lauren and McDonald, 1987) and that are associated with alterations in gill morphology (Al^{3+}: Tietge *et al.,* 1988). Exposure to low environmental pH has been demonstrated to increase the paracellular permeabilities and therefore the net loss of both Na^+ and Cl^- from the fish (McDonald *et al.,* 1991). The increase in permeabilities has been correlated to a 25% decrease in the depth of the tight junctions between CCs and PVCs (Freda *et al.,* 1991). This results in an increase in the net loss of both major ions and necessitates an increase in the mechanisms for ionic uptake irrespective of the changes required to compensate blood acid–base status. Differentiation of these known pathological effects of acute low pH exposure from the physiological/morphological changes required to effect a change in acid–base status is required to compare the acid–base compensatory mechanisms invoked by low pH to those seen during hyperoxia, hypercapnia, or exercise.

Whereas exposure to low pH results in an increase in the requirements for H^+ secretion/base uptake, exposure to elevated environmental pH requires the fish to increase the capacity to either excrete base or take up protons from the water. In rainbow trout (*Oncorhynchus mykiss*) acutely exposed to pH 9.5, branchial $J_{in}^{Na^+}$ and $J_{in}^{Cl^-}$ were severely inhibited. However, after 3 days, there was a 4-fold increase in the fractional area of CCs that was correlated with a return of $J_{in}^{Cl^-}$ to preexposure levels while $J_{in}Na^+$ remained depressed (Wilkie and Wood, 1994). Furthermore,

Wilkie *et al.* (1994) have demonstrated that Lahonton cutthroat trout (*Oncorhynchus clarki henshawi*), a native salmonid species, that is abruptly transferred into the alkaline (pH 9.4) Pyramid Lake in Nevada after spending its juvenile stages in a freshwater environment (pH 8.4) progressively increases the fractional surface area of CCs after this transfer. After 3 days the increase was 4-fold, and by 2 years postmigration, the increase was 20-fold. These increases were accomplished through increases in both the density and surface area of CCs. If the CC is the site of Cl^-/HCO_3^- exchange and hence base excretion, as we propose, then this response would be adaptive and aid in the clearance of base from the fish.

4. Chloride Cell Subtypes

There are discrete inter- and intraspecific differences in the appearance of the CC between species and their responsiveness to variation in acid–base status. The CCs located on the filament epithelium in the brown bullhead catfish (*Ictalurus nebulosus*) appear as highly ornamented pits in the SEM and appear to respond to acid–base disturbances as a relatively homogeneous group (Figs. 1C and 2). The eel *Anguilla rostrata* also appears to have a homogeneous group of CCs but they are not responsive to acid–base disturbances (Fig. 2I), which is consistent with their low and unchanging rates of $J_{in}^{Cl^-}$. Eel CCs are small compared to those of trout, and their surface appears highly ornamented in the SEM.

Attempts have been made to subdivide the CC population in freshwater fish into different distinct subtypes based solely on their morphological appearance and locations. Pisam *et al.* (1987) have distinguished two types of cells in the gills of freshwater guppies (*Lebistes reticulatus*). In their study, α-CCs were lighter staining, elongated cells located at the base of the lamellae and lay in close contact with the capillaries. (Note: although they named their two subtypes α- and β-CCs, there was no attempt to correlate their structures with function or to relate these subtypes to the α- and β-subtypes of mitochondria-rich/intercalated cells found in turtle bladder and the collecting duct, respectively.) The α-CCs had more extensive basolateral infoldings forming an extended network. The vesiculotubular network was less prominent. The β-CCs were darker staining and more ovoid in shape and were located primarily in the interlamellar regions of the gill. The vesiculotubular network was quite prominent in these cells, which were in contact only with the central venous sinus. Pisam *et al.* (1987) suggested that during seawater adaptation, α-CCs transformed into seawater CCs whereas β-CCs degenerated and disappeared. They later found that chronic injections of prolactin, a freshwater-adapting hormone, resulted in the appearance of β-type cells in seawater-adapted tilapia (*Oreochromis niloticus*), suggesting that their identification of a β-type CC is involved in freshwater ionoregulation (Pisam *et al.*, 1993).

Franklin and Davison (1989), Perry and Laurent (1989), Perry *et al.* (1992), and Goss and Perry (1994) have all noted that in salmonids there appear to be varying surface morphologies displayed by the CC. Examples of these morphologies for rainbow trout CCs are shown in Fig. 6. One

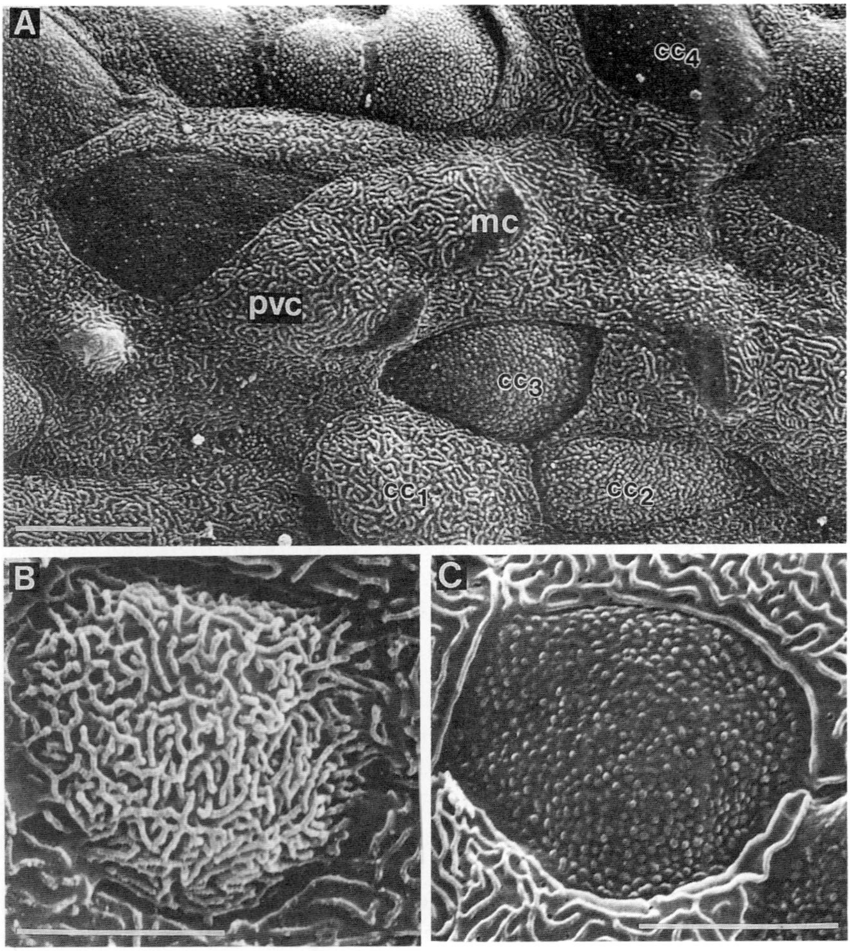

Fig. 6. Representative low- and high-magnification scanning electron microscopy (SEM) photographs of the varying surface morphologies of CCs noted on filament epithelium of rainbow trout (A). One type of CC (CC_1, part B) displayed a dense network of apical microvilli/microridges (A) whereas others (CC_2, part C) displayed varying densities of a network of apical microvilli/microridges (A,C). PVC, pavement cell; MC, mucus cell. (A) bar = 10 μm, (B, C) bar = 5 μm.

type (CC_1) has a dense network of apical membrane microvilli/microridges (Figs. 6A and 6B), whereas others (CC_2) display a network of apical microridges with varying densities (Fig. 6C). However, note that these surface morphologies do not represent two widely different populations, but instead represent a continuum across which an arbitrary division has been placed. Examination of TEM photographs (Fig. 7) from fish displaying disparate CC morphologies reveals noticeable differences in the apical regions. In Fig. 7A, the tubular area is easy to identify. In Fig. 7B, the same area is overcrowded with tubules and the mitochondria have a more disorganized internal structure suggestive of a reduced functional capacity. The reasons for these differences remain in dispute. Although there are differences in staining and site-specific localizations, factors that have been used by others (Pisam *et al.*, 1987, 1993) to distinguish subtypes of CCs, an alternate explanation to the varying staining characteristics of the CC has been put forward by Wendelaar-Bonga and van der Meij (1989). These authors have suggested that the varying degrees of staining noted by Pisam *et al.* (1987) and the variability of the surface structures may be accounted for by differences in the life cycle of the CC. They contend that the variations in staining are due to apoptosis (controlled cell death) of the CC and that the darker-staining cells represent cells that are undergoing degeneration. The potential for subpopulations of mitochondria-rich CCs in freshwater fish has clear implications to our understanding of the mechanisms by which acid–base homeostasis is maintained. The functional significance of these varying morphologies and how their distributions are affected by acid–base disturbances are unclear and represent an area requiring further study.

B. Pavement Cells

In one of the few early studies that investigated the potential involvement of the PC in acid–base regulation (Girard and Payan, 1980), it was demonstrated that Na^+ uptake was performed almost entirely by the cells located on the lamellae. Thus, it was concluded that lamellar PCs, and not CCs, were the cells responsible for Na^+ uptake in freshwater fish. However, these observations were based on the assumption that CCs were found only on the filament epithelia, an observation shown later to be untrue (Laurent *et al.*, 1985). Thereafter, the role of the PC in acid–base compensation remained undefined. Bartels (1989) has shown that certain features in lamprey (*Geotria australis*) PCs are analogous to those found in the granular cells in the frog skin epithelium, a well-documented acid–base-excreting epithelium. This author has suggested that the PC in freshwater lampreys is responsible for Na^+ uptake in a manner similar to that seen in the frog skin.

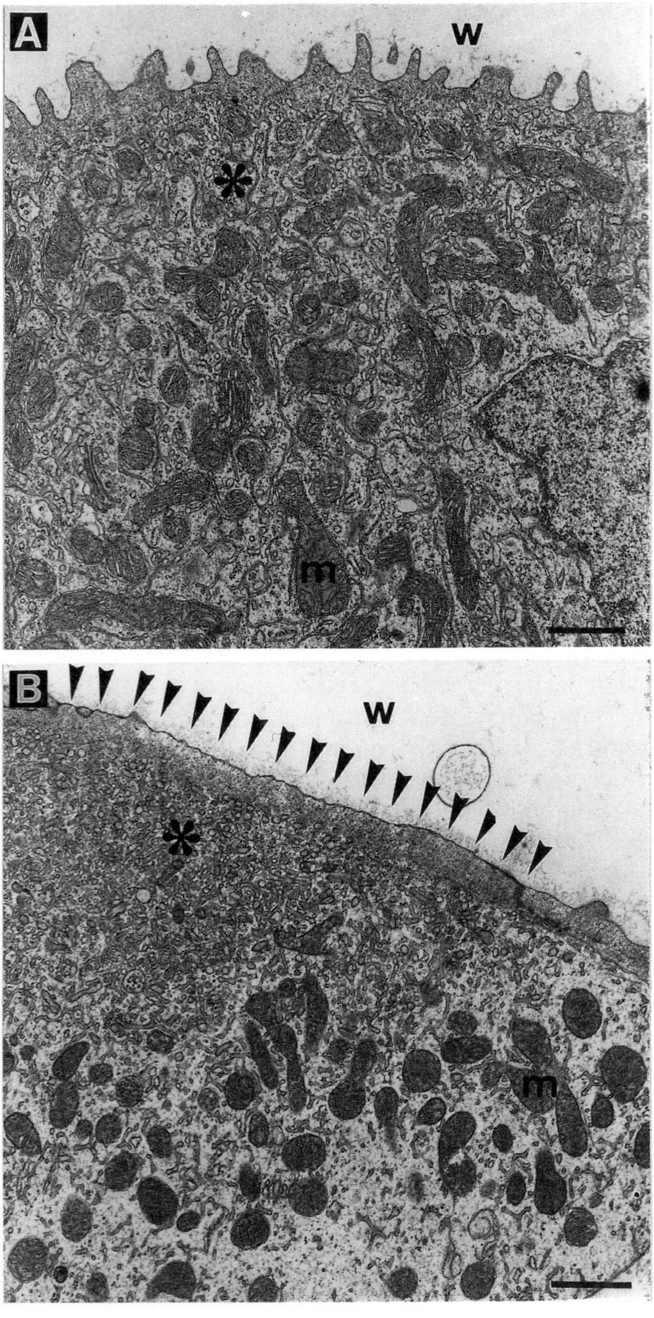

Fig. 7. Representative transmission electron microscope (TEM) photographs of two CCs with varying surface morphologies noted on filament epithelium of rainbow trout (A). One type of CC (A) displayed a dense network of apical microvilli/microridges whereas the

We have presented evidence for morphological alteration of the PC during respiratory acidosis and raised the possibility that the PC may function in Na^+ uptake and in the regulation of acid–base status in freshwater fish (Goss *et al.*, 1992, 1994b). Specifically, in the brown bullhead catfish, exposure to hypercapnia (2% CO_2) resulted in an increase in the microvilli present on the surface of the PC. Exposure to 6 h of hypercapnia resulted in a significant increase in microvilli surface density from 52 ± 1.9 to 64 ± 1.9 intercepts/grid. Figure 8 demonstrates the changes observed that occur during hypercapnia. Increases in microvilli density as a result of exposure to hypercapnia are demonstrated in panels A (control) and B (6 h of hypercapnia). Concomitant with this increase in microvilli surface density, ultrastructural changes were occurring in the PCs as well. Control samples demonstrated PCs characterized by sparse mitochondria and relatively few apical microvilli present (Fig. 8C). However, exposure to hypercapnia resulted in PCs with greater numbers of mitochondria and a concentration of these mitochondria to the subapical portion of the cell (Fig. 8D). The morphological alterations noted here have been associated with increased ion-transporting capacity in other tissues (Madsen and Tisher, 1985). These alterations in PC ultrastructure occur in concert with an increase in net acidic equivalent excretion and elevated rates of Na^+ uptake (see Fig. 1A), although the correspondence between $J_{in}^{Na^+}$ and increase in PC microvilli density is not strictly observed. We suggest that the PC may function in the increased uptake of Na^+ during an acidosis and, possibly, the excretion of acidic equivalents noted during acid–base disturbances. It is noteworthy that we have documented and described vesicles in the PCs of the trout that appear to be identical to the proton pump vesicles of mammalian renal collecting duct (Perry and Laurent, 1993; Laurent *et al.*, 1994). This observation is consistent with the view that the PC is a site of proton excretion and that this process is coupled to inward Na^+ movement via the proton pump/Na^+ channel mechanism (see Lin and Randall, Chapter 9, this volume).

III. FUTURE DIRECTIONS

The freshwater fish gill continues to serve as a useful model for acid–base-regulating epithelia, yet clearly several aspects of the model

---

other displayed no microvilli/microridges in this plane of section (B). Note the lack of microridges in this type of CC (arrowheads) and the absence of mitochondria from the subapical region. W, Water; m, mitochondria. Bar = 1 μm.

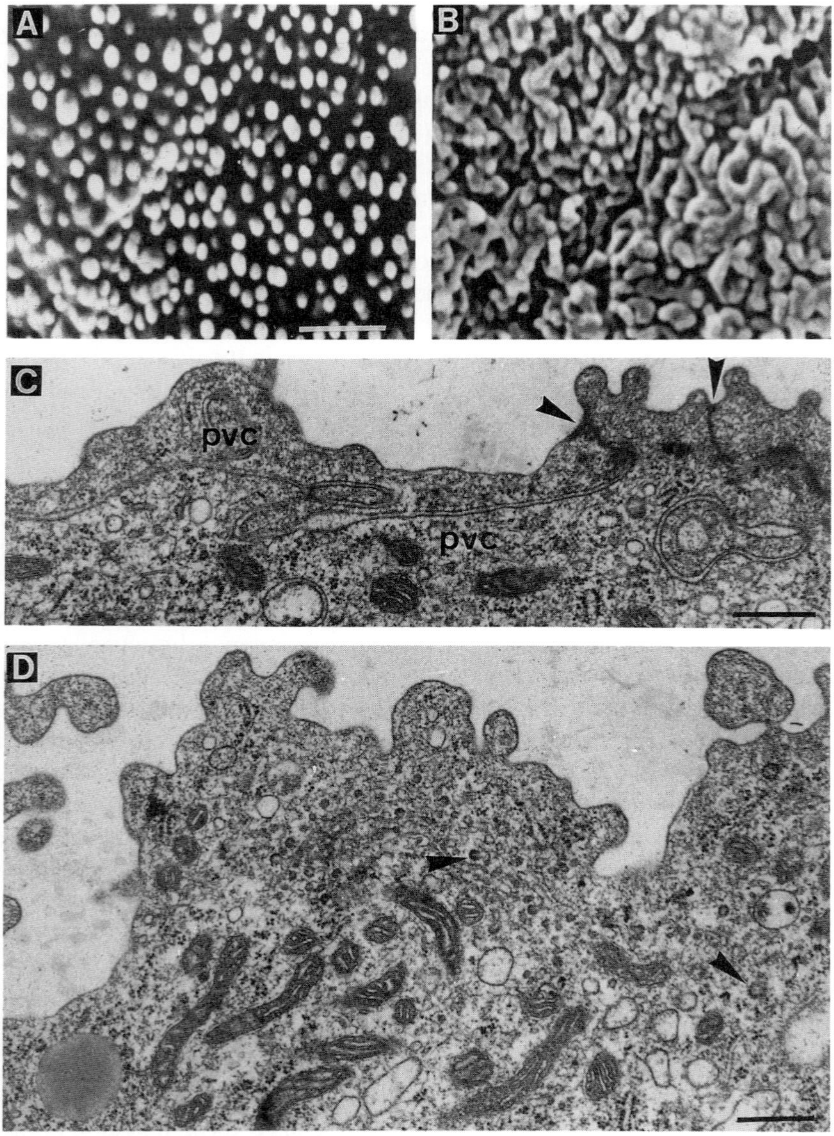

Fig. 8. Representative SEM and high-magnification TEM photographs of pavement cell (PVC) ultrastructure photographs from catfish (*Ictalurus nebulosus*) under control (A) and after 6 h exposure to hypercapnia (B, 2% CO_2 in air). Hypercapnia resulted in an increase in microvilli density on the surface of the cell. In the TEM, under normocapnic conditions, PVC mitochondria were scarce and did not display any particular localization (C). In addition, the PVCs possessed few apical villi. Hypercapnia caused an increase in apical villi and mitochondria that appeared to be localized to the subapical regions of the cell (D). Arrowheads, cell–cell junctions. (A, B) bar = 1 μm, (C, D) bar = 0.5 μm.

remain controversial or unresolved. In particular, future studies should focus on the following four areas.

a. Definitive Localization and Quantification of the Cl^-/HCO_3^- and Na^+/H^+ (H^+ Pump/Na^+ Channel) Mechanisms. Although we have presented evidence for the placement of these transport mechanisms on the CC and PC, respectively, the evidence is largely correlative. Precise localization of the transporters through the use of advanced cellular localization techniques, such as immunocytochemistry, lectin binding, and autoradiography, is required. The applications of these techniques will improve our understanding of the mechanisms involved in acid–base regulation. Furthermore, biochemical and molecular characterization of the mechanisms responsible for activation during short-term and long-term acid–base disturbances should be examined. The effects of alterations in acid–base status on the numbers and affinities of the transporters, the subcellular localization, and the cellular functioning should also be examined to confirm that changes in transport numbers correlate with changes in ion transport.

b. The Mechanism(s) Linking Na^+ Uptake and H^+ Excretion. Conclusive elucidation of the mechanism(s) involved in Na^+ uptake and H^+ excretion is essential, as described earlier by Lin and Randall (Chapter 9, this volume). The two current models (Na^+/H^+ and H^+ pump/Na^+ channel mechanisms; Potts, 1994) should be tested using morphological, immunocytochemical, pharmacological, and/or electrophysiological techniques. A multidisciplinary approach employing these powerful techniques may establish the precise physiological characteristics of Na^+ transport in the freshwater fish gill and then lead to an understanding of some of the methods of regulation. At least four mammalian isoforms of Na^+/H^+ exchange are now known with specific distributions both within tissues and within cells (Wakabayashi *et al.*, 1992), as well as numerous isoforms of the H^+-ATPase subunits (Nelson, 1992). Application of molecular biology techniques such as Northern blot analysis, *in situ* hybridization, and DNA analysis will determine the mechanisms involved in the regulation of acid–base status in freshwater fish, whether these mechanisms involve new isoforms of these transporters, and the degree of homology with other transport systems.

c. Characterization of Varying Surface CC Morphologies. The reasons for the varied external appearance of the gill CCs of salmonids, from highly ornamented with microvilli (CC_1) to possessing few microvilli or "bald" CCs (CC_4), should be examined to determine whether there is more than one cell type present as suggested by some researchers (Pisam *et al.*, 1987; Franklin and Davison, 1989; Franklin, 1990) or whether these varying morphologies simply represent varying stages in the cell cycle as suggested by others (Wendelaar-Bonga and van der Meij, 1989).

d. The Mechanism(s) Responsible for Morphological Adjustments in Gill Epithelia. The mechanisms controlling the alterations in gill morphology remain unclear. The possibilities for regulation of the gill morphology are numerous. It is known that both adrenergic and cholinergic neurons are present in the gill (Bailly *et al.*, 1989) and the former have been implicated in the control of branchial Ca^{2+} uptake (Donald, 1989). However, no studies have examined a role for a neural component in the regulation of blood acid–base status. Experimentally elevated plasma cortisol is known to cause increases in CCFA (Perry and Wood, 1985; Laurent and Perry, 1990), and growth hormone is another humoral agent that has demonstrated the ability to cause dramatic alteration in the gill CC morphology (Madsen, 1990b). However, the role that these hormones play in the regulation of the morphological responses to acid–base disturbances has not been determined. In addition, the possible involvement of other humoral agents (e.g., insulin-like growth factor: McCormick *et al.*, 1991; prolactin: Pisam *et al.*, 1993) has been implicated in fish osmoregulation and the effects of these agents on gill morphology and acid–base regulation should be investigated.

ACKNOWLEDGMENTS

We would like to thank many people for excellent help in the completion of the projects on which this review is based. At CNRS in Strasbourg, France, Dr. Suzanne Dunel-Erb, Claudine Chevalier, Guy Bombarde, and Francois Scheer, and at McMaster University in Hamilton, Canada, Rod Rhem, Steve Munger, and Chris Wood are all gratefully acknowledged. Funding for these projects was provided by NSERC grants to S. Perry and a CNRS grant to P. Laurent. G. Goss was supported by NSERC/OGS postgraduate scholarships.

REFERENCES

Al-Awqati, Q. (1986). Proton-translocating ATPases. *Annu. Rev. Cell Biol.* **2,** 179–199.

Audet, C., and Wood, C. M. (1993). Branchial morphological and endocrine responses of rainbow trout (*Oncorhynchus mykiss*) to a long-term sublethal acid exposure in which acclimation did not occur. *Can. J. Fish. Aquat. Sci.* **50,** 198–209.

Avella, M., and Bornancin, M. (1989). A new analysis of ammonia and sodium transport through the gills of the freshwater rainbow trout (*Salmo gairdneri*). *J. Exp. Biol.* **142,** 155–175.

Avella, M., and Bornancin, M. (1990). Ion fluxes in the gills of freshwater and seawater salmonid fish. *In* "Animal Nutrition and Transport Processes. 2. Transport, Respiration and Excretion: Comparative and Environmental Aspects" (J. P. Truchot and B. Lahlou, eds.), Comp. Physiol., Vol. 6, pp. 1–13. Karger, Basel.

Bailly, Y., Dunel-Erb, S., and Laurent, P. (1989). Neurons controlling the gill vasculature in five species of teleosts. *Cell Tissue Res.* **255,** 567–573.

Bailly, Y., Dunel-Erb, S., and Laurent, P. (1992). The neuroepithelial cells of the fish gill filament: Indolamine-immunocytochemistry and innervation. *Anat. Rec.* **233,** 143–161.

Bartels, H. (1989). Freeze-fracture study of the pavement cell in the lamprey gill epithelium. Analogy of membrane structure with the granular cell in the amphibian urinary bladder. *Biol. Cell.* **66,** 165–171.

Benos, D. J. (1982). Amiloride: A molecular probe of sodium transport in tissues and cells. *Am. J. Physiol.* **243,** C131–C145.

Boutilier, R. G., Iwama, G. K., and Randall, D. J. (1986). The promotion of catecholamine release in rainbow trout, *Salmo gairdneri,* by acute acidosis: Interactions between red cell pH and hemoglobin oxygen carrying capacity. *J. Exp. Biol.* **123,** 145–157.

Brown, D., Gluck, S., and Hartwig, J. (1987). Structure of a novel membrane-containing material in proton secreting epithelial cell and identification as an H^+-ATPase. *J. Cell Biol.* **105,** 1637–1648.

Brown, D., Hirsch, S., and Gluck, S. (1988a). An H^+-ATPase in opposite plasma membrane domains in kidney epithelial cell subpopulations. *Nature (London)* **331,** 622–624.

Brown, D., Hirsch, S., and Gluck, S. (1988b). Localization of a proton-pumping ATPase in rat kidney. *J. Clin. Invest.* **82,** 2114–2126.

Cameron, J. N. (1976). Branchial ion uptake in Arctic grayling: Resting values and the effects of acid–base disturbance. *J. Exp. Biol.* **64,** 711–725.

Cameron, J. N., and Iwama, G. K. (1987). Compensation of progressive hypercapnia in channel catfish and blue crabs. *J. Exp. Biol.* **133,** 183–197.

Cannon, C., van Adelsberg, J., Kelly, S., and Al-Awqati, Q. (1985). Carbon dioxide-induced exocytotic insertion of H^+ pumps in turtle-bladder luminal membrane: Role of cell pH and calcium. *Nature (London)* **314,** 443–446.

Claiborne, J. B., and Heisler, N. (1986). Acid–base regulation and ion transfers in the carp (*Cyprinus carpio*): pH compensation during graded long- and short-term environmental hypercapnia, and the effect of bicarbonate infusion. *J. Exp. Biol.* **126,** 41–62.

Conte, F. P. (1969). Salt secretion. *In* "Fish Physiology" (W. S. Hoar and D. J. Randall, eds.), Vol. 1, pp. 241–283. Academic Press, New York.

Dimberg, K., Hoglund, L. B., Knutsson, P. G., and Ridderstrale, Y. (1981). Histochemical localization of carbonic anhydrase in gill lamellae from young salmon (*Salmo salar* L.) adapted to fresh and salt water. *Acta Physiol. Scand.* **112,** 218–220.

Donald, J. A. (1989). Adrenaline and branchial nerve stimulation inhibit $^{45}Ca^{2+}$ influx into the gills of rainbow trout, *Salmo gairdneri. J. Exp. Biol.* **141,** 441–445.

Doyle, W. L., and Gorecki, D. (1961). The so-called chloride cell of the fish gill. *Physiol. Zool.* **34,** 81–85.

Dunel, S., and Laurent, P. (1980). Functional organization of the gill vasculature in different classes of fish. *In* "Epithelial Transport in the Lower Vertebrates" (B. Lahlou, ed.), pp. 37–58. Cambridge Univ. Press, Cambridge, UK.

Dunel-Erb, S., Bailly, Y., and Laurent, P. (1982). Neuroepithelial cells in fish gill primary lamellae. *J. Appl. Physiol.* **53,** 1342–1353.

Ehrenfeld, J., and Garcia-Romeu, F. (1977). Active hydrogen ion excretion and sodium absorption through isolated frog skin. *Am. J. Physiol.* **233,** F46–F54.

Ehrenfeld, J., Lacoste, I., Garcia-Romeu, F., and Harvey, B. (1990). Interdependence of Na^+ and H^+ transport in frog skin. *In* "Animal Nutrition and Transport Processes.

2. Transport, Respiration and Excretion: Comparative and Environmental Aspects" (J. P. Truchot and B. Lahlou, eds.), Comp. Physiol., Vol. 6, pp. 152–170. Karger, Basel.

Franklin, C. E. (1990). Surface ultrastructural changes in the gills of sockeye salmon (Teleostei: *Oncorhynchus nerka*) during seawater transfer: Comparison of successful and unsuccessful seawater adaptation. *J. Morphol.* **206,** 13–23.

Franklin, C. E., and Davison, W. (1989). SEM observations of morphologically different chloride cells in freshwater-adapted sockeye salmon, *Oncorhynchus nerka. J. Fish Biol.* **34,** 803–804.

Freda, J., Sanchez, D. A., and Bergman, H. L. (1991). Shortening of branchial tight junctions in acid-exposed rainbow trout (*Oncorhynchus mykiss*). *Can. J. Fish. Aquat. Sci.* **48,** 2028–2033.

Fritsche, C., Kleinman, J. G., Bain, J. L. W., Heinen, R. R., and Riley, D. A. (1991). Carbonic anhydrase and proton secretion in turtle bladder mitochondrial-rich cells. *Am. J. Physiol.* **260,** F443–F458.

Girard, J. P., and Payan, P. (1980). Ion exchanges through respiratory and chloride cells in freshwater- and seawater-adapted teleosteans. *Am. J. Physiol.* **7,** R260–R268.

Gluck, S., Kelly, S., and Al-Awqati, Q. (1982). The proton translocating ATPase responsible for urinary acidification. *J. Biol. Chem.* **257,** 9230–9233.

Goss, G. G., Wood, C. M., Laurent, P., and Perry, S. F. (1994a). Morphological responses of the rainbow trout (*Oncorhynchus mykiss*) gill to hyperoxia, base ($NaHCO_3$) and acid (HCl) infusions. *Fish Physiol. Biochem.* **12**(6), 465–477.

Goss, G. G., and Perry, S. F. (1994). Differences in acid–base regulation in rainbow trout (*Oncorhynchus mykiss*) and American eel (*Anguilla rostrata*). *Physiol. Zool.* **67**(2), 381–406.

Goss, G. G., and Wood, C. M. (1990a). Na^+ and Cl^- uptake kinetics, diffusive effluxes, and acidic equivalent fluxes across the gills of rainbow trout: I. Responses to environmental hyperoxia. *J. Exp. Biol.* **152,** 521–547.

Goss, G. G., and Wood, C. M. (1990b). Na^+ and Cl^- uptake kinetics, diffusive effluxes, and acidic equivalent fluxes across the gills of rainbow trout: II. Responses to bicarbonate infusion. *J. Exp. Biol.* **152,** 549–571.

Goss, G. G., and Wood, C. M. (1991). Two-substrate kinetics analysis: A novel approach linking ionic and acid–base transport in freshwater trout. *J. Comp. Physiol., B* **161,** 635–646.

Goss, G. G., Laurent, P., and Perry, S. F. (1992). Gill morphology and acid–base regulation during hypercapnic acidosis in the brown bullhead, *Ictalurus nebulosus. Cell Tissue Res.* **268,** 539–552.

Goss, G. G., Wood, C. M., Laurent, P., and Perry, S. F. (1994a). Morphological responses of the rainbow trout (*Oncorhynchus mykiss*) gill to hyperoxia, base ($NaHCO_3$) and acid (HCl) infusions. *Fish Physiol. Biochem.* **12**(6), 465–477.

Goss, G. G., Laurent, P., and Perry, S. F. (1994b). Gill morphology during hypercapnia in brown bullhead (*I. nebulosus*): Role of chloride cells and pavement cells in acid–base regulation. *J. Fish Biol.* **45,** 705–718.

Handy, R. D. (1989). The ionic composition of rainbow trout body mucus. *Comp. Biochem. Physiol., A* **93,** 571–575.

Harvey, B. J., and Ehrenfeld, J. (1988). Epithelial pH and ion transport regulation by proton pumps and exchangers. *In* "Proton Passage Across Cell Membranes," (G. Bock and J. Marsh, eds.), pp. 139–164. Wiley, Chichester.

Hobe, H., Laurent, P., and McMahon, B. R. (1984). Whole body calcium flux rates in freshwater teleosts as a function of ambient calcium and pH levels: A comparison between the euryhaline trout, *Salmo gairdneri,* and stenohaline bullhead, *Ictalurus nebulosus*. *J. Exp. Biol.* **113,** 237–252.

Hughes, G. (1979). Scanning electron microscopy of the respiratory surfaces of trout gills. *J. Zool.* **188,** 443–453.

Hyde, D. A., and Perry, S. F. (1987). Acid–base and ionic regulation in the American eel (*Anguilla rostrata*) during and after prolonged aerial exposure: Branchial and renal adjustments. *J. Exp. Biol.* **133,** 429–447.

Hyde, D. A., and Perry, S. F. (1989). Differential approaches to blood acid–base regulation during exposure to prolonged hypercapnia in two freshwater teleosts: The rainbow trout (*Salmo gairdneri*) and the American eel (*Anguilla rostrata*). *Physiol. Zool.* **62**(6), 1164–1186.

Kirsch, R. (1972). The kinetics of peripheral exchanges of water and electrolytes in the silver eel (*Anguilla anguilla* L.) in freshwater and seawater. *J. Exp. Biol.* **57,** 489–512.

Kniaz, D., and Arruda, J. A. L. (1990). Adaptation to respiratory acidosis in the turtle bladder. *Proc. Soc. Exp. Biol. Med.* **195,** 84–94.

Kniaz, D., and Arruda, J. A. L. (1991). Adaptation to metabolic alkalosis by the turtle urinary bladder. *Proc. Soc. Exp. Biol. Med.* **196,** 444–450.

Kristensen, P., and Ussing, H. H. (1985). Epithelia organization. *In* "The Kidney: Physiology and Pathophysiology" (W. Seldin and G. Giesbisch, eds.), pp. 173–188. Raven Press, New York.

Lauren, D. J., and McDonald, D. G. (1987). Acclimation to copper by rainbow trout, *Salmo gairdneri:* Physiology. *Can. J. Fish. Aquat. Sci.* **43,** 1488–1496.

Laurent, P. (1984). Gill internal morphology. *In* "Fish Physiology" (W. S. Hoar and D. J. Randall, eds.), Vol. 10A, pp. 73–183. Academic Press, New York.

Laurent, P. (1985). A survey of the morpho-functional relationships between epithelia, vasculature and innervation within the fish gill. *In* "Vertebrate Morphology" (Duncker and Fleischer, eds.), pp. 339–351. Fischer, New York.

Laurent, P. (1989). Gill structure and function: Fish. *In* "Comparative Pulmonary Physiology" (S. C. Wood, ed.), pp. 69–120. Dekker, New York.

Laurent, P., and Dunel, S. (1980). Morphology of gill epithelia in fish. *Am. J. Physiol.* **238,** R147–R159.

Laurent, P., and Perry, S. F. (1990). Effects of cortisol on gill chloride cell morphology and ionic uptake in the freshwater trout, *Salmo gairdneri*. *Cell Tissue Res.* **259,** 429–442.

Laurent, P., Hobe, H., and Dunel-Erb, S. (1985). The role of environmental sodium chloride relative to calcium in gill morphology of freshwater salmonid fish. *Cell Tissue Res.* **240,** 675–692.

Laurent, P., Goss, G. G., and Perry, S. F. (1994). Proton pumps in fish gill pavement cells? *Arch. Int. Physiol. Biochim. Biophys.* **102,** 77–79.

Leino, R. L., and McCormick, J. H. (1984). Morphological and morphometrical changes in chloride cells of the gills of *Pimephales promelas* after chronic exposure to acid water. *Cell Tissue Res.* **236,** 121–128.

Leino, R. L., McCormick, J. H., and Jensen, K. M. (1987). Changes in gill histology of fathead minnows and yellow perch transferred to soft water or acidified water with particular reference to chloride cells. *Cell Tissue Res.* **250,** 389–399.

Madsen, K. M., and Tisher, C. C. (1984). Response of intercalated cells of rat outer medullary collecting duct to chronic metabolic acidosis. *Lab. Invest.* **51,** 268–276.

Madsen, K. M., and Tisher, C. C. (1985). Structure–function relationships in H^+-secreting epithelia. *Fed. Proc., Fed. Am. Soc. Exp. Biol.* **44,** 2704–2709.

Madsen, K. M., and Tisher, C. C. (1986). Structural–functional relationships along the distal nephron. *Am. J. Physiol.* **250,** F1–F5.

Madsen, S. S. (1990a). Cortisol treatment improves the development of hypoosmoregulatory mechanisms in the euryhaline rainbow trout, *Salmo gairdneri*. *Fish Physiol. Biochem.* **8,** 45–52.

Madsen, S. S. (1990b). Enhanced hypoosmoregulatory response to growth hormone after cortisol treatment in immature rainbow trout, *Salmo gairdneri*. *Fish Physiol. Biochem.* **8,** 271–279.

Madsen, S. S. (1990c). Effect of repetitive cortisol and thyroxine injections on chloride cell number and Na^+/K^+-ATPase activity in gills of freshwater acclimated rainbow trout, *Salmo gairdneri*. *Comp. Biochem. Physiol., A* **95,** 171–175.

Maetz, J. (1972). Branchial sodium exchange and ammonia excretion in the goldfish *Carassius auratus*. *J. Exp. Biol.* **58,** 255–275.

McCormick, S. D., Sakamoto, T., Hasewaga, S., and Hirano, T. (1991). Osmoregulatory actions of insulin-like growth factor-1 in rainbow trout (*Oncorhynchus mykiss*). *J. Endocrinol.* **130,** 87–92.

McDonald, D. G., and Prior, E. T. (1988). Branchial mechanisms of ion and acid–base regulation in the freshwater rainbow trout, *Salmo gairdneri*. *Can. J. Zool.* **66,** 2699–2708.

McDonald, D. G., Freda, J., Cavdek, V., Gonzalez, R., and Zia, S. (1991). Interspecific differences in gill morphology of freshwater fish in relation to tolerance of low-pH environments. *Physiol. Zool.* **64**(1), 124–144.

Naon, R., and Mayer-Gostan, N. (1983). Separation by velocity separation of the gill epithelial cells and their ATPase activities in the seawater adapted eel *Anguilla anguilla* L. *Comp. Biochem. Physiol., A* **75,** 541–547.

Nelson, N. (1992). The vacuolar H^+-ATPase—one of the most fundamental ion pumps in nature. *J. Exp. Biol.* **172,** 19–27.

Olson, K. R., and Fromm, P. O. (1973). A scanning electron microscopic study of secondary lamellae and chloride cells of rainbow trout (*Salmo gairdneri*). *Z. Zellforsch. Mikrosk. Anat.* **143,** 439–449.

Perry, S. F., and Goss, G. G. (1994). The effects of experimentally altered gill chloride cell surface area on acid–base regulation in rainbow trout during metabolic alkalosis. *J. Comp. Physiol.* (in press).

Perry, S. F., and Laurent, P. (1989). Adaptational responses of rainbow trout to lowered external NaCl concentration: Contribution of the branchial chloride cell. *J. Exp. Biol.* **147,** 147–168.

Perry, S. F., and Laurent, P. (1993). Environmental effects on fish gill structure and function. *In* "Fish Ecophysiology" (J. C. Rankin and F. B. Jensen, eds.), pp. 231–264. Chapman and Hall, London.

Perry, S. F., and Walsh, P. J. (1989). Metabolism of isolated fish gill cells: Contribution of epithelial chloride cells. *J. Exp. Biol.* **44,** 507–520.

Perry, S. F., and Wood, C. M. (1985). Kinetics of branchial calcium uptake in the rainbow trout: Effects of acclimation to various external calcium levels. *J. Exp. Biol.* **116,** 411–433.

Perry, S. F., Malone, S., and Ewing, D. (1987). Hypercapnic acidosis in the rainbow trout (*Salmo gairdneri*). I. Branchial ion fluxes and blood acid–base status. *Can. J. Zool.* **65,** 888–895.

Perry, S. F., Goss, G. G., and Laurent, P. (1992). The interrelationships between gill chloride cell morphology and ionic uptake in four freshwater teleosts. *Can. J. Zool.* **9,** 1775–1786.

Philpott, C. W. (1966). The use of horseradish peroxidase to demonstrate functional continuity between the plasmalemma and the unique tubular system of the chloride cell. *J. Cell Biol.* **31,** 88A.

Pickering, A. D., Pottinger, T. G., Sumpter, J. P., Carragher, J. F., and Le Bail, P. Y. (1991). Effects of acute and chronic stress on the levels of circulating growth hormone in the rainbow trout, *Oncorhynchus mykiss*. *Gen. Comp. Endocrinol.* **83,** 86–93.

Pisam, M., Caroff, A., and Rambourg, A. (1987). Two types of chloride cells in the gill epithelium of a freshwater-adapted euryhaline fish: *Lebistes reticulatus;* Their modifications during adaptations to seawater. *Am. J. Anat.* **179,** 40–50.

Pisam, M., Auperin, B., Prunet, P., Rentier-Delrue, F., Martial, J., and Rambourg, A. (1993). Effects of prolactin on alpha and beta chloride cells in the gill epithelium of the saltwater adapted tilapia *"Oreochromis niloticus." Anat. Rec.* **235,** 275–284.

Playle, R. C., Goss, G. G., and Wood, C. M. (1989). Physiological disturbances in rainbow trout (*Salmo gairdneri*) during acid and aluminum exposures in soft water of two calcium concentrations. *Can. J. Zool.* **67,** 311–324.

Potts, W. T. W. (1994). Kinetics of sodium uptake in freshwater animals: A comparison of ion-exchange and proton pump hypotheses. *Am. J. Physiol.* **266**(2), R315–R320.

Rahim, S. M., Delaunoy, J.-P., and Laurent, P. (1988). Identification and immunocytochemical localization of two different carbonic anhydrase isoenzymes in teleostean fish erythrocytes and gill epithelia. *Histochemistry* **89,** 451–459.

Rich, A., Dixon, T. E., and Clausen, C. (1990). Changes in membrane conductances and areas associated with bicarbonate secretion in turtle bladder. *J. Membr. Biol.* **113,** 211–219.

Rick, R. (1992). Intracellular ion concentrations in the isolated frog skin epithelium: Evidence for different types of mitochondria-rich cells. *J. Membr. Biol.* **127,** 227–236.

Rosen, S. (1970). Localization of carbonic anhydrase activity in transporting urinary epithelia. *J. Histochem. Cytochem.* **18,** 668–670.

Rosen, S. (1972). Localization of carbonic anhydrase activity in the vertebrate nephron. *Histochem. J.* **4,** 35–48.

Rosen, S., and Freidley, N. J. (1973). Carbonic anhydrase activity in *Rana pipiens* skins: Biochemical and histochemical analysis. *Histochemie* **36,** 1–4.

Sargent, J. R., Thompson, A. J., and Bornancin, M. (1975). Activities and localization of succinic dehydrogenase and Na^+/K^+-activated adenosine triphosphatase in the gills of freshwater and seawater eels (*Anguilla anguilla*). *Comp. Biochem. Physiol., B* **51,** 75–79.

Satlin, L. M., and Schwartz, G. J. (1989). Cellular remodelling of HCO_3^--secreting cells in rabbit renal collecting duct in response to an acidic environment. *J. Cell Biol.* **109,** 1279–1288.

Schuster, V. L., Bonsib, S. M., and Jennings, M. L. (1986). Two types of collecting duct mitochondria-rich (intercalated) cells: Lectin and band 3 cytochemistry. *Am. J. Physiol.* **251,** C347–C355.

Schwartz, G. J., and Al-Awqati, Q. (1985). Carbon dioxide causes exocytosis of vesicles containing H^+ pumps in isolated perfused proximal and collecting tubules. *J. Clin. Invest.* **75,** 1638–1644.

Schwartz, G. J., Barasch, J., and Al-Awqati, Q. (1985). Plasticity of functional epithelial polarity. *Nature* (*London*) **318,** 368–371.

Schwartz, G. J., Satlin, L., and Bergmann, J. E. (1988). Fluorescent characterization of collecting duct cells: A second H^+ secreting type. *Am. J. Physiol.* **255,** F1003–F1014.

Steinmetz, P. R. (1987). Electrogenic proton transport by intercalated cells of tight urinary epithelia. *In* "Proton Passage Across Cell Membranes," (G. Bock and J. Marsh, eds.), pp. 122–138. Wiley, Chichester.

Steinmetz, P. R., Omachi, R. S., and Frazier, H. S. (1967). Independence of hydrogen ion secretion and transport of other electrolytes in turtle bladder. *J. Clin. Invest.* **46,** 1541–1548.

Stetson, D. L., and Steinmetz, P. R. (1985). α and β types of carbonic anhydrase-rich cells in the turtle bladder. *Am. J. Physiol.* **249,** F553–F565.

Stokes, J. B. (1982). Ion transport by the cortical and outer medullary collecting tubule. *Kidney Int.* **22,** 473–484.

Tietge, J. E., Johnson, R. D., and Bergman, H. L. (1988). Morphometric changes in gill secondary lamellae of brook trout (*Salvelinus fontinalis*) after long-term exposure to acid and aluminum. *Can. J. Fish. Aquat. Sci.* **45,** 1643–1648.

Wakabayashi, S., Sardet, C., Farfournoux, P., Counillon, L., Meloche, S., Pages, G., and Pouyssegur, J. (1992). Structure–function of the growth factor-activatable Na^+/H^+ exchanger (NHE1). *Rev. Physiol. Biochem. Pharmacol.* **119,** 157–186.

Weiner, I. D., and Hamm, L. L. (1991). Regulation of Cl^-/HCO_3^- exchange in the rabbit cortical collecting tubule. *J. Clin. Invest.* **87,** 1553–1558.

Wendelaar Bonga, S. E., and van der Meij, C. J. M. (1989). Degeneration and death, by apoptosis and necrosis, of the pavement and chloride cells in the gills of the teleost *Oreochromis mossambicus. Cell Tissue Res.* **255,** 235–243.

Wilkie, M. P., and Wood, C. M. (1994). The effects of extremely alkaline water (pH 9.5) on rainbow trout gill function and morphology. *J. Fish Biol.* **45,** 87–98.

Wilkie, M. P., Wright, P. A., Iwama, G. K., and Wood, C. M. (1994). The physiological adaptations of the Lahonton cutthroat trout (*Oncorhynchus clarki henshawi*), following transfer from well water to the highly alkaline waters of Pyramid lake, Nevada (pH 9.4). *Physiol. Zool.* **67,** 355–380.

Wood, C. M., and Jackson, E. B. (1980). Blood acid–base regulation during environmental hyperoxia in the rainbow trout (*Salmo gairdneri*). *Respir. Physiol.* **42,** 351–372.

Wood, C. M., Wheatley, M., and Höbe, H. (1984). The mechanisms of acid–base and ionoregulation in the freshwater rainbow trout during environmental hyperoxia and subsequent normoxia. III. Branchial exchanges. *Respir. Physiol.* **55,** 175–192.

11

HORMONAL CONTROL OF GILL Na^{+},K^{+}-ATPase AND CHLORIDE CELL FUNCTION

STEPHEN D. McCORMICK

I. INTRODUCTION

The secretion of excess sodium and chloride by teleosts in seawater is carried out by gill chloride cells. These highly specialized cells are characterized by a large, columnar appearance, numerous mitochondria, an extensive tubular system, an apical crypt, and mucosal-serosal exposure (see reviews by Evans *et al.,* 1982; Zadunaisky, 1984; Pisam and Rombourg, 1991). The tubular system is continuous with the basolateral membrane, resulting in a large surface area for the placement of transport proteins. Perhaps the most important of these proteins is Na^+,K^+-ATPase. Also known as the sodium pump, Na^+,K^+-ATPase plays a central role in the salt-secretory function of chloride cells. Present in all animal cells, this energy-dependent, ion-translocating enzyme occurs in high concentrations in most transport epithelia; up to 10^8 sodium pumps may be present in a single chloride cell (Karnaky, 1986)! Na^+,K^+-ATPase participates in ion transport either directly through movement of sodium and potassium across the plasma membrane or indirectly through generation of ionic and electrical gradients.

Regulation of chloride cells and Na^+,K^+-ATPase is critical during movement of fish between fresh water and seawater and within estuaries, and is also important in stenohaline fish under several variable environmental conditions. Since the neuroendocrine system is the primary link between a changing environment and physiological adaptation, the hormonal control of chloride cells and Na^+,K^+-ATPase is critical to euryhalinity. Reviews in this general area have appeared previously (Foskett *et al.,* 1983; Mayer-Gostan *et al.,* 1987) though not recently, and this review will focus primarily on new approaches and recent evidence on the hormonal control of Na^+,K^+-ATPase and chloride cells in teleosts.

II. Na^+,K^+-ATPase AND CHLORIDE CELL FUNCTION

A. Seawater

In seawater the osmoregulatory function of the gill is to secrete excess monovalent ions. Although there are several models for ion movement across the gill (Payan *et al.,* 1984), the model proposed by Silva *et al.* (1977) has the most experimental support (Evans *et al.,* 1982). Na^+,K^+-ATPase, located on the basolateral membrane, creates low intracellular Na^+ and a highly negative charge within the cell. The Na^+ gradient is then used to transport Cl^- into the cell through a $Na^+/K^+/Cl^-$ cotransporter, then Cl^- leaves the cell "downhill" on an electrical gradient through an apical Cl^- channel. Na^+ is transported through a paracellular pathway down its electrical gradient (seawater being more negative than plasma).

In this model, Na^+,K^+-ATPase generates ionic and electrical gradients for use in excretion of Na^+ and Cl^-. Epstein *et al.* (1967) were the first to suggest a role for Na^+,K^+-ATPase in the salt excretory function of the gill, based on their observation of higher levels of gill Na^+,K^+-ATPase activity in seawater-adapted killifish (*Fundulus heteroclitus*). Ouabain, a specific inhibitor of Na^+,K^+-ATPase that binds to the potassium-binding site, strongly inhibits Cl^- excretion and short-circuit current (Silva *et al.*, 1977; Karnaky *et al.*, 1977). The basal location of Na^+,K^+-ATPase was inferred from the greater efficacy of injected ouabain compared to external ouabain exposure (Silva *et al.*, 1977). Cell fractions of dispersed gill tissue that are rich in chloride cells have much higher Na^+,K^+-ATPase activity than other cell fractions (Kamiya, 1972; Sargent *et al.*, 1975). Karnaky *et al.* (1976) were able to visualize high concentration of Na^+,K^+-ATPase in chloride cells through the use of [^{3}H]ouabain autoradiography (Fig. 1). These results indicated that the sodium pump was not located on the apical surface, but rather throughout the cell on the extensive tubular system that is continuous with the basolateral membrane.

The model of Silva *et al.* (1977) offers several predictions of the biochemical properties of chloride cells that have yet to be verified. For example, a bumetanide-sensitive $Na^+/K^+/Cl^-$ cotransporter should be located on the basolateral membrane of chloride cells, and an apical Cl^- channel should be present on the apical membrane of chloride cells. Molecular methods, along with more traditional immunological and pharmacological approaches, should allow exploration of these predictions.

B. Fresh Water

Our current understanding of chloride uptake by the gills in fresh water emphasize Cl^-–HCO_3^- exchange at the apical surface (Avella and Bornancin, 1990). Sodium uptake is accomplished either by apical Na^+–H^+/NH_4^+ exchange or by an apical Na^+ channel that is supported by an electrogenic, apical H^+-ATPase (see Lin and Randall, Chapter 9, this volume). In this model, gill Na^+,K^+-ATPase transports excess intracellular sodium into the blood, and may also provide an electrical gradient for apical transport of sodium and basolateral transport of chloride. However, the relative importance of Na^+,K^+-ATPase in sodium and chloride uptake is unclear. Ouabain has been reported to decrease ion uptake (Epstein *et al.*, 1967) or to have no effect (Kerstetter and Keeler, 1976) in rainbow trout (*Oncorhynchus gairdneri*). Increases in Na^+,K^+-ATPase in a few teleosts upon transfer from seawater to fresh water (discussed in the following) support a role for Na^+,K^+-ATPase in ion uptake. The role of the sodium pump in ion uptake is less certain than its role in ion excretion, and further studies that take account of potential differences among species are needed.

The site of gill ion uptake in teleosts is also uncertain. Foskett and Scheffey (1982) provided direct evidence for chloride cells as the site of chloride excretion in seawater teleosts, but similar direct evidence for

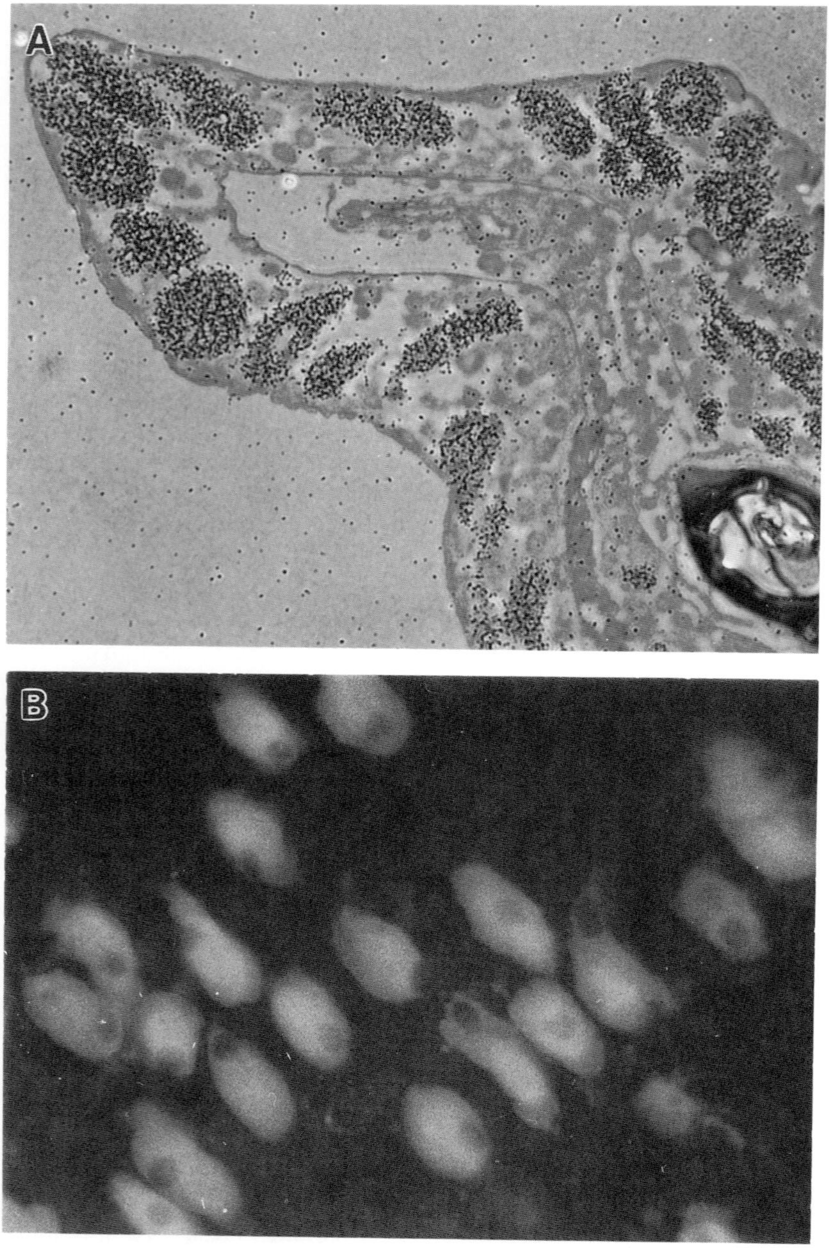

Fig. 1. (A) [^{3}H]Ouabain binding in chloride cells of seawater-adapted killifish. [Reproduced from the *Journal of Cell Biology*, Karnaky *et al.* (1976) **70,** pp. 157–177, by copyright permission of the Rockefeller University Press.] (B) Fluorescent staining of chloride cells in the jawskin of the long-jawed mudsucker following exposure to anthroylouabain. [Reproduced from McCormick (1990a) with permission from Springer-Verlag.] In both preparations, note the high concentration of ouabain binding throughout the cells (but not in the nucleus), indicative of its placement on the extensive tubular system of chloride cells.

chloride cells as the site of ion uptake is lacking. Proliferation in response to low ion content in fresh water is evidence for the role of chloride cells in ion uptake (Laurent and Dunel, 1980; Laurent *et al.,* 1985; Avella *et al.,* 1987; Perry and Laurent, 1989), but there is also evidence for involvement of other cells in the gill (see Goss *et al.,* Chapter 10, this volume). Although variation occurs among species, the appearance of chloride cells in fresh water is distinct from that of the same species in seawater (Pisam *et al.,* 1987; Laurent and Perry, 1991). In fresh water, chloride cells are generally smaller and less columnar, have a less pronounced tubular system, and often do not have apical and serosal contacts. Accessory cells, which interdigitate with seawater chloride cells to form ''leaky'' junctions for passive paracellular sodium efflux, are absent in fresh water (Laurent and Perry, 1991).

III. PROPERTIES OF Na^+,K^+-ATPase

A great deal is known about the structure and biochemistry of Na^+,K^+-ATPase, which is present in all animal cells (see reviews by De Renzis and Bornancin, 1984; Rossier *et al.* 1989; Skou and Esmann, 1992). The protein complex consists of a catalytic α-subunit and a β-subunit; the latter stabilizes folding of the α-subunit, confers K^+ affinities, and is involved in cell–cell interactions (Schmalzing and Gloor, 1994). Both subunits of Na^+,K^+-ATPase have been sequenced from one fish species to date, the Pacific electric ray (*Torpedo californica*), which contains an electric organ rich in Na^+,K^+-ATPase. The α-subunit consists of 1022 amino acids and has a calculated molecular mass of 112,000 and six to eight membrane-spanning domains (Kawakami *et al.,* 1985). The β-subunit is a 35-kDa glycoprotein of 300 amino acids with a single membrane-spanning domain (Noguchi *et al.,* 1986). Its amino acid sequence is less highly conserved among vertebrates (60%) than the α-subunit (Rossier *et al.,* 1989). The α-subunit has also been fully sequenced in one teleost, the white sucker (*Catostomus commersoni*), and consists of 1027 amino acids and a predicted eight membrane-spanning domains (Schörock *et al.,* 1991). There is substantial sequence similarity with other vertebrate Na^+,K^+-ATPase α-subunits (74–98%). A partial sequence (450 base pairs) of the gene encoding the rainbow trout α-subunit also shows high sequence similarity (Kisen *et al.,* 1994). Three isoforms of the α-subunit and four isoforms of the β-subunit have been isolated in mammals; these isoforms are present in varying proportions in different organs (Skou and Esmann, 1992).

The stoichiometry of Na^+,K^+-ATPase appears to be similar in all animals investigated (Skou and Esmann, 1992). Three internal sodium ions are transported outward in exchange for two potassium ions. The sodium pump is therefore electrogenic, creating a potential difference across the cell membrane. Each translocation of ions requires the hydrolysis of ATP and is accompanied by a conformational change. Binding sites for ions, ATP, and inhibitors are all contained on the α-subunit.

IV. METHODS

A. Quantitation and Localization of Na^+,K^+-ATPase

Measurement of Na^+,K^+-ATPase is most frequently accomplished by measurement of V_{max} in a microsomal preparation or crude homogenate and expressed as activity per unit wet weight or protein content. The breakdown of ATP is detected (most often by measuring inorganic phosphate or ADP) in the presence and absence of ouabain. Such biochemical measurements are, by their nature, method dependent. Efforts should be made to optimize the conditions for expression of enzyme activity, including tissue preparation, ion concentrations, pH, ouabain, detergent, and temperature. Several methods specific for measurement of gill Na^+,K^+-ATPase activity of teleosts have been presented (Johnson *et al.,* 1977; Zaugg, 1982; Mayer-Gostan and Lemaire, 1991), including a "microassay" suitable for nonlethal biopsies and organ cultured tissues (McCormick, 1993). These biochemical assays are an approximation of the total number of sodium pumps and rapid activation or deactivation (e.g., phosphorylation) may not be detected by these methods.

[^{3}H]Ouabain binding to gill homogenates and whole tissue has also been used to measure the number of sodium pumps. In all cases examined to date, gill ouabain binding and Na^+,K^+-ATPase activity are strongly correlated (Sargent and Thomson, 1974; Stagg and Shuttleworth, 1982). B_{max} values for teleost gills range from 3 to 20 pmol/mg dry weight in the three species examined (Hossler *et al.,* 1979; Stagg and Shuttleworth, 1982; McCormick and Bern, 1989). Since chloride cells usually represent less than 10% of the gill, the sodium pump content of individual chloride cells is much higher (Karnaky, 1986). Hootman and Ernst (1988) outline a method for using [^{3}H]ouabain to measure turnover (*in situ* activity) and sodium pump numbers in intact cells.

Autoradiographic methods using [^{3}H]ouabain have localized Na^+,K^+-ATPase to chloride cells and provided information on their intracellular location (Fig. 1). A fluorescent ouabain analog, anthroylouabain, has been recently used to examine Na^+,K^+-ATPase-rich chloride cells in the opercular membrane (Fig. 1). This probe has the advantage of ouabain specificity and increased fluorescence upon binding (Fortes, 1977). A major disadvantage is the high autofluorescence in the wavelengths necessary to view anthracene. Nonetheless, in tilapia this stain distinguishes the freshwater from the seawater form of the chloride cell. Histochemical methods have also been used to localize Na^+,K^+-ATPase in fish gills and are reviewed in De Renzis and Bornancin (1984).

Molecular methods have recently been applied to the measurement of Na^+,K^+-ATPase gene expression. Using a 450-base pair fragment of the α-subunit gene, Northern blot analysis was used by Kisen *et al.* (1994) to detect a prominent mRNA of 3.7 kilobases in rainbow trout gill. This probe was found to be useful for measuring gill Na^+,K^+-ATPase gene expression in several other teleosts. Northern blots can also be used to measure gill Na^+,K^+-ATPase mRNA in brown trout (*Salmo trutta*) using cDNA of the *Xenopus* α-subunit (S. Madsen, personal communication). These studies revealed that gill Na^+,K^+-ATPase α-subunit mRNA levels are higher in fish in seawater than in fish in fresh water; increases occur within 1 day of transfer to seawater in brown trout.

B. Morphology and Function of Chloride Cells

The distinguishing characteristics of chloride cells, especially numerous mitochondria and an extensive tubular system, permit their identification from other cells in the gill and opercular membrane at the light microscope level. The vital mitochondrial stain DASPEI (2-*p*-dimethylaminostyrylethylpyridiniumiodide) has been used widely on opercular membranes (Karnaky *et al.*, 1984). Such vital stains are rarely used on the gill because of tissue thickness and complex structure, although gill dispersions are possible (Perry and Walsh, 1989). Champy-Maillet's fixative (a 0.2% osmic acid with saturated zinc powder and 25% mg/ml iodine) stains plasma membranes and the extensive tubular system, rendering the entire chloride cell black (e.g., Avella *et al.*, 1987). Great care must be taken in the examination of chloride cells using postfixation staining methods such as hematoxylin and eosin, as positive identification of chloride cells can be problematic. Surprisingly, immunohistochemistry has not been widely used in studies of chloride cells. Currently available methods for protein isolation should permit production of antibodies specific for teleost Na^+,K^+-ATPase, transport proteins, or mitochondrial enzymes.

A variety of methods can be used to probe chloride cell function. Methods for direct and indirect measurement of ion movements have been widely used and are reviewed in detail elsewhere in this volume (see Marshall, Chapter 1). Methods for examining chloride cell function in skin and opercular membranes, particularly the use of Ussing chambers to examine ion transport and electrophysiology, have been reviewed (McCormick, 1994). Although the vibrating probe has been used successfully to localize chloride secretion to chloride cells (Foskett and Scheffey, 1982), it has not yet been used to examine regulation of chloride cell function. Perhaps the most powerful new tools for exploring chloride cell physiology are ion-sensitive fluorescent dyes. These have already proven useful in basic studies of ion transport (see Chapters 1 and 8, this volume), and should be equally useful in regulatory (endocrine) studies of ion uptake and secretion. To date, these methods have been used only with chloride cells in the opercular membrane, though confocal laser microscopy may permit their use in gill tissue.

C. Organ and Cell Culture

In vitro methods are useful for determining the direct action of hormones and other agents on function of cells and tissues. Recent advances in gill respiratory cell culture (see Pärt and Bergstrom, Chapter 8, this volume) and rectal gland cell culture for elasmobranchs (see Valentich *et al.,* Chapter 7, this volume) illustrate the contributions that such methods can make to our understanding of transport physiology. Unfortunately, a method for long-term culture of chloride cells is not yet available. Isolated chloride cells have relatively poor viability in culture using methods suitable for other gill cells (Peter Pärt, personal communication). It is unclear whether chloride cell preparations would possess the transport capabilities of intact tissue, particularly if accessory cells are required for ion secretion. Nonetheless, chloride cell cultures would be very useful, if only to examine regulation of transport proteins such as Na^+,K^+-ATPase.

Long-term organ cultures of primary gill filaments and opercular membranes have been used in several studies of the hormonal regulation of Na^+,K^+-ATPase (McCormick and Bern, 1989; McCormick, 1990b; McCormick *et al.,* 1991a; Madsen and Bern, 1993; see McCormick, 1994, for detailed methods). Though more difficult to prepare, the opercular membrane culture has the advantage of being more accessible than gills for morphological and transport studies (e.g., Ussing chamber or ion-sensitive dye) following hormone exposure. In preparing culture media, physiological levels of pH, ions, and gases appropriate for the species being studied should be used. In many cases this means only a slight modification of existing commercial culture media. Minimal Essential Medium (MEM) has been widely used and apparently supports a variety of teleost cell types equally or better than media with more additives (Fernandez *et al.,* 1993). Gassing for physiological ranges of pH (7.7–8.0),

pCO_2 (1–4 mm Hg), and HCO_3^- (3–10 m*M*) (Heisler, 1993) can theoretically be achieved by using 0.3–0.5% CO_2 gas. One percent CO_2 is commercially available and has been used successfully in gill organ culture (S. D. McCormick, unpublished). Saturation with 5% CO_2, widely used in fish cell and organ culture, results in supraphysiological levels (for teleosts) of CO_2 and HCO_3^-, though this is apparently not detrimental to growth of cells or maintenance of tissue in culture (McCormick, 1990b; Fernandez *et al.*, 1993). The health and responsiveness of cells in culture are the most important criteria, and these can be judged through vital stains, histological appearance, and biochemical analyses.

V. ENVIRONMENTAL AND DEVELOPMENTAL REGULATION

An excellent review has been written on the environmental influences on chloride cells (Laurent and Perry, 1991), though a similar overview on gill Na^+,K^+-ATPase is lacking. Since space limitations preclude an exhaustive review, only heuristic and recent research will be presented here.

The most widely recognized and investigated environmental determinant of gill Na^+,K^+-ATPase and chloride cell development is salinity. For almost all teleosts examined to date, which includes several dozen species, increasing salinity results in higher levels of gill or opercular Na^+,K^+-ATPase activity (Kirschner, 1980; De Renzis and Bornancin, 1984). This is true both for comparisons within species and, generally, when comparing freshwater and marine forms. Despite this generalization, it is enlightening to examine the exceptions. No difference in gill Na^+,K^+-ATPase activity was found between freshwater- and seawater-adapted flounder (*Platichthys flesus;* Stagg and Shuttleworth, 1982) and striped bass (*Morone saxatilis;* Madsen *et al.*, 1994). Gill Na^+,K^+-ATPase activity has been found to be higher in fresh water in sea bass (*Dicentrarchus labrax*), thick-lipped mullet (*Chelon labrosus*), and Australian bass (*Macquaria novemaculeata;* Lasserre, 1971; Langdon, 1987). These fish are generally considered to be of marine origin. One possible explanation, which lacks experimental support, is that these fish maintain higher gill permeability and ion fluxes than other fish in fresh water and therefore require higher levels of gill Na^+,K^+-ATPase.

Exposure of fish to low ion content in fresh water results in a dramatic increase in chloride cell density, particularly on the secondary lamellae, which is presumably involved in ion uptake. Surprisingly, we know little of the impact of ion-poor water on gill Na^+,K^+-ATPase. In Atlantic salmon

(*Salmo salar*), gill Na^+,K^+-ATPase activity does not increase in ion-poor water (S. D. McCormick, unpublished).

Gill Na^+,K^+-ATPase activity generally increases following acclimation to low temperature, but there is contradictory evidence (even in the same species) that makes interpretations difficult (see De Renzis and Bornancin, 1984). In addition to methodological considerations, interpretation of temperature effects is made even more difficult by potential interaction of physiological demands on the gills, such as gas exchange and ion transport.

Although moderate changes in pressure have no effect on gill Na^+,K^+-ATPase activity *in vitro,* changes in pressure that might be experienced in the deep ocean result in rapid deactivation of gill Na^+,K^+-ATPase (Gibbs and Somero, 1989). This effect of pressure, which is presumably the result of changes in membrane fluidity and the structure of the enzyme, is less pronounced in fish living at greater depths. Deep-living species generally have lower gill Na^+,K^+-ATPase activity, which may reflect the lower activity and metabolic rate of these species (Gibbs and Somero, 1990).

Diet and nutrition may also play a role in regulation of chloride cell function. Food restriction in tilapia reduces chloride cell number and gill Na^+,K^+-ATPase activity, but this apparently has no effect on salinity tolerance (Kultz and Jurss, 1991). Diets high in NaCl can increase gill Na^+,K^+-ATPase activity and chloride cell numbers in rainbow trout, but this is not a universal finding among teleosts (Salman and Eddy, 1987).

Size has a potential but largely unexamined influence on chloride cell development. Gibbs and Somero (1990) found that total gill Na^+,K^+-ATPase (activity per fish) was related to size; this result was largely related to allometric growth of the gill rather than to changes within the gill. There is great variation of the scaling coefficient among species, indicating that gill Na^+,K^+-ATPase may increase or decrease with increasing size. No size relationship between protein-specific gill Na^+,K^+-ATPase activity and size was found in brook trout (*Salvelinus fontinalis*) weighing from 2 to 380 g (McCormick and Naiman, 1984a). Chloride cells on skin and yolk sac of small embryonic and larval fishes, present prior to gill development, may be important sites for ion exchange in fresh water and seawater (Ayson *et al.,* 1994).

Developmental events play a crucial role in regulating chloride cells in some diadromous fishes. In both downstream-migrating juvenile salmonids (known as smolts) and maturing "silver" European eels (*Anguilla anguilla*), chloride cell numbers and gill Na^+,K^+-ATPase activity increase prior to exposure to seawater (Zaugg and McLain, 1970; Thomson and Sargent, 1977; see reviews by McCormick and Saunders, 1987; Hoar, 1988). These preparatory adaptations result in increased salinity tolerance of the migrants. In anadromous salmonids, this developmental event is size dependent and cued by increasing daylength (Saunders and Henderson, 1970; Duston and Saunders, 1990). Some of the endocrine factors discussed in Section VI are likely to be responsive not only to salinity but also to environmental and developmental influences.

VI. HORMONAL REGULATION

A. Cortisol

Cortisol was the first hormone shown to stimulate gill Na^+,K^+-ATPase activity (Pickford *et al.,* 1970b); following this work with killifish, American eel (*Anguilla rostrata*), tilapia (*Oreochromis mossambicus*), and several salmonids have been shown to respond similarly (Epstein *et al.,* 1971; Richman and Zaugg, 1987; Björnsson *et al.,* 1987; Dange, 1986; Madsen, 1990c; Bisbal and Specker, 1991). Stenohaline fish have not been sufficiently examined, and it would be of some interest to determine if stenohalinity is in part due to an inability to respond to cortisol (or other hormones). Hypophysectomy reduces gill Na^+,K^+-ATPase activity in American eel, killifish, and coho salmon (*Oncorhynchus kisutch*), which can be at least partially restored by cortisol treatment (Pickford *et al.,* 1970b; Butler and Carmichael, 1972; Björnsson *et al.,* 1987; Richman and Zaugg, 1987). This effect is presumably through the loss of pituitary ACTH (a cortisol secretagogue), though other pituitary hormones are involved in regulating circulating cortisol and gill Na^+,K^+-ATPase (see the following).

Cortisol treatment also affects the morphology and development of chloride cells (Fig. 2). Exogenous cortisol causes increased chloride cell numbers in American eel, tilapia, brown trout (*Salmo trutta*), rainbow trout, and coho salmon (Foskett *et al.,* 1981; Madsen, 1990a,c; Doyle and Epstein, 1993; Richman and Zaugg, 1987). Chloride cell size is less often measured, but was found to increase following cortisol treatment in brown trout (Madsen, 1990c). Doyle and Epstein (1972) reported increased development of the tubular system in chloride cells of the American eel after cortisol injection. As might be expected from these effects on gill Na^+,K^+-ATPase activity and chloride cells, cortisol treatment also increases salinity tolerance in eel, brown trout, and Atlantic salmon (Epstein *et al.,* 1971; Madsen, 1990c; Bisbal and Specker, 1991).

In vitro stimulation of gill and opercular membrane Na^+,K^+-ATPase by cortisol has been demonstrated in coho salmon and tilapia, respectively (Fig. 3; McCormick and Bern, 1989; McCormick, 1990b), indicating its direct effect on these tissues. Using anthroylouabain, increased gill

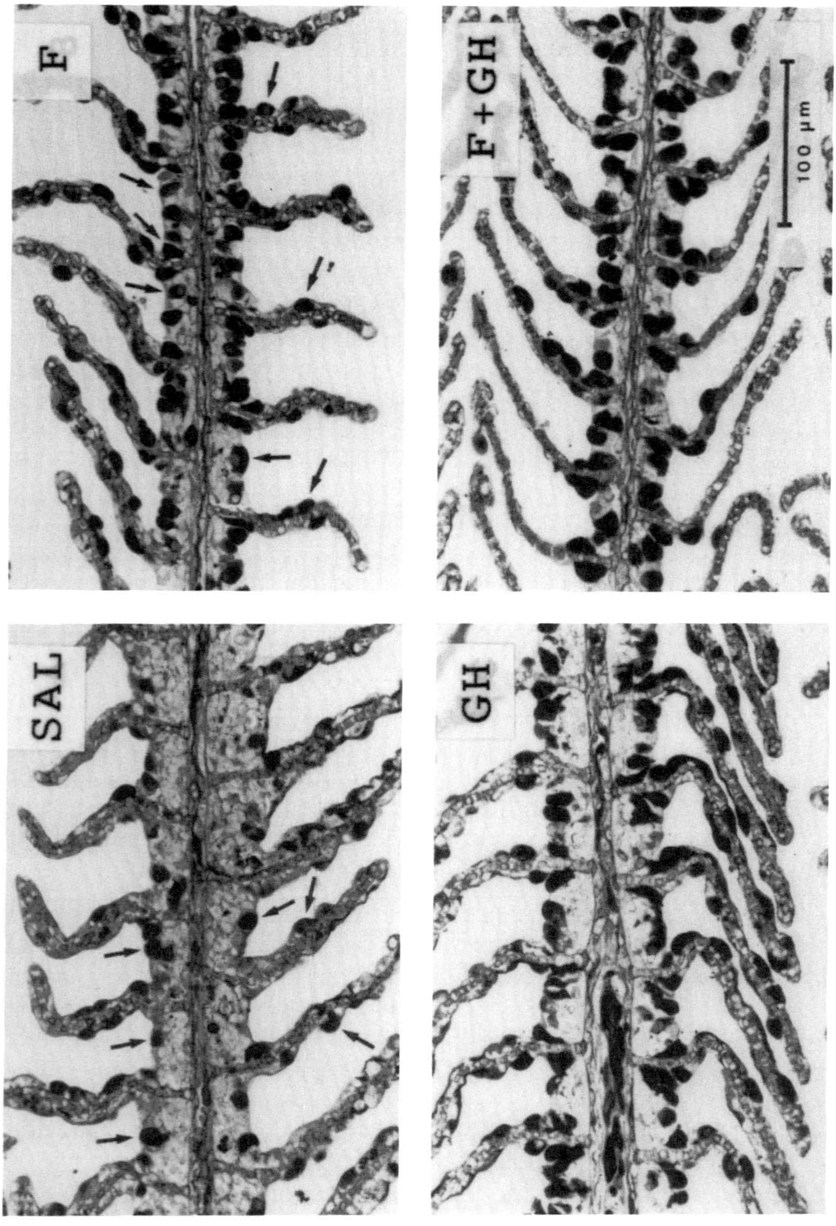

Fig. 2. The effect of cortisol (F) and growth hormone (GH) treatments on gill chloride cells of brown trout (SAL = saline-injected). Tissue was fixed in Champy-Maillet's fixative, which stains the extensive tubular system of chloride cells. Note that cortisol and growth hormone each increase cell number and cell size. When injected together, cell size is increased slightly (but not significantly) over treatment with either hormone alone. In the same experiment, cortisol and growth hormone acted additively to increase gill Na^+,K^+-ATPase activity. [Reproduced from Madsen (1990c) with permission from Academic Press.]

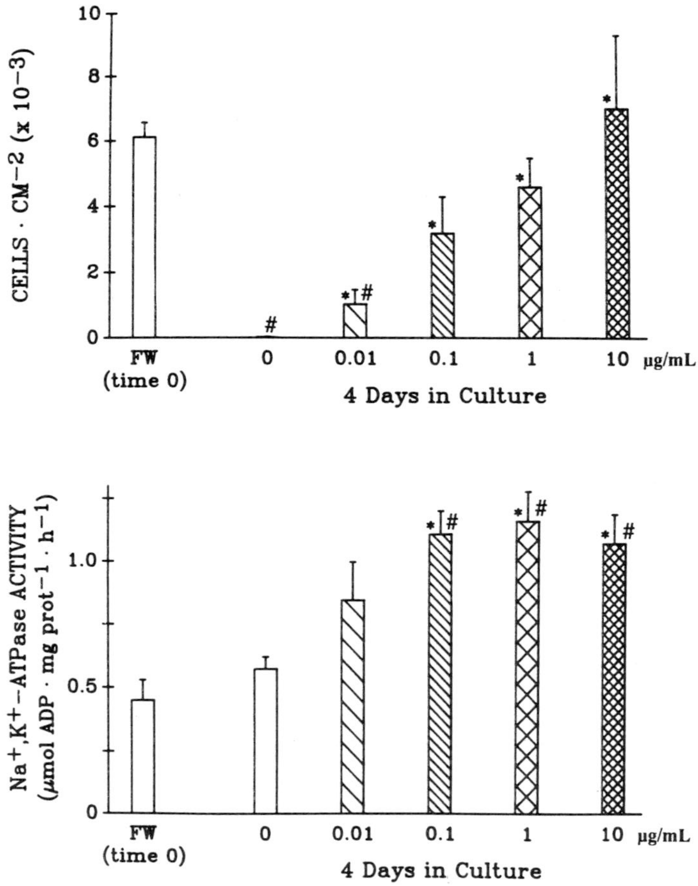

Fig. 3. Effect of cortisol *in vitro* on chloride cell number and Na^+,K^+-ATPase activity in opercular membrane of freshwater-adapted tilapia. Upper: In the absence of cortisol, chloride cell numbers decrease dramatically, and are increased in a dose-dependent manner by cortisol. Lower: Cortisol increases Na^+,K^+-ATPase activity to levels higher than seen initially. [*, indicates significant difference from control; #, indicates significant difference from initial (Time 0).] Cell height and the number of cells spanning the opercular membrane are also increased. [Reproduced from McCormick (1990b) with permission from The American Physiological Society.]

Na^+,K^+-ATPase was shown to be primarily a response of chloride cells. Cortisol also increased cell height and the proportion of chloride cells spanning the opercular membrane, but chloride cell numbers did not increase. From this evidence it was concluded that cortisol directly causes differentiation of the seawater form of the chloride cell. These *in vitro* findings contrast with the *in vivo* effect of cortisol in tilapia, where chloride cell density increases, but chloride cell size does not (Foskett *et al.*, 1981).

These differences suggest that other endocrine factors, stimulated by or acting with cortisol *in vivo,* are involved in chloride cell proliferation in tilapia. Electrophysiological properties of the opercular membrane are also not affected by cortisol *in vivo* (Foskett *et al.,* 1981), indicating that other hormones (or salinity itself) are required for activating sodium and chloride secretion.

Developmental differences in the responsiveness of gill Na^+,K^+-ATPase activity to cortisol *in vivo* and *in vitro* have been found in coho and Atlantic salmon (McCormick *et al.,* 1991a). Maximum responsiveness occurs when normal increases in gill Na^+,K^+-ATPase activity begin in early spring. As in the parr–smolt transformation itself, cortisol responsiveness can be altered by photoperiod. Changes in responsiveness may be due in large part to changes in cortisol receptors, which have been found in gill tissue of several euryhaline species (Sandor *et al.,* 1984; Chakraborti *et al.,* 1987; Maule and Schreck, 1990). In coho salmon and anadromous rainbow trout, gill cytosolic cortisol receptors are most numerous from January to March (Shrimpton *et al.,* 1994; McLeese *et al.,* 1994), the time of maximum *in vitro* responsiveness. Cortisol receptors may also be one avenue by which growth hormone interacts with cortisol (see the following). Weisbart *et al.* (1987) found that cytosolic cortisol receptors decrease and nuclear receptors increase following exposure of brook trout to seawater. This response is similar to that seen following injection of cortisol, suggesting that translocation of the cortisol receptor to the nucleus is involved in seawater acclimation. However, the location of cortisol receptors in gill tissue and their importance in chloride cell differentiation and/or proliferation have yet to be established. Corticoid receptor antagonists such as RU-486 (Baulieu, 1989) may prove to be useful in this regard.

Reported inabilities of cortisol treatment to stimulate gill Na^+,K^+-ATPase activity in salmonids and tilapia (Langdon *et al.,* 1984; Hegab and Hanke, 1984; Redding *et al.,* 1984) may in part relate to developmental differences in responsiveness to cortisol. The time course and method of administration of cortisol may also alter its effectiveness. Specker *et al.* (1994) have developed a vegetable oil-based implant that results in prolonged elevation of physiological levels of cortisol.

Often referred to as a "seawater-adapting hormone," it is interesting to find that cortisol may be more than permissively involved in ion uptake. Laurent and Perry (1990) have shown that cortisol treatment of freshwater rainbow trout results in increased influx of Na^+ and Cl^-, and increased number and apical surface area of gill chloride cells. Although it has yet to be demonstrated that cortisol increases net influx of Na^+ and Cl^-, this combination of ion transport and morphological evidence is particularly powerful, and similar results were obtained for European eel, bullhead catfish (*Ictalurus nebulosus*), and tilapia (Perry *et al.*, 1992). Cortisol also increases gill H^+-ATPase, which has been implicated in Na^+ uptake in coho salmon in fresh water (Lin and Randall, 1993). This potential dual effect of cortisol raises several fundamental questions in our understanding of chloride cell function. Does cortisol simultaneously increase both freshwater and seawater forms of the chloride cell? Or are chloride cells "bipolar," with the same cells able to switch rapidly from ion uptake to ion secretion? Answers to these questions must await the development of methods that will permit the *functional* differentiation of chloride cells.

A possible rapid action (minutes to hours) of cortisol on ion transport should not be discounted. Cortisol increases quickly following exposure to seawater (and other stressors) (Schreck, 1981). Mineralocorticoids in amphibians and mammals rapidly increase sodium retention (Minuth *et al.*, 1987). Forrest *et al.* (1973) found increased sodium efflux in American eels treated for 2 days with cortisol and then transferred to seawater. This effect occurred prior to an increase in gill Na^+,K^+-ATPase activity, which occurred after 14 days of cortisol treatment.

B. Growth Hormone

Sakamoto *et al.* (1993) present evidence for the role of growth hormone (GH) in seawater acclimation of salmonids, including the effects of exogenous GH, changes in circulating GH, and metabolic clearance of GH and GH receptors. Exogenous GH increases salinity tolerance in Atlantic salmon, sockeye salmon (*Oncorhynchus nerka*), and rainbow and brown trout; this effect was found to be independent of the hormone's influence on body size (Komourdjian *et al.*, 1976; Clarke *et al.*, 1977; Bolton *et al.*, 1987; Madsen 1990c). Treatment with GH for more than 1 week results in increased gill Na^+,K^+-ATPase activity and chloride cells in all of these species (Boeuf *et al.*, 1990; Madsen, 1990b,c). Growth hormone partially restores the decrease in gill Na^+,K^+-ATPase activity and salinity tolerance following hypophysectomy of coho salmon (Björnsson *et al.*, 1987; Richman *et al.*, 1987). As with cortisol in eels, growth hormone treatment for 2 days can be shown to improve ion regulatory performance of rainbow trout and Atlantic salmon in seawater, prior to a detectable increase in gill Na^+,K^+-ATPase activity (Collie *et al.*, 1989; McCormick *et al.*, 1991b). The short-term and long-term effects of cortisol and growth hormone may involve different mechanisms (such as protein activation, protein synthesis, mitogenesis, and hemodynamics), though much remains to be investigated in this regard.

Almost all the published work on the role of growth hormone in teleost hypoosmoregulation has been conducted with salmonids. However, Flik *et al.* (1993) found that GH treatment doubled chloride cells density in tilapia opercular membrane. GH also increases the salinity tolerance of hypophysectomized tilapia (T. Sakamoto and E. G. Grau, personal communication), suggesting that a significant role of GH in salt secretion may prove to be common among euryhaline fishes.

There is a strong interaction between growth hormone and cortisol in the regulation of salt secretion. GH and cortisol act in synergy to increase gill Na^+,K^+-ATPase activity and salinity tolerance in brown trout, rainbow trout, and Atlantic salmon (Madsen, 1990b,c). Findings of an increase in gill cortisol receptors following growth hormone treatment in coho salmon (Shrimpton *et al.*, 1995) suggest at least one pathway for this synergy. Although growth hormone receptors have been found in gill tissue of several salmonids (Gray *et al.*, 1990; Yao *et al.*, 1991; Sakamoto and Hirano, 1991), there is currently no evidence for a direct action of GH on Na^+,K^+-ATPase activity (McCormick *et al.*, 1991a). GH may also act through the interrenal to affect cortisol release. Young (1988) found that both *in vivo* and *in vitro* GH treatment increased responsiveness of the coho salmon interrenal to ACTH. However, an effect of GH on circulating levels of cortisol has yet to be demonstrated. In addition to its effect on cortisol receptors and cortisol secretion, at least one other avenue for the action of growth hormone is through its major influence on insulin-like growth factor I (IGF-I).

C. Insulin-like Growth Factor I

IGF-I is a 70-amino acid polypeptide that is produced primarily in the liver but also in several other tissues. Growth hormone is the most important secretogogue, and the physiological activity of IGF-I is controlled by several binding proteins, which like IGF-I itself are under complex endocrine and nutritional control. The cDNA of IGF-I has been isolated from five salmonids (Duan *et al.*, 1994), and the deduced amino acid sequence has an 80% similarity with mammalian IGF-I. Other aspects of the structure and actions of IGF-I in ectothermic vertebrates can be found in Bern *et al.* (1991).

IGF-I mRNA increases in gill and kidney of rainbow trout following seawater exposure (Sakamoto and Hirano, 1993). As with growth hormone, increases in salinity tolerance have been observed within 48 hr of injection of IGF-I in rainbow trout and Atlantic salmon (McCormick *et al.*, 1991b, S. D. McCormick, unpublished). It is not known whether this short-term effect involves the well-known mitogenic activity of IGF-I or some other mechanism of action. In other vertebrates, IGF-I has varied osmoregulatory actions, including rapid effects on glomerular filtration, direct stimulation of sodium transport in isolated epithelia, and mitogenesis of renal cells (Hammerman *et al.,* 1993).

IGF-I may also have a role in the long-term action of growth hormone and cortisol in stimulating gill Na^+,K^+-ATPase activity and chloride cells. Treatment of brown trout with cortisol, GH, or IGF-I increases gill Na^+,K^+-ATPase α-subunit mRNA (S. Madsen, personal communication). IGF-I treatment also increases responsiveness of gill Na^+,K^+-ATPase to cortisol in Atlantic salmon (S. D. McCormick, unpublished). IGF-I increases Na^+,K^+-ATPase in gill organ culture of coho salmon if the fish are pretreated with GH (Madsen and Bern, 1993). It appears, however, that IGF-I cannot carry out all the long-term effects of GH. Treatment of Atlantic salmon with GH for 2 weeks results in two-fold increases in gill Na^+,K^+-ATPase activity, but the same period of treatment with IGF-I has no effect (S. D. McCormick, unpublished). Although the mitogenic actions commonly associated with IGF-I suggest that this hormone may play a role in GH-mediated increases in numbers (or differentiation) of chloride cells, this remains to be established. Some of the action of IGF-I may be through paracrine or autocrine pathways (local hormone production); in rainbow trout, IGF-I mRNA levels in the gill are highest in chloride cells (T. Sakamoto and S. Hyodo, personal communication).

D. Prolactin

Since the seminal findings of Pickford and Phillips (1959) on the role of prolactin in maintaining ion homeostasis in freshwater killifish, prolactin has been shown to have a role in ion uptake in most, if not all, teleosts (see Hirano, 1986, for a review of prolactin in fish). In euryhaline fish, plasma prolactin levels decrease rapidly following transfer from fresh water to seawater, and increase upon "reverse" transfer. Surprisingly few investigations have examined the biochemical and morphological effects of prolactin on fish osmoregulatory organs. Prolactin decreases gill Na^+,K^+-ATPase activity and increases kidney Na^+,K^+-ATPase activity in hypophysectomized, freshwater killifish (Pickford *et al.,* 1970a). In seawater tilapia there was no effect of prolactin on gill Na^+,K^+-ATPase activity in spite of a profound effect on plasma ions and chloride cells (Herndon *et al.,* 1991). Madsen and Bern (1992) reported that prolactin injections reduce gill Na^+,K^+-ATPase activity in freshwater rainbow trout and antagonize the action of GH in increasing salinity tolerance and gill Na^+,K^+-ATPase activity. In contrast, Atlantic salmon implanted with prolactin show increased gill Na^+,K^+-ATPase activity, with no antagonism of the actions of GH (Boeuf *et al.,* 1994). This confusing picture of the effects of prolactin

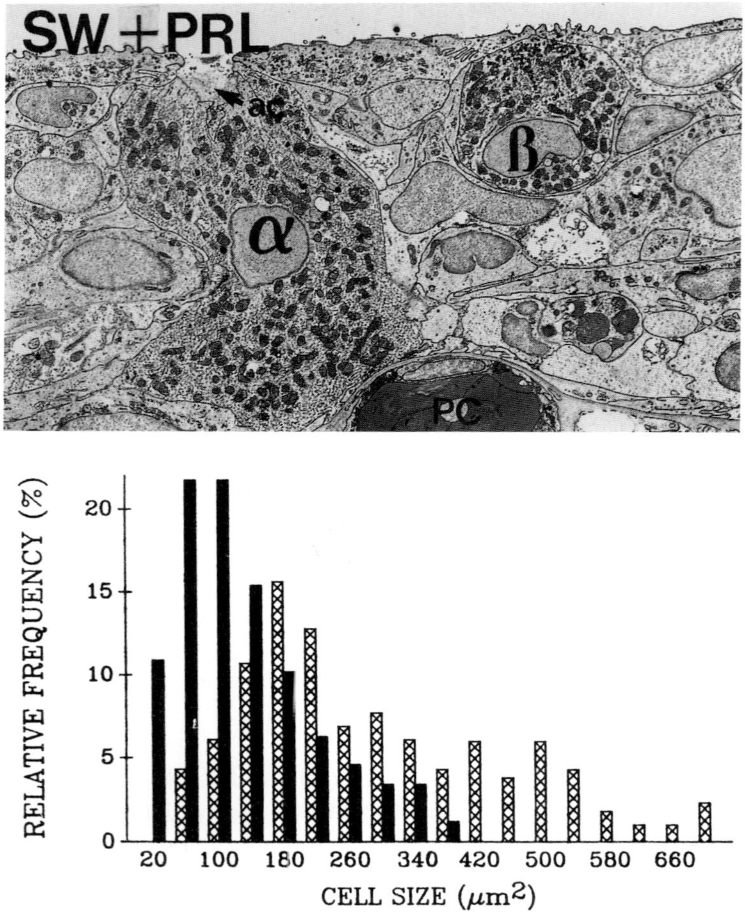

Fig. 4. The effects of prolactin (PRL) on chloride cell size and morphology in tilapia. Upper: α- and β-cells (the latter being the seawater form of the chloride cell) of the Nile tilapia (ac = apical crypt). Prolactin injection in seawater-adapted Nile tilapia caused the appearance of α-cells and decreased size of β-cells. [Reproduced from Pisam, M., Auperin, B., Prunet, P., Rentierdelrue, F., Martial, J., and Rambourg, A. (1993). Effects of prolactin on alpha and beta chloride cells in the gill epithelium of the saltwater adapted tilapia *Oreochromis niloticus*. *Anat. Rec.* **235,** 275–284. Copyright © (1993) by Wiley–Liss, Inc., a division of John Wiley & Sons, Inc. Reprinted by permission of John Wiley & Sons, Inc.] Lower: In Mozambique tilapia, prolactin (solid bars) treatment causes a dramatic shift in chloride cell size compared to saline (hatched bars) treatment. The vital mitochondrial stain DASPEI was used to identify chloride cells in this study. [Reproduced from Herndon *et al.* (1991) with permission from Academic Press.] The size distribution of chloride cells in these studies underscores the need to view them as a heterogeneous population of cells with varying degrees of development.

may stem from the use of heterologous hormones, species differences or developmental differences.

Although limited, current information indicates that prolactin strongly affects chloride cell morphology. Based on reductions in opercular membrane conductance and short-circuit current following prolactin treatment of seawater-adapted tilapia, Foskett *et al.* (1982) postulated that prolactin reduces chloride cell numbers and active transport in the remaining chloride cells. Herndon *et al.* (1991) found that prolactin treatment of seawater-adapted tilapia resulted in a dramatic reduction in chloride cell size without changing chloride cell density. Chloride cell height and proportion of cells spanning the opercular membrane were reduced, suggesting that these cells were effectively "removed" as chloride secretory cells (Fig. 4). Working with the closely related but stenohaline Nile tilapia (*Oreochromis niloticus*), Pisam *et al.* (1993) found that prolactin injection caused the appearance of the smaller "β-chloride cells" that were previously absent from these fish in seawater. The remaining "α-chloride cells" were also reduced in size and had shallower apical crypts and reduced tubular systems (Fig. 4).

These effects of prolactin on chloride cells raise several interesting questions. Is decreased cell size the result of prolactin acting on existing (secretory) chloride cells, or are undeveloped cells affected? How rapid do morphological changes occur and are they sufficient to explain the effects of prolactin on ion regulation? The possible mode(s) of action of prolactin have not been examined. Prolactin receptors have been found in gill tissue (Dauder *et al.*, 1990; Prunet and Auperin, 1994), but as yet there is no evidence for direct action of prolactin on the gills or opercular membrane and regulation of prolactin receptors has not been investigated. Investigations into other possible endocrine factors involved in the prolactin response and interaction of prolactin with other endocrine systems are required.

E. Thyroid Hormones

The reported effects of thyroid hormones in teleost hypoosmoregulation are contradictory. Thyroxine (T_4) or 3,5′,3′-triiodo-L-thyronine (T_3) incorporated in food can improve salinity tolerance of coho and Atlantic salmon (Fagerlund *et al.*, 1980; Refstie, 1982; Saunders *et al.*, 1985). It is unclear whether this is an effect on osmoregulation per se, since the effect apparently depends on prior increases in body size, and salinity tolerance in salmonids is size dependent (McCormick and Naiman, 1984b). Studies on chum and amago salmon have shown no effect of T_4 or thiourea, an inhibitor of T_4 production, on salinity tolerance (Miwa and Inui, 1983;

Iwata *et al.*, 1987) despite the fact that these treatments were effective in altering circulating levels of thyroid hormones and body silvering (a morphological change that is part of the parr–smolt transformation). The possible influence of thyroid hormones on salinity tolerance in nonsalmonids has not been examined.

Thyroxine has no apparent positive effect by itself on gill Na^+,K^+-ATPase activity in tilapia or amago, coho, and Atlantic salmon (Miwa and Inui, 1985; Saunders *et al.*, 1985; Björnsson *et al.*, 1987; Dange, 1986). In rainbow trout, T_4 injections had no effect on gill Na^+,K^+-ATPase activity (Madsen, 1990a), whereas T_3 immersion decreased activity (Omeljaniuk and Eales, 1986). In contrast to these studies, Madsen and Korsgaard (1989) found that multiple injections of thyroxine could advance increases in gill Na^+,K^+-ATPase activity and chloride cell numbers that occur during the parr–smolt transformation of Atlantic salmon.

Greater agreement in the literature is found regarding the interaction of thyroid hormones with other endocrine axes. Thyroxine increases the capacity of cortisol to increase gill Na^+,K^+-ATPase activity in tilapia (Dange, 1986). In amago salmon, T_4 and GH in combination were capable of elevating gill Na^+,K^+-ATPase, whereas each hormone alone had no effect. The most convincing evidence for a role of thyroid hormones in seawater acclimation utilized an inhibitor of 5′-monodeiodination of T_4 (Lebel and Leloup, 1992; Leloup and Lebel, 1993). In these studies, rainbow and brown trout have been found to require conversion of T_4 to T_3 for normal and GH-stimulated acclimation to seawater. From these results the authors suggest that the action of growth hormone on hypoosmoregulatory mechanisms is through its increase of T_4 to T_3 conversion (de Luze *et al.*, 1989). However, this seems to be an incomplete explanation given the equivocal effect of T_3 on salinity tolerance and its general inability to increase gill Na^+,K^+-ATPase activity. An alternative explanation is that T_3 is necessary for the peripheral action of GH (or IGF-I), possibly through regulation of receptors or binding proteins. Other indirect pathways for thyroid hormone action include their ability to increase pituitary GH production (Moav and McKeown, 1992) and interrenal sensitivity to ACTH (Young and Lin, 1988). Thyroid hormone receptors have been found in liver, kidney, and gill of rainbow trout, brown trout, and European eel (Bres and Eales, 1988; Lebel and Leloup, 1989), but their endocrine regulation and response to salinity have not been examined.

F. Sex Steroids

Normal sexual maturation and treatment with exogenous sex steroids have been shown to have a negative effect on the ability of several salmonids to adapt to seawater (McCormick and Naiman, 1985; Ikuta *et al.*, 1987; Lundqvist *et al.*, 1989; Schmitz and Mayer, 1993). Repeated injections of 17β-estradiol result in decreased gill chloride cell density and Na^+,K^+-ATPase activity in Atlantic salmon (Madsen and Korsgaard, 1989). The mode of action of sex steroids, particularly whether they are acting directly on the gill or through other endocrine systems, is currently unknown. It is also unclear whether the action of sex steroids on hypoosmoregulation is peculiar to the anadromous life history of salmonids in which sexual maturation occurs only in fresh water; research on other fishes is clearly warranted.

G. Rapid Activation

Little work has been done on the rapid activation of Na^+,K^+-ATPase. Although it can be assumed that any of the hormones involved in rapid changes in ion efflux will involve stimulation of Na^+,K^+-ATPase, it is not clear whether these will be direct (stimulation of the pump itself) or indirect (through increased substrate availability provided by stimulation or inhibition of other transporters or channels). Important advances in this area will come from methods that permit rapid, simultaneous measurement of ion transport and morphological or biochemical changes in chloride cells. Ion-sensitive dyes and fluorescent probes may be particularly valuable in this area.

VII. SUMMARY AND PROSPECTUS

Any summary of the endocrine control of physiological function in fishes must confront the diversity and evolutionary history of this large group and the sometimes contradictory results in the literature. Nevertheless, a hypothetical model of endocrine regulation of Na^+,K^+-ATPase and chloride secretory cells will apply to some species (Fig. 5). Evidence to date indicates that cortisol (F), growth hormone (GH), insulin-like growth factor I (IGF-I), and thyroid hormones (T_4 and T_3) increase Na^+,K^+-ATPase and/or promote the differentiation of chloride secretory cells. Cortisol has been shown to have direct action on Na^+,K^+-ATPase and chloride cells, and cortisol receptors have been found in gill tissue. Growth hormone acts in several ways: by increasing gill cortisol receptors, by increasing the sensitivity of the interrenal to adrenocorticotropic hormone (ACTH), and by production of IGF-I. Growth hormone receptors have been found in both gill and kidney. Although IGF-I is known to increase salinity tolerance and Na^+,K^+-ATPase activity, its mode of action is

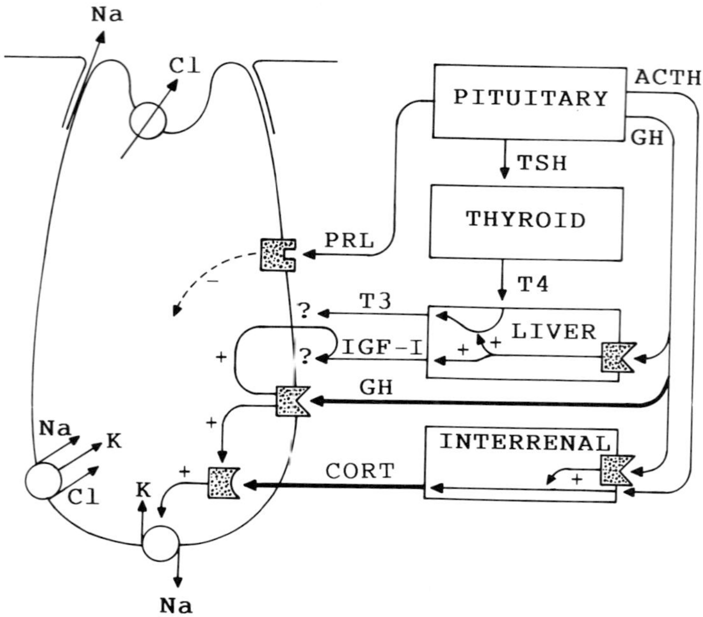

Fig. 5. Model of endocrine regulation of Na^+,K^+-ATPase and chloride secretory cells. Evidence to date indicates that cortisol (CORT), growth hormone (GH), insulin-like growth factor I (IGF-I), and thyroid hormones (T_4 and T_3) increase Na^+,K^+-ATPase and/or promote the differentiation of chloride secretory cells, whereas prolactin (PRL) is inhibitory. Adrenocorticotropic hormone (ACTH) and thyroid-stimulating hormone (TSH) are pituitary hormones involved in stimulating secretion of cortisol and thyroid hormones, respectively. See text for details.

unknown and receptors in the gill have not been demonstrated. The degree to which the GH/IGF-I axis is important in nonsalmonids requires clarification. Thyroid hormones promote chloride cell development, possibly through their interaction with the GH/IGF-I axis, though this has yet to be fully explored. Prolactin inhibits the development of the chloride secretory cell and promotes the development of the freshwater form of the chloride cell. Prolactin's effects on gill Na^+,K^+-ATPase are equivocal, and although prolactin receptors have been found in the gill, its mode of action is unclear. The universality of this model is suspect, and particularly may not apply to those species in which increased Na^+,K^+-ATPase is associated with exposure to fresh water rather than seawater.

Recent and upcoming advances promote optimism about the solution of the questions posed in this review. The advance of molecular methods will allow for the greater use of homologous hormones and radioimmunoassays and the detection of changes in receptor gene expression. Isolation of the cDNA for other ATPases and cotransport proteins of chloride cells (in fresh water and seawater) will be powerful tools for determining the functional attributes of chloride cells and their endocrine control. Continued development of methods for organ and cell culture will advance our understanding of direct hormone actions. Combining the new methods with classical physiology, histology, and endocrinology will be especially useful to our understanding of chloride cell regulation.

ACKNOWLEDGMENTS

I thank Mark Shrimpton and Joe Zydlewski for their helpful comments and discussion. Howard Bern, Tetsuye Hirano, Trevor Shuttleworth, and an anonymous reviewer made many helpful comments in review. I am grateful to Victoria McCormick for her rendition of Fig. 5.

REFERENCES

Avella, M., and Bornancin, M. (1990). Ion fluxes in the gills of freshwater and seawater salmonid fish. *In* "Animal Nutrition and Transport Processes. 2. Transport, Respiration and Excretion: Comparative and Environmental Aspects" (J. P. Truchet and B. Lahlou, eds.), Comp. Physiol. Vol. 6, pp. 1–13. Karger, Basel.

Avella, M., Masoni, A., Bornancin, M., and Mayer-Gostan, N. (1987). Gill morphology and sodium influx in the rainbow trout (*Salmo gairdneri*) acclimated to artificial freshwater environments. *J. Exp. Zool.* **241,** 159–169.

Ayson, F. G., Kaneko, T., Hasegawa, S., and Hirano, T. (1994). Development of mitochondrion-rich cells in the yolk-sac membrane of embryos and larvae of tilapia, *Oreochromis mossambicus,* in fresh water and seawater. *J. Exp. Zool.* **270,** 129–135.

Baulieu, E. E. (1989). Contragestion and other clinical applications of RU486, an antiprogesterone at the receptor. *Science* **245,** 1351–1357.

Bern, H. A., McCormick, S. D., Kelley, K. M., Gray, E. S., Nishioka, R. S., Madsen, S. S., and Tsai, P. I. (1991). Insulin-like growth factors "under water": Role in growth and function of fish and other poikilothermic vertebrates. *In* " Modern Concepts of Insulin-like Growth Factors" (E. M. Spencer, ed.), pp. 85–96. Elsevier, New York.

Bisbal, G. A., and Specker, J. L. (1991). Cortisol stimulates hypo-osmoregulatory ability in Atlantic salmon, *Salmo salar* L. *J. Fish Biol.* **39,** 421–432.

Björnsson, B. T., Yamauchi, K., Nishioka, R. S., Deftos, L. J., and Bern, H. A. (1987). Effects of hypophysectomy and subsequent hormonal replacement therapy on hormonal and osmoregulatory status of coho salmon, *Oncorhynchus kisutch. Gen. Comp. Endocrinol.* **68,** 421–430.

Boeuf, G., Prunet, P., and Le Bail, P.-Y. (1990). Is growth hormone treatment able to stimulate the smoltification in the Atlantic salmon? *Seances Acad. Sci.* **310,** 75–80.

Boeuf, G., Marc, A. M., Prunet, P., Le Bail, P.-Y., and Smal, J. (1994). Stimulation of the parr–smolt transformation by hormonal treatment in Atlantic salmon (*Salmo salar* L.). *Aquaculture* **121,** 195–208.

Bolton, J. P., Collie, N. L., Kawauchi, H., and Hirano, T. (1987). Osmoregulatory actions of growth hormone in rainbow trout (*Salmo gairdneri*). *J. Endocrinol.* **112,** 63–68.

Bres, O., and Eales, J. G. (1988). High-affinity, limited-capacity triiodothyronine-binding sites in nuclei from various tissues of the rainbow trout (*Salmo gairdneri*). *Gen. Comp. Endocrinol.* **69,** 71–79.

Butler, D. G., and Carmichael, F. J. (1972). (Na^+-K^+)-ATPase activity in eel (*Anguilla rostrata*) gill in relation to change in environmental salinity: Role of adrenocortical steroids. *Gen. Comp. Endocrinol.* **19,** 421–427.

Chakraborti, P. K., Weisbart, M., and Chakraborti, A. (1987). The presence of corticosteroid receptor activity in the gills of the brook trout, *Salvelinus fontinalis*. *Gen. Comp. Endocrinol.* **66,** 323–332.

Clarke, W. C., Farmer, S. W., and Hartwell, K. M. (1977). Effects of teleost pituitary growth hormone on growth of *Tilapia mossambica* and on growth and seawater adaptation of sockeye salmon (*Oncorhynchus nerka*). *Gen. Comp. Endocrinol.* **33,** 174–178.

Collie, N. L., Bolton, J. P., Kawauchi, H., and Hirano, T. (1989). Survival of salmonids in seawater and the time-frame of growth hormone action. *Fish Physiol. Biochem.* **7,** 315–321.

Dange, A. D. (1986). Branchial Na^+,K^+-ATPase activity in freshwater or saltwater acclimated tilapia *Oreochromis* (*Sarotherodon*) *mossambicus*: Effects of cortisol and thyroxine. *Gen. Comp. Endocrinol.* **62,** 341–343.

Dauder, S., Young, G., Hass, L., and Bern, H. A. (1990). Prolactin receptors in liver, kidney, and gill of the tilapia (*Oreochromis mossambicus*): Characterization and effect of salinity on specific binding of iodinated ovine prolactin. *Gen. Comp. Endocrinol.* **77,** 368–377.

de Luze, A., Leloup, J., Papkoff, H., Kikuyama, S., and Kawauchi, H. (1989). Effects of vertebrate prolactins and growth hormones on thyroxine 5′-monodeiodination in the eel (*Anguilla anguilla*): A potential bioassay for growth hormone. *Gen. Comp. Endocrinol.* **73,** 186–193.

De Renzis, G., and Bornancin, M. (1984). Ion transport and gill ATPases. *In* "Fish Physiology" (W. S. Hoar and D. J. Randall, eds.), Vol. 10B, pp. 65–104.

Doyle, W. L., and Epstein, F. H. (1972). Effects of cortisol treatment and osmotic adaptation on the chloride cells in the eel, *Anguilla rostrata*. *Cytobiologie* **6,** 58–73.

Duan, C., Duguay, S. J., Swanson, P., Dickhoff, W. W., and Plisetskaya, E. (1994). Tissue-specific expression of insulin-like growth factor I messenger ribonucleic acids in salmonids: Developmental, hormonal, and nutritional regulation. *In* "Perspectives in Comparative Endocrinology" (K. G. Davey, R. E. Peter, and S. S. Tobe, eds.), pp. 365–372. National Research Council of Canada, Ottawa.

Duston, J., and Saunders, R. L. (1990). The entrainment role of photoperiod on hypoosmoregulatory and growth-related aspects of smolting in Atlantic salmon (*Salmo salar*). *Can. J. Zool.* **68,** 707–715.

Epstein, F. H., Katz, A. I., and Pickford, G. E. (1967). Sodium- and potassium-activated adenosine triphosphatase of gills: Role in adaptation of teleosts to salt water. *Science* **156,** 1245–1247.

Epstein, F. H., Cynamon, M., and McKay, W. (1971). Endocrine control of Na-K-ATPase and seawater adaptation in *Anguilla rostrata*. *Gen. Comp. Endocrinol.* **16,** 323–328.

Evans, D. H., Claiborne, J. B., Farmer, L., Mallery, C., and Krasny, E. J. (1982). Fish gill ionic transport methods and models. *Biol. Bull.* (*Woods Hole, Mass.*) **163,** 108–130.

Fagerlund, U. H. M., Higgs, D. A., McBride, J. R., Plotnikoff, M. D., and Dosanjh, B. S. (1980). The potential for using anabolic hormones 17a-methyltestosterone and (or) 3,5,3'-triiodo-L-thyronine in the fresh water rearing of coho salmon (*Oncorhynchus kisutch*) and the effects on subsequent seawater performance. *Can. J. Zool.* **58,** 1424–1432.

Fernandez, R. D., Yoshimizu, M., Ezura, Y., and Kimura, T. (1993). Comparative growth response of fish cell lines in different media, temperatures, and sodium chloride concentrations. *Gyobyo Kenkyu* **28,** 27–34.

Flik, G., Atsma, W., Fenwick, J. C., Rentierdelrue, F., Smal, J., and Bonga, S. E. W. (1993). Homologous recombinant growth hormone and calcium metabolism in the tilapia, *Oreochromis mossambicus,* adapted to fresh water. *J. Exp. Biol.* **185,** 107–119.

Forrest, J. N., Cohen, A. D., Schon, D. A., and Epstein, F. H. (1973). Na^+ transport and Na-K-ATPase in gills during adaptation to seawater: Effects of cortisol. *Am. J. Physiol.* **224,** 709–713.

Fortes, P. A. G. (1977). Anthroylouabain: A specific fluorescent probe for the cardiac glycoside receptor. *Biochemistry* **16,** 531–540.

Foskett, J. K., and Scheffey, C. (1982). The chloride cell: Definitive identification as the salt-secretory cell in teleosts. *Science* **215,** 164–166.

Foskett, J. K., Logsdon, C. D., Turner, T., Machen, T. E., and Bern, H. A. (1981). Differentiation of the chloride extrusion mechanism during seawater adaptation of a teleost fish, the cichlid *Sarotherodon mossambicus. J. Exp. Biol.* **93,** 209–224.

Foskett, J. K., Machen, T. E., and Bern, H. A. (1982). Chloride secretion and conductance of teleost opercular membrane: Effects of prolactin. *Am. J. Physiol.* **242,** R380–R389.

Foskett, J. K., Bern, H. A., Machen, T. E., and Conner, M. (1983). Chloride cells and the hormonal control of teleost fish osmoregulation. *J. Exp. Biol.* **106,** 255–281.

Gibbs, A., and Somero, G. N. (1989). Pressure adaptation of Na^+/K^+-ATPase in gills of marine teleosts. *J. Exp. Biol.* **143,** 475–492.

Gibbs, A., and Somero, G. N. (1990). Na^+-K^+-Adenosine triphosphatase activities in gills of marine teleost fishes: Changes with depth, size and locomotory activity level. *Mar. Biol. (Berlin)* **106,** 315–321.

Gray, E. S., Young, G., and Bern, H. A. (1990). Radioreceptor assay for growth hormone in coho salmon (*Oncorhynchus kisutch*) and its application to the study of stunting. *J. Exp. Biol.* **256,** 290–296.

Hammerman, M. R., Oshea, M., and Miller, S. B. (1993). Role of growth factors in regulation of renal growth. *Annu. Rev. Physiol.* **55,** 305–321.

Hegab, S. A., and Hanke, W. (1984). The significance of cortisol for osmoregulation in carp (*Cyprinus carpio*) and tilapia (*Sarotherodon mossambicus*). *Gen. Comp. Endocrinol.* **54,** 409–417.

Heisler, N. (1993). Acid–base regulation. *In* "The Physiology of Fishes" (D. H. Evans, ed.), pp. 343–378. CRC Press, Boca Raton, FL.

Herndon, T. M., McCormick, S. D., and Bern, H. A. (1991). Effects of prolactin on chloride cells in opercular membrane of seawater-adapted tilapia. *Gen. Comp. Endocrinol.* **83,** 283–289.

Hirano, T. (1986). The spectrum of prolactin action in teleosts. *In* "Comparative Endocrinology: Developments and Directions" (C. L. Ralph, ed.), pp. 53–74. A. R. Liss, New York.

Hoar, W. S. (1988). The physiology of smolting salmonids. *In* "Fish Physiology" (W. S. Hoar and D. Randall, eds.), Vol. 11B, pp. 275–343. Academic Press, Orlando, FL.

Hootman, S. W., and Ernst, S. A. (1988). Estimation of Na^+,K^+-pump numbers and turnover in intact cells with [^{3}H]ouabain. *In* "Methods in Enzymology" (S. Fleischer and B. Fleischer, eds.), Vol. 156. pp. 213–228. Academic Press, Orlando, FL.

Hossler, F. E., Ruby, J. R., and McIlwain, T. D. (1979). The gill arch of the mullet, *Mugil cephalus*. II. Modification in surface ultrastructure and Na,K-ATPase content during adaptation to various salinities. *J. Exp. Zool* **208,** 399–405.

Ikuta, K., Aida, K., Okumoto, N., and Hanyu, I. (1987). Effects of sex steroids on the smoltification of masu salmon, *Oncorhynchus masou. Gen. Comp. Endocrinol.* **65,** 99–110.

Iwata, M., Komatsu, S., Hasegawa, S., Ogasawara, T., and Hirano, T. (1987). Inconsistent effect of thyroid hormone alteration on seawater adaptability of fry of chum salmon *Oncorhynchus keta. Nippon Suisan Gakkaishi* **53,** 1969–1973.

Johnson, S. L., Ewing, R. D., and Lichatowich, J. A. (1977). Characterization of gill (Na + K)-activated adensine triphosphatase from chinook salmon, *Oncorhynchus tshawytscha. J. Exp. Zool.* **199,** 345–354.

Kamiya, M. (1972). Sodium–potassium-activated adenosine triphosphatase in isolated chloride cells from eel gills. *Comp. Biochem. Physiol., B* **43,** 611–617.

Karnaky, K. J. (1986). Structure and function of the chloride cell of *Fundulus heteroclitus* and other teleosts. *Am. Zool.* **26,** 209–224.

Karnaky, K. J., Kinter, L. B., Kinter, W. B., and Stirling, C. E. (1976). Teleost chloride cell. II. Autoradiographic localization of gill Na,K-ATPase in killifish *Fundulus heteroclitus* adapted to low and high salinity environments. *J. Cell Biol.* **70,** 157–177.

Karnaky, K. J., Degnan, K. J., and Zadunaisky, J. A. (1977). Chloride transport across isolated opercular epithelium of killifish: A membrane rich in chloride cells. *Science* **195,** 203–205.

Karnaky, K. J., Degnan, K. J., Garretson, L. T., and Zadunaisky, J. A. (1984). Identification and quantification of mitochondria-rich cells in transporting cells in transporting epithelia. *Am. J. Physiol.* **246,** R770–R775.

Kawakami, K., Noguchi, S., Noda, M., Takahashi, H., Toshiko, O., Kawamura, M., Nojima, H., Nagano, K., Hirose, T., Inayama, S., Hayashida, H., Miyata, T., and Numa, S. (1985). Primary structure of the a-subunit of *Torpedo californica* (Na^+ + K^+)-ATPase deduced from cDNA sequence. *Nature (London)* **316,** 733–736.

Kerstetter, T. H., and Keeler, M. (1976). On the interaction of NH_4^+ and Na^+ fluxes in the isolated trout gill. *J. Exp. Biol.* **64,** 517–527.

Kirschner, L. B. (1980). Comparison of vertebrate salt-excreting organs. *Am. J. Physiol.* **7,** R219–R223.

Kisen, G., Gallais, C., Auperin, B., Klungland, H., Sandra, O., Prunet, P., and Andersen, O. (1994). Northern blot analysis of the Na^+,K^+-ATPase α-subunit in salmonids. *Comp. Biochem. Physiol., B* **107,** 255–259.

Komourdjian, M. P., Saunders, R. L., and Fenwick, J. C. (1976). The effect of porcine somatotropin on growth and survival in seawater of Atlantic salmon (*Salmo salar*) parr. *Can. J. Zool.* **54,** 531–535.

Kultz, D., and Jurss, K. (1991). Acclimation of chloride cells and Na/K-ATPase to energy deficiency in tilapia (*Oreochromis mossambicus*). *Zool. Jahrb., Abt. Allg. Zool. Physiol.* **95,** 39–50.

Langdon, J. S. (1987). Active osmoregulation in the Australian bass, *Macquaria novemaculeata* (Steindachner), and the golden perch, *Macquaria ambigua* (Richardson) (Percichthyidae). *Aust. J. Mar. Freshwater Res.* **38,** 771–776.

Langdon, J. S., Thorpe, J. E., and Roberts, R. J. (1984). Effects of cortisol and ACTH on gill Na^+/K^+-ATPase, SDH and chloride cells in juvenile Atlantic salmon, *Salmo salar. Comp. Biochem. Physiol., A* **77,** 9–12.

Lasserre, P. (1971). Increase of (Na^+ + K^+)-dependent ATPase activity in gills and kidneys of two marine teleosts, *Crenimugil labrosus* (Risso, 1826) and *Dicentrarchus labrax* (Linnaeus, 1758), during adaptation to fresh water. *Life Sci.* **10,** 113–119.

Laurent, P., and Dunel, S. (1980). Morphology of gill epithelia in fish. *Am. J. Physiol.* **7,** R147–R159.

Laurent, P., and Perry, S. F. (1990). Effects of cortisol on gill chloride cell morphology and ionic uptake in the freshwater trout, *Salmo gairdneri*. *Cell Tissue Res.* **259,** 429–442.

Laurent, P., and Perry, S. F. (1991). Environmental effects on fish gill morphology. *Physiol. Zool.* **64,** 4–25.

Laurent, P., Hobe, H., and Dunel-Erb, S. (1985). The role of environmental sodium chloride relative to calcium in gill morphology of freshwater salmonid fish. *Cell Tissue Res.* **240,** 675–692.

Lebel, J. M., and Leloup, J. (1989). Triiodothyronine binding to putative solubilized nuclear thyroid hormone receptors in liver and gill of the brown trout (*Salmo trutta*) and the European eel (*Anguilla anguilla*). *Gen. Comp. Endocrinol.* **75,** 301–309.

Lebel, J. M., and Leloup, J. (1992). Triiodothyronine is required for the acclimation to seawater of the brown trout (*Salmo trutta*) and rainbow trout (*Oncorhynchus mykiss*). *C. R. Seances Acad. Sci., Ser. 3* **314,** 461–468.

Leloup, J., and Lebel, J. M. (1993). Triiodothyronine is necessary for the action of growth hormone in acclimation to seawater of brown (*Salmo trutta*) and rainbow trout (*Oncorhynchus mykiss*). *Fish Physiol. Biochem.* **11,** 165–173.

Lin, H., and Randall, D. J. (1993). H^+-ATPase activity in crude homogenates of fish gill tissue: Inhibitor sensitivity and environmental and hormonal regulation. *J. Exp. Biol.* **180,** 163–174.

Lundqvist, H., Borg, B., and Berglund, I. (1989). Androgens impair seawater adaptability in smolting Baltic salmon (*Salmo salar*). *Can. J. Zool.* **67,** 1733–1736.

Madsen, S. S. (1990a). Effect of repetitive cortisol and thyroxine injections on chloride cell number and Na^+/K^+-ATPase activity in gill of freshwater acclimated rainbow trout, *Salmo gairdneri*. *Comp. Biochem. Physiol., A* **95,** 171–176.

Madsen, S. S. (1990b). Enhanced hypoosmoregulatory response to growth hormone after cortisol treatment in immature rainbow trout, *Salmo gairdneri*. *Fish Physiol. Biochem.* **8,** 271–279.

Madsen, S. S. (1990c). The role of cortisol and growth hormone in seawater adaptation and development of hypoosmoregulatory mechanisms in sea trout parr (*Salmo trutta trutta*). *Gen. Comp. Endocrinol.* **79,** 1–11.

Madsen, S. S., and Bern, H. A. (1992). Antagonism of prolactin and growth hormone: Impact on seawater adaptation in two salmonids, *Salmo trutta* and *Oncorhynchus mykiss*. *Zool. Sci.* **9,** 775–784.

Madsen, S. S., and Bern, H. A. (1993). *In vitro* effects of insulin-like growth factor-I on gill Na^+,K^+-ATPase in coho salmon, *Oncorhynchus kisutch*. *J. Endocrinol.* **138,** 23–30.

Madsen, S. S., and Korsgaard, B. (1989). Time-course effects of repetitive oestradiol-17β and thyroxine injections on the natural spring smolting of Atlantic salmon, *Salmo salar* L. *J. Fish Biol.* **35,** 119–128.

Madsen, S. S., McCormick, S. D., Young, G., Endersen, J. S., Nishioka, R. S., and Bern, H. A. (1994). Physiology of seawater acclimation in the striped bass, *Morone saxatilis* (Walbaum). *Fish Physiol. Biochem.* **13,** 1–11.

Maule, A. G., and Schreck, C. B. (1990). Glucocorticoid receptors in leukocytes and gill of juvenile coho salmon (*Oncorhynchus kisutch*). *Gen. Comp. Endocrinol.* **77,** 448–455.

Mayer-Gostan, N., and Lemaire, S. (1991). Measurements of fish gill ATPases using microplates. *Comp. Biochem. Physiol., B* **98,** 323–326.

Mayer-Gostan, N., Wendelaar Bonga, S. E., and Balm, P. H. M. (1987). Mechanisms of hormone actions on gill transport. *In* "Vertebrate Endocrinology: Fundamentals and Biomedical Implications" (P. K. T. Pang and M. P. Schreibman, eds.), Vol. 2, pp. 211–238. Academic Press, Boston.

McCormick, S. D. (1990a). Fluorescent labelling of Na^+,K^+-ATPase in intact cells by use of a fluorescent derivative of ouabain: Salinity and teleost chloride cells. *Cell Tissue Res.* **260,** 529–533.

McCormick, S. D. (1990b). Cortisol directly stimulates differentiation of chloride cells in tilapia opercular membrane. *Am. J. Physiol.* **259,** R857–R863.

McCormick, S. D. (1993). Methods for non-lethal gill biopsy and measurement of Na^+,K^+-ATPase activity. *Can. J. Fish. Aquat. Sci.* **50,** 656–658.

McCormick, S. D. (1994). Opercular membranes and skin. *In* "Biochemistry and Molecular Biology of Fishes" (P. W. Hochachka and T. P. Mommsen, eds.), Vol. 3, pp. 231–238. Elsevier, Amsterdam.

McCormick, S. D., and Bern, H. A. (1989). *In vitro* stimulation of Na^+,K^+-ATPase activity and ouabain binding by cortisol in coho salmon gill. *Am. J. Physiol.* **256,** R707–R715.

McCormick, S. D., and Naiman, R. J. (1984a). Osmoregulation in the brook trout, *Salvelinus fontinalis*. I. Diel, photoperiod and growth related physiological changes in freshwater. *Comp. Biochem. Physiol., A* **79,** 7–16.

McCormick, S. D., and Naiman, R. J. (1984b). Osmoregulation in the brook trout, *Salvelinus fontinalis*. II. Effects of size, age and photoperiod on seawater survival and ionic regulation. *Comp. Biochem. Physiol., A* **79,** 17–28.

McCormick, S. D., and Naiman, R. J. (1985). Hypoosmoregulation in an anadromous teleost: Influence of sex and maturation. *J. Exp. Zool.* **234,** 193–198.

McCormick, S. D., and Saunders, R. L. (1987). Preparatory physiological adaptations for marine life in salmonids: Osmoregulation, growth and metabolism. *Am. Fish. Soc. Symp.* **1,** 211–229.

McCormick, S. D., Dickhoff, W. W., Duston, J., Nishioka, R. S., and Bern, H. A. (1991a). Developmental differences in the responsiveness of gill Na^+,K^+-ATPase to cortisol in salmonids. *Gen. Comp. Endocrinol.* **84,** 308–317.

McCormick, S. D., Sakamoto, T., Hasegawa, S., and Hirano, T. (1991b). Osmoregulatory action of insulin-like growth factor I in rainbow trout (*Oncorhynchus kisutch*). *J. Endocrinol.* **130,** 87–92.

McLeese, J. M., Johnsson, J., Huntley, F. M., Clarke, W. C., and Weisbart, M. (1994). Seasonal changes in osmoregulation, cortisol, and cortisol receptor activity in the gills of parr/smolt of steelhead trout and steelhead rainbow trout hybrids, *Oncorhynchus mykiss*. *Gen. Comp. Endocrinol.* **93,** 103–113.

Minuth, W. W., Steckelings, U., and Gross, P. (1987). Complex physiological and biochemical action of aldosterone in toad urinary bladder and mammalian renal collecting duct cells. *Renal Physiol.* **10,** 297–310.

Miwa, S., and Inui, Y. (1983). Effects of thyroxine ant thiourea on the parr–smolt transformation of Amago salmon (*Oncorhynchus kisutch*). *Bull. Natl. Res. Inst. Aquacult. (Jpn.)* **4,** 41–52.

Miwa, S., and Inui, Y. (1985). Effects of L-thyroxine and ovine growth hormone on smoltification of amago salmon (*Oncorhynchus kisutch*). *Gen. Comp. Endocrinol.* **58,** 436–442.

Moav, B., and McKeown, B. A. (1992). Thyroid hormone increases transcription of growth hormone mRNA in rainbow trout pituitary. *Horm. Metab. Res.* **24,** 10–14.

Noguchi, S., Noda, M., Takahashi, H., Kasakami, K., Ohta, T., Nagano, K., Hirose, T., Inayama, S., Kawamura, M., and Numa, S. (1986). Primary structure of the B-subunit of *Torpedo californica* (Na^+ + K^+)-ATPase deduced from the cDNA sequence. *FEBS Lett.* **196,** 315–320.

Omelanjiuk, R. J., and Eales, J. G. (1986). The effect of 3,5,3′-triiodo-L-thyronine on gill Na^+/K^+-ATPase of rainbow trout (*Salmo gairdneri*), in fresh water. *Comp. Biochem. Physiol., A* **84,** 427–429.

Payan, P., Girard, J. P., and Mayer-Gostan, N. (1984). Branchial ion movements in teleosts: The roles of respiratory and chloride cells. *In* "Fish Physiology" (W. S. Hoar and D. J. Randall, eds.), Vol. 10B, pp. 39–63. Academic Press, New York.

Perry, S. F., and Laurent, P. (1989). Adaptational responses of rainbow trout to lowered external NaCl concentration: Contribution of the branchial chloride cell. *J. Exp. Biol.* **147,** 147–168.

Perry, S. F., and Walsh, P. J. (1989). Metabolism of isolated fish gill cells: Contribution of epithelial chloride cells. *J. Exp. Biol.* **144,** 507–520.

Perry, S. F., Goss, G. G., and Laurent, P. (1992). The interrelationships between gill chloride cell morphology and ionic uptake in four freshwater teleosts. *Can. J. Zool.* **70,** 1775–1786.

Pickford, G. E., and Phillips, J. G. (1959). Prolactin, a factor promoting survival of hypophysectomized killifish in freshwater. *Science* **130,** 454–455.

Pickford, G. E., Griffith, R. W., Torretti, J., Hendlez, E., and Epstein, F. H. (1970a). Branchial reduction and renal stimulation of Na^+,K^+-ATPase by prolactin in hypophysectomized killifish in freshwater. *Nature (London)* **228,** 378–379.

Pickford, G. E., Pang, P. K. T., Weinstein, E., Torreti, J., Hendlez, E., and Epstein, F. H. (1970b). The response of hypophysectomized cyprinodont, *Fundulus heteroclitus,* to replacement therapy with cortisol: Effects on blood serum and sodium–potassium activated adenosine triphosphatase in the gills, kidney and intestinal mucosa. *Gen. Comp. Endocrinol.* **14,** 524–534.

Pisam, M., and Rombourg, A. (1991). Mitochondria-rich cells in the gill epithelium of teleost fishes: An ultrastructural approach. *Int. Rev. Cytol.* **130,** 191–232.

Pisam, M., Caroff, A., and Rambourg, A. (1987). Two types of chloride cells in the gill epithelium of a freshwater-adapted euryhaline fish: *Lebistes reticulatus;* Their modifications during adaptation to saltwater. *Am. J. Anat.* **179,** 40–50.

Pisam, M., Auperin, B., Prunet, P., Rentierdelrue, F., Martial, J., and Rambourg, A. (1993). Effects of prolactin on alpha and beta chloride cells in the gill epithelium of the saltwater adapted tilapia *Oreochromis niloticus. Anat. Rec.* **235,** 275–284.

Prunet, P., and Auperin B. (1994). Prolactin receptors. *In* "Fish Physiology" (W. S. Hoar and D. Randall, eds.), Vol. 13, pp. 367–391. Academic Press, New York.

Redding, J. M., Schreck, C. B., Birks, E. K., and Ewing, R. D. (1984). Cortisol and its effects on plasma thyroid hormone and electrolyte concentrations in fresh water and during seawater acclimation in yearling coho salmon, *Oncorhynchus kisutch. Gen. Comp. Endocrinol.* **56,** 146–155.

Refstie, T. (1982). The effect of feeding thyroid hormones on saltwater tolerance and growth rates of Atlantic salmon. *Can. J. Zool.* **60,** 2706–2712.

Richman, N. H., and Zaugg, W. S. (1987). Effects on cortisol and growth hormone on osmoregulation in pre- and desmoltified coho salmon (*Oncorhynchus kisutch*). *Gen. Comp. Endocrinol.* **65,** 189–198.

Richman, N. H., Nishioka, R. S., Young, G., and Bern, H. A. (1987). Effects of cortisol and growth hormone replacement on osmoregulation in hypophysectomized coho salmon *Oncorhynchus kisutch. Gen. Comp. Endocrinol.* **67,** 194–201.

Rossier, B. C., Geering, K., and Kraehenbuhl, J. P. (1989). Regulation of the sodium pump: How and why? *Trends Biochem. Sci.* **12,** 483–487.

Sakamoto, T., and Hirano, T. (1991). Growth hormone receptors in the liver and osmoregulatory organs of rainbow trout: Characterization and dynamics during adaptation to seawater. *J. Endocrinol.* **130,** 425–433.

Sakamoto, T., and Hirano, T. (1993). Expression of insulin-like growth factor-I gene in osmoregulatory organs during seawater adaptation of the salmonid fish: Possible mode of osmoregulatory action of growth hormone. *Proc. Natl. Acad. Sci. U.S.A.* **90,** 1912–1916.

Sakamoto, T., McCormick, S. D., and Hirano, T. (1993). Osmoregulatory actions of growth hormone and its mode of action in salmonids: A review. *Fish Physiol. Biochem.* **11,** 155–164.

Salman, N. A., and Eddy, F. B. (1987). Response of chloride cell numbers and gill Na^+/K^+-ATPase activity of freshwater rainbow trout (*Salmo gairdneri* Richardson) to salt feeding. *Aquaculture* **61,** 41–48.

Sandor, T., DiBattista, J. A., and Mehdi, A. Z. (1984). Glucocorticoid receptors in the gill tissue of fish. *Gen. Comp. Endocrinol.* **53,** 353–364.

Sargent, J. R., and Thomson, A. J. (1974). The nature and properties of the inducible sodium-plus-potassium ion-dependent adenosine triphosphatase in the gills of eels (*Anguilla anguilla*) adapted to fresh water and sea water. *Biochem. J.* **144,** 69–75.

Sargent, J. R., Thomson, A. J., and Bornancin, M. (1975). Activities and localization of succinic dehydrogenase and Na^+/K^+-activated adenosine triphosphatase in the gills of fresh water and sea water eels (*Anguilla anguilla*). *Comp. Biochem. Physiol., B* **51,** 75–79.

Saunders, R. L., and Henderson, E. B. (1970). Influence of photoperiod on smolt development and growth of Atlantic salmon (*Salmo salar*). *J. Fish. Res. Board Can.* **27,** 1295–1311.

Saunders, R. L., McCormick, S. D., Henderson, E. B., Eales, J. G., and Johnston, C. E. (1985). The effect of orally administered 3,5,3′-triiodo-L-thyronine on growth and salinity tolerance of Atlantic salmon (*Salmo salar*). *Aquaculture* **45,** 143–156.

Schmalzing, G., and Gloor, S. (1994). Na^+/K^+-pump beta-subunits: Structure and functions. *Cell. Physiol. Biochem.* **4,** 96–114.

Schmitz, M., and Mayer, I. (1993). Effect of androgens on seawater adaptation in Arctic char, *Salvelinus alpinus*. *Fish Physiol. Biochem.* **12,** 11–20.

Schonrock, C., Morley, S. D., Okawara, Y., Lederis, K., and Richter, D. (1991). Sodium and potassium ATPase of the teleost fish *Catostomus commersoni*—sequence, protein structure and evolutionary conservation of the alpha-subunit. *Biol. Chem. Hoppeseyler.* **372,** 279–286.

Schreck, C. B. (1981). Stress and compensation in teleostean fishes: Response to social and physical factors. *In* "Stress and Fish" (A. D. Pickering, ed.), pp. 295–321. Academic Press, New York.

Shrimpton, J. M., Bernier, N. J., and Randall, D. J. (1994). Changes in cortisol dynamics in wild and hatchery reared juvenile coho salmon. *Can. J. Fish Aquat. Sci.* **51,** 2179–2187.

Shrimpton, J. M., Devlin, R. H., Donaldson, E. M., Mclean, E., and Randall, D. J. (1995). Increases in gill corticosteroid receptor abundance and saltwater tolerance in juvenile coho salmon (*Oncorhynchus kisutch*) treated with growth hormone and placental lactogen. *Gen. Comp. Endocrinol.* (in press).

Silva, P., Solomon, R., Spokes, K., and Epstein, F. H. (1977). Ouabain inhibition of gill Na-K-ATPase: Relationship to active chloride transport. *J. Exp. Zool.* **199,** 419–426.

Skou, J. C., and Esmann, M. (1992). The Na,K-ATPase. *J. Bioenerg. Biomembr.* **24,** 249–261.

Specker, J. L., Portesi, D. M., Cornell, S. C., and Veillette, P. A. (1994). Methodology for implanting cortisol in Atlantic salmon and effects of chronically elevated cortisol on osmoregulatory physiology. *Aquaculture* **121,** 181–193.

Stagg, R. M., and Shuttleworth, T. J. (1982). Na^+,K^+-ATPase, ouabain binding and ouabain-sensitive oxygen consumption in gill from *Platichthys flesus* adapted to seawater and freshwater. *J. Cell. Physiol.* **147,** 93–99.

Thomson, A. J., and Sargent, J. R. (1977). Changes in the levels of chloride cells and

(Na^+ + K^+)-dependent ATPase in the gills of yellow and silver eels adapting to seawater. *J. Exp. Zool.* **200,** 33–40.

Weisbart, M., Chakraborti, P. K., Gallivan, G., and Eales, J. G. (1987). Dynamics of cortisol receptor activity in the gills of the brook trout, *Salvelinus fontinalis,* during seawater adaptation. *Gen. Comp. Endocrinol.* **68,** 440–448.

Yao, K., Niu, P.-D., Le Gac, F., and Le Bail, P.-Y. (1991). Presence of specific growth hormone binding sites in rainbow trout (*Oncorhynchus mykiss*) tissue: Characterization of the hepatic receptor. *Gen. Comp. Endocrinol.* **81,** 72–82.

Young, G. (1988). Enhanced response of the interrenal of coho salmon (*Oncorhynchus kisutch*) to ACTH after growth hormone treatment *in vivo* and *in vitro*. *Gen. Comp. Endocrinol.* **71,** 85–92.

Young, G., and Lin, R. (1988). Response of the interrenal to adrenocorticotropic hormone after short-term thyroxine treatment of coho salmon (*Oncorhynchus kisutch*). *J. Exp. Zool.* **245,** 53–58.

Zadunaisky, J. A. (1984). The chloride cell: The active transport of chloride and the paracellular pathway. *In* "Fish Physiology" (W. S. Hoar and D. Randall, eds.), Vol. 11B, pp. 275–343. Academic Press, New York.

Zaugg, W. S. (1982). A simplified preparation for adenosine triphosphatase determination in gill tissue. *Can. J. Fish. Aquat. Sci.* **39,** 215–217.

Zaugg, W. S., and McLain, L. R. (1970). Adenosine triphosphatase activity in gills of salmonids: Seasonal variations and salt water influences in coho salmon, *Oncorhynchus kisutch*. *Comp. Biochem. Physiol.* **35,** 587–596.

12

CALCIUM TRANSPORT PROCESSES IN FISHES

GERT FLIK, PIETER M. VERBOST,
AND SJOERD E. WENDELAAR BONGA

I. INTRODUCTION

Vertebrates are dependent on calcium for the formation of the skeleton and for many cellular functions. This requires the regulation within narrow limits of the calcium concentration, in particular ionic calcium, of the intracellular (resting [Ca^{2+}]: 10^{-7} mol liter^{-1}) and intercellular fluids ([Ca^{2+}]: about 1.3×10^{-3} mol liter^{-1}). Changes in the intercellular [Ca^{2+}] levels beyond the physiological limits disturb neural, muscular, and cardiovascular functions, and may result in tetany (hypocalcemia), lethargy (hypercalcemia), and eventually death of the animals. Studies on eels that are easily stanniectomized (removal of the major source of hypocalcemic hormone in fish) indicate that fish, more so than higher vertebrates, may survive extreme hypercalcemia for considerable time (Hanssen *et al.*, 1989), although all researchers know that such stanniectomized fish become lethargic.

Complicated regulatory systems are involved in controlling Ca^{2+} levels. These systems seem to be more complex in the terrestrial than in the aquatic vertebrates. On the land, calcium is taken up from the food, essentially an unreliable calcium source. This, and the episodic character of feeding, may have been the driving force for the evolution of bone as a calcium reservoir, and for the development of an intricate system of bone cells and blood vessels effecting both deposition and mobilization of bone minerals. The reservoir function of the bone enables terrestrial vertebrates to survive periods of dietary calcium shortage or high calcium demands, such as during rapid growth or female reproduction. Parathyroid hormone (PTH), calcitonin, and 1,25-dihydroxyvitamin D_3 are the primary hormones controlling these processes, whereas hormones such as prolactin, growth hormone, and estrogens perform minor roles. PTH is the dominant hormone in terrestrial animals for maintenance of the homeostasis of the body fluids with respect to calcium, not only under normal conditions, but also during growth spurts, yolk deposition, avian egg shell formation, and calcification of the embryonic skeleton of the mammalian offspring (Wendelaar Bonga and Pang, 1992).

With respect to the availability of calcium, the situation in fishes is fundamentally different from that of the terrestrial vertebrates. The differences concern the availability of calcium and the accessibility of the external calcium sources. In addition to the food, aquatic animals have access to the ionic calcium present in the water. This may occur in concentrations well above those of the intercellular fluids, as in seawater (around 10 mmol liter^{-1}), or in concentrations similar to or well below these internal levels, as in fresh water (0.1–3 mmol liter^{-1}). Under both conditions, fishes are able to extract calcium from the water. In contrast to land animals, which usually have to economize on calcium and may suffer from calcium shortage caused by inadequate dietary availability or poor functioning of the regulatory mechanisms (a not uncommon phenomenon during aging), fishes normally are facing a surplus of calcium. This is due to their continuous access to the water, in combination with efficient uptake mechanisms that enable them to maintain a positive calcium balance, unless the water calcium concentration is very low or when toxic pollutants are interfering with calcium uptake (see Section V). This favorable situation of fishes with respect to calcium may explain why in fish bone the cellular and vascular elements are missing, which are essential for bone to function as a readily available calcium reservoir, and why PTH is missing in fishes and amphibians with a typical aquatic way of life.

In fishes, only a PTH-gene-related peptide seems to be expressed in the pituitary gland (Fraser *et al.,* 1991). This hypercalcemic hormone, present in and indispensable for all terrestrial vertebrates, must have evolved during the water-to-land transition in early vertebrate evolution in conjunction with the development of bone as a calcium reservoir (Pang, 1973; Pang *et al.*, 1980, 1989; Wendelaar Bonga and Pang, 1992). Interestingly, in teleost fishes the hormone dominating the calcium homeostasis of the blood plasma, stanniocalcin, has a hypocalcemic function. This indicates that reduction rather than increase of plasma calcium concentration is the more important response for maintaining plasma calcium levels in freshwater as well as seawater teleost fishes. Stanniocalcin is produced in the Stannius bodies, organs unique to teleostean and holostean fishes. It is a fast-acting hormone that effectively reduces the active calcium uptake in the gills (see reviews by Hirano, 1989; Wagner and Friesen, 1989; Verbost *et al.*, 1993a) as well as the intestine (Sundell *et al.*, 1992a) in freshwater and seawater fishes.

This chapter will focus on the membrane mechanisms of transepithelial calcium transport in the principal organ effecting the uptake of calcium from the environment, the gills, but also the intestine. Because these organs are in intimate contact with the environment, their limiting epithelia are the first to be affected by water pollutants and toxic food contaminants. In particular, some heavy metals are known to interact with calcium uptake mechanisms. Such interactions are briefly dealt with at the end of this chapter.

II. CALCIUM TRANSPORT IN GILLS AND INTESTINE

A. Gills

Freshwater fishes show a substantial exchange of calcium with the ambient water. The major organs involved are the gills, the intestine, and the kidneys, with the gills being the most important organs (Fenwick, 1989). In early literature, the view prevailed that calcium uptake via the gills of freshwater fishes was a passive process. The arguments for this assumption were that there was no appreciable difference in concentration of calcium in blood (blood plasma total calcium around 3 mmol liter^{-1}, free Ca^{2+} around 1.5 mmol liter^{-1}) and water, and only a small potential difference, mostly favorable for passive influx of Ca^{2+} over the branchial epithelium (typically around 5 mV negative inside relative to the water). However, the equilibrium potential for Ca^{2+} (calculated according to the Nernst equation and applying the water and plasma Ca^{2+} *activities*) is always more negative than the transepithelial potential measured in fresh water (reported values range from around 0 to -16 mV at neutral pH; Maetz and Bornancin, 1975; McWilliams and Potts, 1978; Perry and Wood,

1985; Perry and Flik, 1988). Thus, in the epithelium of the gills of freshwater fishes, the electrochemical gradient for Ca^{2+} (driving the passive flux) is generally directed outwardly. Although the branchial epithelium of freshwater fishes is qualified as "tight" (Wendelaar Bonga *et al.*, 1983; Hirano, 1986), a substantial passive efflux of calcium has been reported. The foregoing considerations have stimulated research on active calcium uptake in teleost fishes, and it is now well accepted that the branchial Ca^{2+} uptake in freshwater fishes mainly occurs transcellularly (see review by Fenwick, 1989), most likely through specialized ion-transporting cells, referred to as chloride cells, mitochondria-rich cells, or ionocytes (Mayer Gostan *et al.*, 1983; Flik *et al.*, 1984a,b; Perry and Wood, 1985; Perry and Flik, 1988; Perry *et al.*, 1992); in this chapter we will subsequently refer to these cells as ionocytes.

Seawater fishes maintain a high, inside-positive, transepithelial potential across the branchial epithelium (Maetz and Bornancin, 1975; Young *et al.*, 1988) that is higher than the equilibrium potential for Ca^{2+} and, therefore, a passive influx of Ca^{2+} from the seawater is also unlikely (Verbost *et al.*, 1994). Together with the effective and inhibitory control of the active uptake of calcium via the transcellular route, this is an important mechanism to prevent hypercalcemia in seawater.

In addition to the gills, several nonbranchial areas of the skin may be implicated in calcium uptake. Although early reports on a role of the fins have not been confirmed, other studies have shown that the ionocytes present in the skin likely account for the active calcium uptake across the integument, as was demonstrated for skin preparations of the rainbow trout (Marshall *et al.*, 1992), the Nile tilapia (McCormick *et al.*, 1992), and the killifish (Burghardt, 1993) under *in vitro* conditions. Given the evidence that the Ca^{2+} influx of the gills is most likely also mediated by ionocytes, the contribution of extrabranchial areas of the skin may be proportional to the relative numbers of the ionocytes present in these areas. In larvae of many species (Hwang, 1989), ionocytes occur throughout the skin, but in juveniles they reduce in numbers in the extrabranchial areas, and in general extrabranchial ionocytes are very scarce or absent in adults of most species, with the exception of the inner branchial epithelium of a few (e.g., tilapia, killifish), where substantial numbers may be present (Foskett and Scheffey, 1982). In fishes exposed to stressors, the ionocytes may increase outside as well as inside the branchial area, as we have shown for the dorsal skin of carp (Iger and Wendelaar Bonga, 1994). The contribution of extrabranchial areas may therefore be significant, although one should realize that this conclusion derives from data obtained with *in vitro* preparations. In tilapia the opercular membranes contain about 15% of the total Na^+/K^+-ATPase activity—an enzymatic marker of ionocytes—of the gills (Wendelaar Bonga *et al.*, 1990). Marshall *et al.*, 1995 have concluded that up to 46% of the total calcium uptake in killifish could occur via opercular skin.

The study of branchial calcium transport mechanisms started in the 1970s (e.g., Fenwick, 1976, 1979; Ma and Copp, 1978). In particular, the wider availability and application of tracers for Ca^{2+} (^{45}Ca and ^{47}Ca) and detection equipment contributed significantly to this development. Research focused on the mechanisms of Ca^{2+} transport in the gills of euryhaline fishes (mainly eel, tilapia, and trout) kept in fresh water. These studies led to the identification of Ca^{2+}-ATPase (Flik *et al.*, 1985a,b) and Na^+/Ca^{2+} exchange (Flik *et al.*, 1990, 1993) that account for all or most of the active Ca^{2+} uptake. These and other biochemical studies, in combination with flux studies, have resulted in a model for branchial calcium uptake that is now widely accepted. The model was based on the assumption—and this was recently confirmed (G. Flik, unpublished observation)—that the intracellular Ca^{2+} concentration of the ionocytes is in the submicromolar range, similar to that of other cells. The model implies that Ca^{2+} enters the ionocytes down the electrochemical gradient across the apical membrane, is transported via the cytoplasm, and finally is extruded into the blood across the basolateral membrane, against a steep electrochemical gradient. The available evidence for this model will be discussed in Section III.

B. Intestine

Under normal, nonstressed conditions, the drinking rate of freshwater fishes is very low, and the contribution of the intestine to calcium uptake is restricted to dietary calcium. Although several studies have indicated that 70% or more of the total calcium uptake of freshwater fishes originates from the water (Berg, 1968; Ogino and Takeda, 1978; Flik *et al.*, 1985b), the ratio between calcium gained from ambient water and from food, and thus of the contribution of gills and intestine to calcium uptake, is variable and depends on the calcium concentration of water and food. In low-calcium water the relative contribution of the food increases, whereas feeding low-calcium diets stimulates branchial uptake (Ichii and Mugiya, 1983). However, whereas a total lack of dietary calcium can be completely compensated by branchial uptake, very low calcium concentrations of the water induce hypocalcemia and impair growth (Ichii and Mugiya, 1983; Flik *et al.*, 1986b).

In seawater fishes, the role of dietary calcium in calcium metabolism is probably negligible. Nevertheless, the relative contribution of the intestine to total calcium uptake seems to be at least as important as in freshwater fishes, mainly because of the high drinking rate of seawater fishes, which is necessary to compensate osmotic water losses (Fletcher, 1978; Björnsson and Nilsson, 1985; Sundell and Björnsson, 1988). The intestinal uptake of calcium has been verified in *in vitro* experiments with everted or right-side-out gut sacs and a significant net uptake from the luminal side has been demonstrated for both freshwater and seawater fishes (Flik *et al.*, 1982; Fenwick, 1984; Collie and Hirano, 1984). Using an elegant technique involving perfusion of the vasculature and lumen of the isolated cod intestine, Sundell and Björnsson (1988) determined unidirectional calcium influxes and effluxes. The results revealed a chlorpromazine-sensitive, and thus possibly Ca^{2+}-ATPase-mediated, component, as well as a passive diffusional Ca^{2+}-influx component, in line with the notion that the intestinal epithelium is a leaky epithelium. Other biochemical investigations have revealed Ca^{2+}-ATPase and Na^+/Ca^{2+} exchange activities in the gut of tilapia, with characteristics similar to those of the gills, and likely with a similar function (Flik *et al.*, 1990). It is possible that the current model for explaining branchial Ca^{2+} uptake is also applicable to the intestinal epithelium, although the relative contribution of the calcium-extruding mechanisms and the endocrine control of the transport process may show differences. For the gills, stanniocalcin is the dominant—and inhibitory—factor, with other hormones such as prolactin and cortisol in accessory, stimulatory roles (see review by Wendelaar Bonga and Pang, 1992). For the intestine, the role of stanniocalcin, although undeniable (Takagi *et al.*, 1985; Sundell *et al.*, 1992a), is less prominent, in favor of vitamin D_3 metabolites (Flik *et al.*, 1982; Fenwick *et al.*, 1984; Sundell and Björnsson, 1990).

III. Ca^{2+} TRANSPORT IN THE GILLS OF TELEOST FISHES

A. Transport across the Apical Plasma Membrane

For the transcellular uptake of Ca^{2+} in the gills, the apical membrane of the ionocyte forms the primary barrier for Ca^{2+} between the water and the fish. Taking into consideration the physiological conditions—millimolar concentrations of Ca^{2+} outside and submicromolar concentrations in the cytosol—the importance of the apical membrane becomes fully apparent in keeping Ca^{2+} out to maintain the physiological intracellular Ca^{2+} concentration. Prolonged elevation of cytosolic calcium can be toxic, as it can act on endonucleases in the nucleus to trigger programmed cell death (Berridge, 1994). At the same time, the apical membrane must be permeable to some degree to let Ca^{2+} through for transcellular Ca^{2+} uptake. The logical conclusion is that most if not all of the control over this uptake is effected at the apical membrane. Indeed, data obtained so far confirm this hypothesis.

Stanniocalcin—the predominant hypocalcemic hormone in teleost fishes (Pang *et al.*, 1980; Wendelaar Bonga and Pang, 1986)—has no effect on basolateral ATP-dependent Ca^{2+} transport (Verbost *et al.*, 1993b), whereas it does reduce the permeability for Ca^{2+} of the apical membrane (Verbost *et al.*, 1989; Fig. 1). The Ca^{2+} pump activity is considered to reflect the whole-body uptake of Ca^{2+} (Flik *et al.*, 1984a, 1985a). An

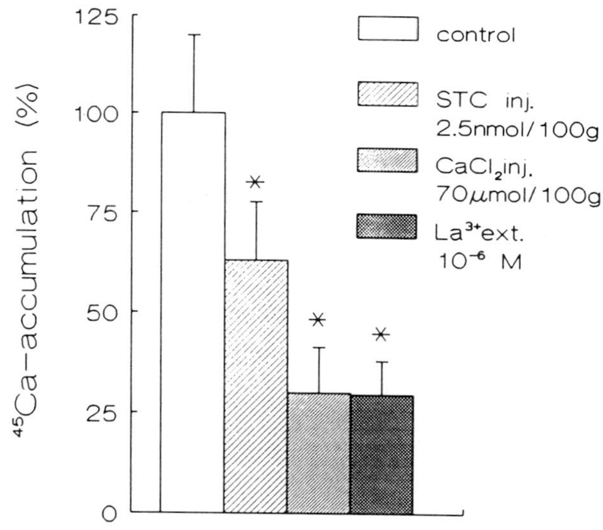

Fig. 1. $^{45}Ca^{2+}$ accumulation in rainbow trout (*Oncorhynchus mykiss*) branchial epithelium after manipulation of apical membrane permeability to Ca^{2+}. Fishes were exposed for 1 h to $^{45}Ca^{2+}$ (1 MBq liter $^{-1}$) in artificial fresh water containing 0.7 mmol liter^{-1} Ca, quickly anesthetized in bicarbonate-buffered (pH 7.4) methane sulfonate salt, and injected with 5000 IU Na heparin. The gills were rapidly cleared from blood by perfusion with saline via the ventral aorta. Subsequently the gill arches were excised and the branchial epithelium was carefully collected by scraping with a microscope slide. Aliquots of branchial epithelium were digested in H_2O_2 and analyzed for $^{45}Ca^{2+}$. Control fish received saline injections, trout STC was injected at 2.5 nmol per 100 g, and $CaCl_2$ was injected at 70 μmol per 100 g $CaCl_2^-$ injection releases endogenous stanniocalcin (Verbost *et al.*, 1993b). One group was exposed to 1 μmol liter^{-1} $LaCl_3$ during tracer exposure. The $^{45}Ca^{2+}$ accumulation in the control was set to 100%. All three treatments decreased the accumulation of $^{45}Ca^{2+}$. Mean values ± standard error of the mean are given for six fishes per treatment. The results indicate that Ca^{2+} entry in the branchial epithelium is controlled by stanniocalcin, exerting its effects on a lanthanum-sensitive pathway. (*, $P < 0.01$, compared to controls.)

argument against apical control seems to be the way in which prolactin acts. Prolactin has hypercalcemic activities that coincide with activation of the basolateral ATP-dependent Ca^{2+} transport (Flik *et al.*, 1984b, 1986a; see also Section III,B). However, the effects of prolactin require days to develop, as opposed to those of stanniocalcin, which become evident within minutes (Lafeber *et al.*, 1988). It could well be that the increase in Ca^{2+} pump capacity following prolactin treatment is not causally related to the hypercalcemia since there appears to be an enormous overcapacity of the basolateral Ca^{2+} pump system (Verbost *et al.*, 1993b).

On the other hand, it is likely that there are several ways to control the transcellular uptake. For instance, stimulation of transcellular Ca^{2+} uptake by 1,25-dihydroxyvitamin D_3 in intestine of higher vertebrates is accompanied by changes in apical Ca^{2+} transport (Rasmussen and Barret, 1984; Kaune *et al.*, 1992), induction of calcium binding proteins in the cytosol (Wassermann and Fullmer, 1983), and stimulation of basolateral Ca^{2+} transport (Ghijsen *et al.*, 1982; Ghijsen and Van Os, 1982). Since 1,25-dihydroxyvitamin D_3 is a steroid hormone, many of its effects will result from transcriptional events following transfer to the nucleus and binding to a cytosolic steroid receptor. Certainly the mechanisms of control are completely different from those of the peptide hormones stanniocalcin and prolactin, which are hydrophilic and unable to pass the cell membrane. The example shows, however, that all steps involved in the passage of Ca^{2+} through the cell may be influenced to achieve a change in transport rate and that, therefore, all of them should be studied. Although a hypercalcemic control does not seem to be the first necessity for fishes (Wendelaar Bonga and Pang, 1992), it is known that in teleost fishes 1,25-dihydroxyvitamin D_3 has a hypercalcemic effect, in both freshwater (Swarup and Srivastav, 1982; Fenwick, 1984; Fenwick *et al.*, 1984; Swarup *et al.*, 1991) and marine fishes (Sundell *et al.*, 1992b, 1993). Vitamin D metabolites were shown to have a stimulatory effect on the intestinal Ca^{2+} uptake in fishes (Chartier *et al.*, 1979; Flik *et al.*, 1982; Fenwick *et al.*, 1984; Sundell *et al.*, 1992b), but the precise mechanism of action is not known at present.

1. Channel Typing

Direct analysis of Ca^{2+} transport across the branchial apical membrane has not been realized, because no one has yet been able to isolate this membrane. The methods known for the isolation of intestinal or renal brush border membranes (Klaren *et al.*, 1993; Hennessen *et al.*, 1993) do not yield a purified apical membrane fraction from gill cells (our unpublished observation). For future research, patch-clamp analysis applied to the apical membrane is the most promising technique to show the presence of the Ca^{2+} channels and to define the type of channel, as shown by work on primary cultures of the rabbit distal convoluted tubule (Poncet *et al.*, 1992). The latter study showed the presence of lanthanum-sensitive Ca^{2+} channels in the apical membrane that likely represented the apical uptake pathway for transepithelial Ca^{2+} transport. In primary cultures, polarized cells always bind with their basolateral side to the substrate and thus expose the apical membrane, ready for clamping.

Indirect analyses (studies with intact fishes, in which whole-body uptake is equated with branchial uptake of Ca^{2+}) of branchial apical Ca^{2+} transport have provided many indications that Ca^{2+} enters the gills via voltage-independent Ca^{2+} channels. Interestingly, results from toxicological studies on the effects of heavy metals on Ca^{2+} transport have contributed substantially to the current model. The concept of a voltage-independent channel is derived from work showing that blockers of L-type channels had no effect on whole-body Ca^{2+} influx, whereas cobalt (Co^{2+}) and lanthanum (La^{3+}) reduced Ca^{2+} influx significantly in both whole-body uptake studies and the trout isolated-head preparation (Perry and Flik, 1988; Verbost *et al.*, 1987a, 1989). A surprising but so far consistent finding is that La^{3+} (at concentrations of 10^{-6}–10^{-5} mol liter^{-1}) has no effect on transepithelial Ca^{2+} influx when measured on isolated epithelia in an Ussing chamber (skin of trout covering the cleithral bone: Marshall *et al.*, 1992; killifish opercular membrane: Burghardt, 1993). This contrasts with the idea that La^{3+} reduces Ca^{2+} influx by occupying Ca^{2+} binding sites. The explanation for these apparently conflicting results is unknown at the moment. However, the implication—that Ca^{2+} influx in isolated membranes differs from that in intact tissues—deserves attention in future research.

A distinct argument in the search for the channel type involved in the regulated Ca^{2+} influx is that channel permeability is reduced by stanniocalcin (Verbost *et al.*, 1989, 1993a,b). Stanniocalcin, as a glycopeptide, has to convey its message via a second messenger system after binding to a receptor in the plasma membrane, and this defines the channel as a second messenger-operated Ca^{2+} channel (SMOC).

From tracer flux studies it was concluded that cadmium (Cd^{2+}) can enter the gill cells via the Ca^{2+} channels that are involved in the influx of Ca^{2+} (Verbost *et al.*, 1989). The permeability of Ca^{2+} channels for Cd^{2+} is not specific for the fish gill. For instance, Cd^{2+} entry in canine kidney epithelial cells is mediated, at least in part, by Ca^{2+} channels (Flanagan and Friedman, 1991). This opens possibilities to study these Ca^{2+} channels. Also, since Ca^{2+}, unlike Cd^{2+}, is quickly extruded after entering the epithelium (Cd^{2+} will first occupy intracellular binding sites for which it has higher affinity than Ca^{2+}), Cd^{2+} can be a useful tool to study the channels.

This fact was used in a Cd-tracer uptake study on cultured branchial cells from rainbow trout (Block and Pärt, 1992), from which they concluded that the respiratory (pavement) cells do not contain the La^{3+}-sensitive channels that are permeable to (both Cd^{2+} and) Ca^{2+}. This conclusion is consistent with a predominant role for the ionocytes in the uptake of Ca^{2+} and Cd^{2+} from the water.

2. Channel Regulation

The current debate focuses more on the mechanism of regulation of the channels rather than on the type of Ca^{2+} channel. On the basis of work on higher vertebrates, two conceivable pathways are regulation by (de-)phosphorylation of channels present in the apical membrane and by changing the number of channels. An example of regulation through activation of channels already resident in the apical membrane is that of vasopressin, which increases apical Na^{+} permeability in amphibian renal A6 cells by phosphorylation (Oh *et al.*, 1993). PTH was found to increase intracellular Ca^{2+} in mouse distal tubule cells through insertion of Ca^{2+} channels in the apical membrane (Bacskai and Friedman, 1990).

A completely different means of regulation is that of changing the size of the membrane surface in contact with the water. This property seems unique for the ionocytes in fish gill, as opposed to intestinal or renal cells in fishes or higher vertebrates, and is usually found in association with an increase in ionocyte numbers. In most species the apical membrane area becomes enlarged by forming an apical crypt (Pisam *et al.*, 1987, 1988; Fig. 2A). In salmonids this is usually effected by protrusion of this membrane, for instance, in ion-deficient or acidified water (Laurent *et al.*, 1985; Fig. 2B). Such protrusion could also result from uncovering of ionocyte apical plasma membrane by receding pavement cells (see Goss *et al.*, Chapter 10, this volume). Morphometrical analyses strongly indicate that a positive relation exists between calcium uptake and the exposed surface area of ionocytes in freshwater fishes (Perry *et al.*, 1992).

To investigate the messenger system involved in the control of the channels, two approaches are indicated. One could take a hormone that is known to affect Ca^{2+} influx and look for its effect on second messenger levels in the gill cells, or one could change second messenger levels and

---

Fig. 2. Representative transmission electron micrographs of ionocytes in branchial lamellar epithelium from (A) rainbow trout (*Oncorhynchus mykiss*; ×6175) and (B) tilapia (*Oreochromis mossambicus;* ×8550). The ionocyte of the trout protrudes with its apical part between pavement cells (p); b, blood space. The apical area of the ionocyte in the tilapia is covered by cell processes of pavement cells (p); the outer membrane invaginates to form an apical pit (a).

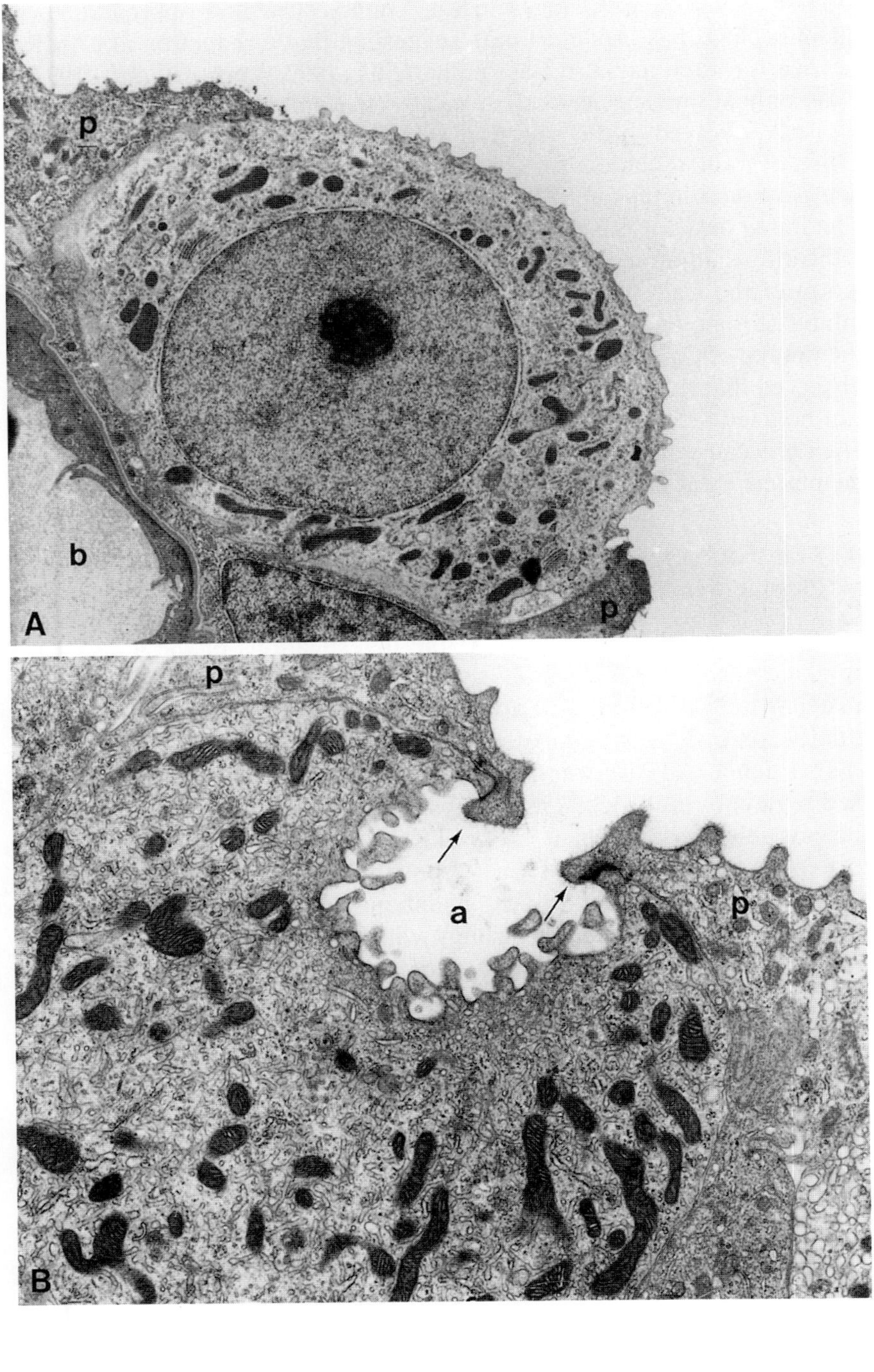
p
b
A
p
p
a
p
B

look for subsequent changes in Ca^{2+} influx. The first approach, using stanniocalcin, has produced data suggesting that a reduction in cAMP is needed for a reduction in Ca^{2+} influx (Flik, 1990; Verbost *et al.*, 1993b). Data on just one species (eel) suggested that inositol trisphosphate (IP_3) was not involved in the action of stanniocalcin (Verbost *et al.*, 1993b). This does not exclude a role for Ca^{2+} coming from pools other than IP_3-sensitive ones in the control of the Ca^{2+} channels in the apical membrane. The initial data for the second approach came from a toxicological study indicating that intracellular Ca^{2+} could indeed function in a feedback regulation of the Ca^{2+} channels. Accumulation of Cd^{2+}, which is known to inhibit storage and extrusion of Ca^{2+} (Verbost *et al.*, 1987a,b), reduced the uptake of Ca^{2+} in the gill cells (Verbost *et al.*, 1989). It has since been observed that thapsigargin (a specific inhibitor of the ER–Ca^{2+} pump that can be used to increase intracellular Ca^{2+}; Kijima *et al.*, 1991; Ghosh *et al.*, 1991) caused a 60 to 70% reduction in Ca^{2+} influx in killifish opercular membrane (Marshall *et al.*, 1995).

B. Transport across the Basolateral Plasma Membrane

As mentioned in Section II,A, the investigations of the active Ca^{2+} uptake mechanisms in fish gills started in the 1970s with the search for an energized Ca^{2+} pump. In early studies, a Ca^{2+}-ATPase with a very low affinity for Ca^{2+} was postulated to provide the driving force. Later on, this particular ATPase was defined as a nonspecific phosphatase activity (see review by Fenwick, 1989). Subsequently, attention was focused on the possible presence of a Ca^{2+}-ATPase with characteristics sufficient to explain the extrusion of Ca^{2+} from the ionocytes against the steep electrochemical gradient for Ca^{2+} maintained over the basolateral plasma membrane of these cells, in analogy with the model proposed for Ca^{2+} transport in rat intestine (Ghijsen *et al.*, 1982).

1. Pump Characterization

A series of specific criteria were advanced to positively identify the calcium pump in the basolateral plasma membrane of the gills of fishes. First, the location of the calcium pump in the basolateral plasma membrane of an ion-transporting cell implies that the pump is colocated with markers for this membrane compartment. The Na^+/K^+-ATPase activity is such a marker (Karnaky *et al.*, 1976) and the fortunate situation is that this enzyme is more abundantly present in the ionocytes with their elaborately invaginated basolateral plasma membrane. Therefore, a membrane isolation procedure that yields membranes that are specifically purified with respect to Na^+/K^+-ATPase is likely to yield a membrane preparation that is enriched in basolateral plasma membranes, and more in particular plasma membranes of the ionocytes of the branchial epithelium. Procedures have been described that allow the isolation of membrane fractions that are essentially devoid of membranes from erythrocytes (that are always, even after perfusion of the gills with saline, present in scrapings of branchial epithelium collected at the start of an isolation procedure), or subcellular membranes from endoplasmic reticulum, Golgi apparatus, or mitochondria (see review by Flik and Verbost, 1993). This appears to be a prerequisite as Ca^{2+} pumps are present in all membranes that face the cytosol (Grover and Khan, 1992), and it is only the plasma membrane calcium pump that mediates extrusion for epithelial calcium uptake.

Kinetic criteria have been advanced to identify the plasma membrane calcium pump. A prerequisite for the analysis of enzymes with high affinity for Ca^{2+} is the use of Ca^{2+}-buffering substances in the assay media. As a rule of thumb one should use Ca^{2+} buffers in assay media when Ca^{2+} concentrations below 50 μmol liter^{-1} are required. The principle behind the use of calcium buffers is not different from that of pH buffers. However, the chemistry relating to Ca^{2+} buffering in physiological media is rather complex. One has to take into account ionic strength, pH, and temperature of the medium to calculate free-Ca^{2+} levels for complex combinations of ligands (e.g., ATP, EGTA, HEDTA, and NTA) and metals (Ca and Mg). The published computer program "Chelator" (Schoenmakers *et al.*, 1992b) can provide these for the researcher in perfect agreement with actual (measured) concentrations.

A major problem encountered in the study of fish branchial plasma membranes is the exuberant phosphatase activity mentioned earlier (Flik *et al.*, 1983): it is from this pool of phosphatases that one has to identify the Ca^{2+}-transporting enzyme, which hydrolyzes only a small percentage (typically less than 5% in our preparations) of the ATP available. The release of phosphate from ATP is taken as a measure of the calcium pump activity *in vitro*. Specific ATP preference, submicromolar affinity for Ca^{2+}, and Michaelis–Menten kinetics of phosphatase activity have been advanced as criteria to discriminate a high-affinity Ca^{2+}-ATPase in eel gill plasma membranes from nonspecific phosphatases that consume chelates of ATP and Ca^{2+} or Mg^{2+} (Flik *et al.*, 1983, 1984a). The discrimination of the actual transporting ATPase is essential but may be difficult. A commendable study on "pumping and non-pumping" ATPase activity was published by Lin and Russell (1988), who elegantly showed that rat liver plasma membrane contains two Ca^{2+}-ATPase activities with comparable kinetics, one of which is the molecular correlate of a calcium pump, the other an ecto-ATPase involved in regulating extracellular levels of adenosine nucleotides. It follows that in addition to copurification with Na^+/K^+-ATPase and proper Ca^{2+} kinetics, more criteria should be advanced to identify a calcium pump involved in epithelial transport than ATP-dependent and Ca^{2+}-stimulated phosphate release.

In studies with resealed plasma membrane vesicles, in which the same criteria were applied as mentioned here for the phosphatase assays, the activity of the calcium pump was assayed as ATP-driven $^{45}Ca^{2+}$ accumulation into the vesicular space (Fig. 3). This "vectorial" assay unambiguously revealed the activity of a calcium pump (Flik *et al.*, 1985a). The plasma membrane-associated calcium pump of gills proved to be calmodulin dependent (Flik *et al.*, 1985c; Perry and Flik, 1988), a characteristic of plasma membrane calcium pumps in general (Grover and Khan, 1992). In the plasma membrane preparations used for these studies, thapsigargin did not affect kinetics of the Ca^{2+} pump activity in gill plasma membrane preparations, and this observation confirmed the purity of this preparation. It seems justified then to state that fish gill plasma membranes contain a P-type (Strehler, 1991) calmodulin-dependent, high-affinity Ca^{2+}-ATPase, which is the driving force for Ca^{2+} transport in the tissue.

In two other osmoregulatory organs in fishes, namely, the intestine of

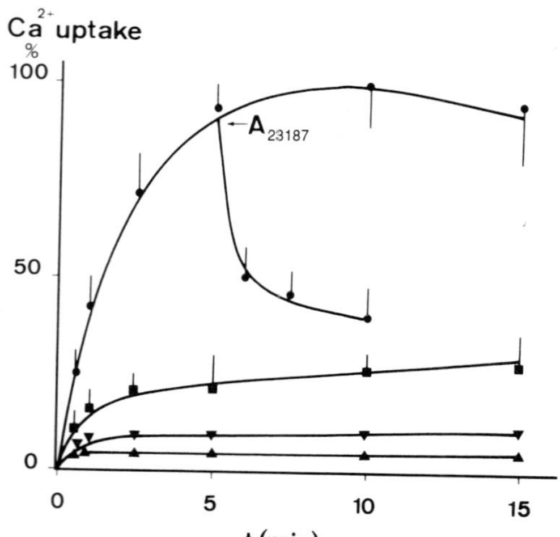

Fig. 3. Ca^{2+} transport in eel (*Anguilla rostrata* LeSueur) gill plasma membrane vesicles. ATP (circles), but not AMP (triangles) or ADP (inverted triangles), stimulates $^{45}Ca^{2+}$ accumulation (determined under V_{max} conditions, 1 μmol liter^{-1} Ca^{2+}) in resealed vesicles. Addition of the Ca^{2+} ionophore A23187 largely inhibits (squares) or reverses (arrow) accumulation stimulated by ATP. The release of Ca^{2+} from the vesicles by the ionophore indicates that ATP-driven Ca^{2+} transport creates an outwardly directed concentration gradient for Ca^{2+} in these vesicles. (Data taken from Flik *et al.*, 1985c.)

tilapia (Flik *et al.*, 1990; Schoenmakers *et al.*, 1993) and the kidney of the swordtail (Doneen, 1993), evidence has been provided for a high-affinity Ca^{2+}-ATPase activity that may be equated with the calcium pump of the respective tissues. These organs, with their particular physiological roles in calcium homeostasis, provide interesting models to study the involvement of Ca^{2+} transport mechanisms in epithelial Ca^{2+} transport.

2. Regulation

Two hormones that exert hypercalcemic actions in fishes, namely, prolactin in eel (Flik *et al.*, 1984b, 1989) and tilapia (Flik *et al.*, 1986b, 1994) and cortisol in trout (Flik and Perry, 1989), also enhanced calcium pumping activity in the plasma membrane fraction of the gills. Indeed, in prolactin- and cortisol-treated fishes the branchial calcium influx from the water increased (Flik and Perry, 1989; Flik *et al.*, 1994), and it thus appears that the calcium pumping capacity of the branchial epithelium is adjusted to the calcium uptake from the water under control of these hormones. Observations on tilapia indicate that the branchial ionocyte density increases in a dose dependent manner, and in parallel with the branchial Ca^{2+} influx, when fishes are treated with prolactin. Therefore, we assume that in prolactin-treated fishes, new populations of ionocytes become active. Moreover, it appears that the relative density of the Ca^{2+}-ATPase relative to the Na^+/K^+-ATPase increases upon prolactin treatment, suggesting that the expression of the calcium pump in fish gills is enhanced by specific control of this pituitary hormone.

3. Alternative Ca^{2+} Transport

The Na^+/Ca^{2+} exchanger is a second mechanism for energized transport of Ca^{2+} across plasma membranes (Carafoli, 1987). This carrier appears to be particularly important in excitable tissues, where Na^+ and Ca^{2+} countercurrents underlie the events related to the generation of action potentials. Yet, this carrier has also been demonstrated in nonexcitable cells and thus it may serve different functions. The carrier exchanges Na^+ for Ca^{2+} and it may participate in Ca^{2+} extrusion (e.g., in squid giant axon, where Na^+/K^+-ATPase activity maintains an inward sodium gradient as driving force; Baker *et al.*, 1969) as well as in Na^+ extrusion (e.g., in dog erythrocytes, which lack Na^+/K^+-ATPase, but where an ATP-driven Ca^{2+} extrusion pump maintains an inward Ca^{2+} gradient as driving force; Taylor, 1989). However, in rat renal and intestinal tissue, no clear function in Ca^{2+} extrusion could be attributed to the exchanger (Van Heeswijk *et al.*, 1984; Ghijsen *et al.*, 1983). Interestingly, in the intestine of tilapia, Ca^{2+} transport from mucosa to serosa appears to be largely dependent on a powerful Na^+/Ca^{2+} exchange activity (Flik *et al.*, 1990; Schoenmakers *et al.*, 1993). Subsequently, this exchange carrier was also demonstrated in branchial plasma membranes (Flik *et al.*, 1993; Flik and Verbost, 1993; Verbost *et al.*, 1994). The kinetic parameters of the exchanger have been compared to those of the ATP-driven calcium pump, and this comparison led us to conclude that in the gills the ATPase may play a more pronounced role in active Ca^{2+} transport in branchial epithelium (Verbost *et al.*, 1994): we calculated a twofold higher activity of the ATPase than of the exchanger at prevailing cytosolic Ca^{2+} levels. Considering the higher turnover for Na^+ in seawater fish gills (Maetz, 1974; Evans, 1979), one may have predicted a more pronounced role for the exchanger in calcium handling in seawater fish gills. However, the relative densities of the calcium transporters do not differ in tilapia well acclimated to either fresh water or seawater. Better understanding of the involvement and role of the exchanger in calcium transport in branchial and opercular epithelium awaits physiological studies.

Attention should be paid to a possible control by atrial natriuretic factor (ANF), as this peptide and its second messenger cyclic GMP have been shown to stimulate the exchanger in rat smooth muscle cells (Furukawa *et al.*, 1991). Also, catecholamines are known to enhance Na^+/Ca^{2+} exchange activity (Khoyi *et al.*, 1991), and the gills of fishes are an important target for catecholamines (e.g., Perry *et al.*, 1984). Thus, with the demonstration of Na^+/Ca^{2+} exchange in gills, its control by catecholamines should be considered, and the fish model deserves future research in this field.

IV. TRANSPORT IN PERMEABILIZED CELLS

The plasma membrane of cells can be permeabilized with saponin, which forms pores of 8–10 nm in cholesterol-rich membranes: the lipophilic parts of saponin interact with cholesterol, resulting in a ring structure with a central hydrophilic channel (Wakasugi *et al.*, 1982). Membranes with low cholesterol levels, like the membranes of ER and mitochondria, are hardly permeabilized by the saponin treatment (Burgess *et al.*, 1983). Thus permeabilization of isolated cells results in an intact intracellular compartment that is continuous with the surrounding medium and easily accessible for probing. This condition offers the possibility of measuring the Ca^{2+}-buffering characteristics of the mitochondria and ER by suspending the leaky cells in Ca^{2+}-buffered media supplemented with cytosolic K^+ and Na^+ concentrations.

Calcium stores have an important function in maintaining low intracellular Ca^{2+} concentrations (in conjunction with the Ca^{2+} pumps in the plasma membrane). Furthermore, the ER functions in intracellular signaling as a reservoir for Ca^{2+} that is mobilized in response to a messenger (IP_3, ryanodine, Ca^{2+}; Rasmussen and Barret, 1984; Berridge, 1993). The generally accepted concept is that all cells equipped with mitochondria and ER regulate their Ca^{2+} concentration in a comparable way; the mitochondria function as a Ca^{2+} sink when the Ca^{2+} levels surpass levels of 0.5 μmol liter^{-1}, whereas the ER is able to function as a fine regulator thanks to the high affinity of its Ca^{2+} pumps (Van Os *et al.*, 1988). In agreement with this notion, fish gill cells were found to have an IP_3-releasable Ca^{2+} pool (Flik *et al.*, 1993) that is thapsigargin-sensitive and has a high affinity for Ca^{2+} (it takes up Ca^{2+} in an ATP-dependent way in a medium containing 0.1 μmol liter^{-1} Ca^{2+}); a ruthenium red-sensitive pool was demonstrated that has a low affinity for Ca^{2+} (mitochondrial uptake starts at concentrations higher than 0.3 μmol liter^{-1}). The inhibitor of mitochondrial Ca^{2+} uptake, ruthenium red, has no effect on the uptake measured at 0.1 μmol liter^{-1} Ca^{2+}, but almost completely inhibits uptake at 1.0 μmol liter^{-1} Ca^{2+} (P. M. Verbost, unpublished results).

From the foregoing one may conclude that with respect to cellular calcium homeostasis, the gill cells are much like other cells. Also, their intracellular Ca^{2+} concentration is maintained at low resting values around 0.1 μmol liter^{-1} as in other cells. Many cellular events are regulated by fluctuations in free Ca^{2+} triggered by receptor activation on the cell surface (Berridge, 1994). To date it has not been determined which fish hormones act via controlling the release of Ca^{2+} from the stores in the gill cells, although there is little doubt that there will be some. The topic of Ca^{2+} as second messenger needs a lot of attention at the level of the stores themselves; for instance, how are they refilled after a Ca^{2+} discharge (Putney and Bird, 1993)? A potentially important related finding is that endogenous ouabain—as an adrenal cortical hormone or a paracrine hormone—could play a role in the refilling of the stores and consequently increase the potential responsiveness of the cells (Blaustein, 1993). If the underlying mechanism (inhibition of Na^+/K^+-ATPase and subsequent uptake of Ca^{2+} mediated by Na^+/Ca^{2+} exchange) also functions in fish gills, it could cause differences in Ca^{2+} influx or responsiveness to calcium-regulatory hormones. A summary of Ca^{2+} transport mechanisms in branchial ionocytes is given in Fig. 4.

V. INTERACTION OF CADMIUM WITH TRANSCELLULAR CALCIUM TRANSPORT

Cd^{2+} interferes in a specific fashion with Ca^{2+} transport mechanisms and this is generally explained by the many similarities between Ca^{2+} and Cd^{2+}. Cd is positioned between zinc and mercury in group IIb of the

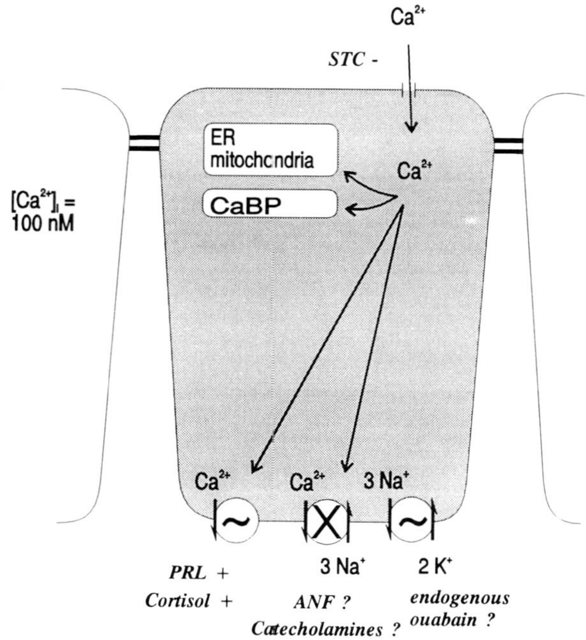

Fig. 4. Diagram of Ca^{2+} transport mechanisms in the ionocyte of the gills and their possible modes of regulation. Water and blood calcium concentrations are in the millimolar range. Ca^{2+} from the water enters the cell passively down an electrochemical gradient through stanniocalcin (STC)-controlled Ca^{2+} channels. In the cytosol, as yet undefined calcium-binding proteins (CaBP) and sequestering of Ca^{2+} in endoplasmic reticulum (ER) and mitochondria maintain cytosolic Ca^{2+} levels ($Ca^{2+}{}_i$) around 100 nmol liter^{-1}. In the basolateral plasma membrane, two Ca^{2+} pumps may extrude Ca^{2+} to the blood, a high-affinity Ca^{2+}-ATPase and a Na^+/Ca^{2+} exchanger (dependent on the Na^+ gradient maintained by the Na^+/K^+-ATPase). The calcitropic hormones cortisol (in trout) and prolactin (PRL; in tilapia and eel) determine the relative abundance of the ATPases and by doing so control the calcium transport capacity of the gills. Catecholamines and atrial natriuretic factor (ANF), through its second messenger cGMP, should be considered as potential regulators of the exchanger. Endogenous ouabain may influence cellular calcium homeostasis by controlling Na^+/K^+-ATPase activity and hence Na^+-dependent Ca^{2+} extrusion via the exchanger.

periodic table. It has an atomic mass of 112.4 and an ionic radius of 97 pm (Ca^{2+} = 99 pm). Compared to Ca, Cd is more polarizable, because of its larger number of electrons and more polarizing as a result of the organization of its electron shells. Important for the ion activity in solution are the stability constants of complexes with the metal ions. In particular, nitrogen- and sulfur-group-containing ligands will form strong complexes with Cd and more so when the groups are deprotonated (Aylett, 1979).

Cd has become widely spread in our environment (in water as well as air) as a result of zinc smelting and electroplating industries. Cd is considered a nonessential element for life and nanomolar concentrations are reported as toxic (Verbost *et al.*, 1987a). In a chemical sense, the cadmium ion has a very mobile character (comparable to Ca^{2+}) and this may explain the toxicity of cadmium as well as its omnipresence.

In particular, waterborne Cd is toxic for fishes (McDonald and Wood, 1993). A characteristic of Cd poisoning is that sublethal levels induce hypocalcemia, in all likelihood through an inhibition of branchial Ca^{2+} uptake (Verbost *et al.*, 1987a). The interaction of cadmium with Ca^{2+} transport through the branchial epithelium may occur at three distinct levels: the entry step, the transfer through the cytosol, and the extrusion step.

a. The Entry Step. In Section III,A, it was discussed that Cd enters the branchial epithelium via a stanniocalcin-regulated Ca^{2+} pathway, probably a SMOC-type Ca^{2+} channel. Therefore, this topic will not be further discussed here. Interestingly, an alternative mode of entry for Cd in cells was presented by Lou and coworkers (1991), who showed that monovalent cation complexes of Cd, $Cd(OH)(HCO_3)_2^-$, and $Cd(OH)(HCO_3)Cl^-$ enter human red blood cells via a HCO_3^-/Cl^- exchanger. A comparable exchanger has also been demonstrated in fish gills (Perry *et al.*, 1984), but a possible involvement of this pathway in Cd uptake has not received attention to our knowledge. Whatever the mechanism, Cd easily and rapidly enters the branchial epithelium of fishes (Verbost *et al.*, 1989), and this is when problems arise. Reported levels of waterborne Cd will never inhibit Ca^{2+} entry by competition at the apical entry step.

b. Transfer through the Cytosol. In the cytosol, Cd may bind to a variety of high-affinity Ca^{2+} receptors, such as Ca^{2+} binding proteins and Ca^{2+} sites on ATPases, and to sulfhydryl groups as in metallothioneins. Actual measurements of cytosolic Cd levels are not technically feasible at present, but we assume by deduction that the levels of ionized Cd^{2+} will normally be very low in the cytosol. The effects of sequestering mechanisms (binding to proteins, extrusion to intracellular compartments) on the passage of Cd through the cytosol have received little attention so far. Transepithelial Cd^{2+} flux (typical values around 50 nmol h^{-1} per kg^{-1} fish at 100 nmol $liter^{-1}$ waterborne Cd) is readily measured in gills (Pärt and Svanberg, 1981; Verbost *et al.*, 1989), and this means that significant cytosolic Cd traffic occurs. However, the exact nature of the cytosolic shuttle mechanisms for Ca^{2+} as well as Cd^{2+} in fish ionocytes are unknown.

c. The Extrusion Step. Substantial evidence has been presented that the ATP-driven calcium pump in the gills of fishes is inhibited by Cd (Verbost *et al.*, 1988). In experiments with plasma membrane vesicles it was demonstrated that the ATPase does not detectably translocate Cd^{2+}. The high-affinity binding of Cd^{2+} to the Ca^{2+} transport site of the calcium pump most likely renders the ATPase inactive. Yet the Cd flux through the epithelium is too high to be explained by diffusion, and therefore the transport via some other carrier protein was anticipated. Evidence for carrier-mediated Cd^{2+} extrusion across fish plasma membranes was provided by Schoenmakers and coworkers (1992a), working on the Ca^{2+} carriers in the plasma membrane of the enterocyte of tilapia. Cd^{2+} was shown to inhibit Ca^{2+} transport over the basolateral plasma membrane mediated by ATPase as well as Na^+/Ca^{2+} exchange activity. However, the exchanger transported Cd^{2+} at a significant rate in its Ca^{2+}/Ca^{2+} mode, and this mechanism may form a basis for transcellular Cd^{2+} flux. However, this mechanism does not appear to be sufficient to guarantee cellular calcium homeostasis, as exposure of fishes to waterborne Cd invariably increases apoptosis of ionocytes (Pratap and Wendelaar Bonga, 1993), regulated cell death that is initiated by increasing cytosolic Ca^{2+} levels (Berridge, 1994).

REFERENCES

Aylett, B. J. (1979). The chemistry and bioinorganic chemistry of cadmium. *In* "The Chemistry, Biochemistry and Biology of Cadmium" (M. Webb, ed.), pp. 1–44. Elsevier, Amsterdam.

Bacskai, B. J., and Friedman, P. A. (1990). Activation of latent Ca^{2+} channels in epithelial cells are activated by parathyroid hormone. *Nature* (*London*) **347**, 388–391.

Baker, P. F., Blaustein, M. P., Hodgkin, A. L., and Steinhardt, R. A. (1969). The influence of calcium on sodium efflux in squid axons. *J. Physiol.* (*London*) **200**, 431–458.

Berg, A. (1968). Studies on the metabolism of calcium and strontium in freshwater fish. Relative contribution of direct and intestinal absorption. *Mem. Ist. Ital. Idrobiol. Dott. Marco de Marchi* **23**, 161–169.

Berridge, M. (1993). Intracellular channels and calcium spiking. *Biomed. Res.* **14**(Suppl. 2), 21–27.

Berridge, M. (1994). The biology and medicine of calcium signalling. *Mol. Cell. Endocrinol.* **98**, 119–124.

Björnsson, B. T., and Nilsson, S. (1985). Renal and extrarenal excretion of calcium in the marine teleost, *Gadus morhua*. *Am. J. Physiol.* **248**, R18–R22.

Blaustein, M. P. (1993). Physiological effects of endogenous ouabain—Control of intracellular Ca^{2+} stores and cell responsiveness. *Am. J. Physiol.* **264**, C1367–C1387.

Block, M., and Pärt, P. (1992). Uptake of ^{109}Cd by cultured gill epithelial cells from rainbow trout (*Oncorhynchus mykiss*). *Aquat. Toxicol.* **23**, 137–151.

Burgess, G. M., McKinney, J. S., Fabiato, A. A., Leslie, A., and Putney, J. W. (1983). Calcium pools in saponin-permeabilized guinea pig hepatocytes. *J. Biol. Chem.* **258,** 15336–15345.

Burghardt, J. S. (1993). Ca^{2+} transport (*in vitro*) in the opercular epithelium of freshwater adapted *Fundulus heteroclitus*. Honours B.Sc. Thesis, St. Francis Xavier University, Antigonish, Nova Scotia, Canada.

Carafoli, E. (1987). Intracellular calcium homeostasis. *Annu. Rev. Biochem.* **56,** 395–433.

Chartier, M.-M., Milet, C., Martelly, E., Lopez, E., and Warrot, S. (1979). Stimulation par le vitamin D_3 et le 1,25-dihydroxyvitamine D_3 de l'absorption intestinale du calcium chez l'anguille (*Anguilla anguilla* L.) *J. Physiol.* (*Paris*) **75,** 275–282.

Collie, N. L., and Hirano, T. (1984). Effects of stanniectomy on intestinal calcium transport in the Japanese eel. *Zool. Sci.* **1,** 966.

Doneen, B. A. (1993). High-affinity Ca^{2+}-Mg^{2+}-ATPase in kidney of euryhaline *Gillichthys mirabilis:* Kinetics, subcellular distribution and effects of salinity. *Comp. Biochem. Physiol.* **106,** 719–728.

Evans, D. H. (1979). Osmotic and ionic regulation by freshwater and marine fishes. *In* "Environmental Physiology of Fishes" (M. A. Ali, ed.), pp. 93–122. Plenum, New York.

Fenwick, J. C. (1976). Effects of stanniectomy on calcium activated adenosinetriphosphate activity in the gills of freshwater adapted North American eels, *Anguilla rostrata* LeSueur. *Gen. Comp. Endocrinol.* **29,** 383--387.

Fenwick, J. C. (1979). Ca^{2+}-activated adenosinetriphosphatase activity in the gills of freshwater- and seawater-adapted eels (*Anguilla rostrata*). *Comp. Biochem. Physiol., B* **62,** 67–70.

Fenwick, J. C. (1984). Effect of vitamin D_3 (cholecalciferol) on plasma calcium and intestinal ^{45}calcium absorption in goldfish, *Carassius auratus* L. *Can. J. Zool.* **62,** 34–36.

Fenwick, J. C. (1989). Calcium exchange across fish gills. *In* "Vertebrate Endocrinology: Fundamentals and Biomedical Implications" (P. K. T. Pang and M. P. Schreibman, eds.), Vol. 3, pp. 319–342. Academic Press, San Diego, CA.

Fenwick, J. C., Smith, K., Smith, J., and Flik, G. (1984). Effects of various vitamin D analogs on plasma calcium and phosphorus and intestinal calcium absorption in fed and unfed American eels, *Anguilla rostrata*. *Gen. Comp. Endocrinol.* **55,** 398–404.

Flanagan, J. L., and Friedman, P. A. (1991). Parathyroid hormone-stimulated cadmium accumulation in Madin-Darby canine kidney cells. *Toxicol. Appl. Pharmacol.* **109,** 241–250.

Fletcher, C. R. (1978). Osmotic and ionic regulation in the cod (*Gadus callarias* L.). I. Water balance. *J. Comp. Physiol.* **124,** 149–155.

Flik, G. (1990). Hypocalcin physiology. *In* "Progress in Comparative Endocrinology" (A. Epple, C. G. Scanes, and H. H. Stetson, eds.), pp. 578–585. Wiley-Liss, New York.

Flik, G., and Perry, S. F. (1989). Cortisol stimulates whole body calcium uptake and the branchial calcium pump in freshwater rainbow trout. *J. Endocrinol.* **120,** 75–82.

Flik, G., and Verbost, P. M. (1993). Calcium transport in fish gills and intestine. *J. Exp. Biol.* **184,** 17–29.

Flik, G., Reijntjens, F. M. J., Stikkelbroeck, J., and Fenwick, J. C. (1982). 1,25-Vit. D_3 and calcium transport in the gut of tilapia (*Sarotherodon mossambicus*). *J. Endocrinol.* **94,** 40.

Flik, G., Wendelaar Bonga S. E., and Fenwick, J. C. (1983). Ca^{2+}-dependent phosphatase and Ca^{2+}-dependent ATPase activity in plasma membranes of eel gill epithelium. I. Identification of Ca^{2+}-activated ATPase activities with non-specific phosphatase activities. *Comp. Biochem. Physiol. B* **76,** 745–754.

Flik, G., Wendelaar Bonga, S. E., and Fenwick, J. C. (1984a). Ca^{2+}-dependent phosphatase and Ca^{2+}-dependent ATPase activities in plasma membranes of eel gill epithelium. II. Evidence for transport high-affinity Ca^{2+}-ATPase. *Comp. Biochem. Physiol.* **79,** 9–16.

Flik, G., Wendelaar Bonga, S. E., and Fenwick, J. C. (1984b). Ca^{2+}-dependent phosphatase and Ca^{2+}-dependent ATPase activities in plasma membranes of eel gill epithelium. III. Stimulation of branchial high-affinity Ca^{2+}-ATPase activity during prolactin-induced hypercalcemia in American eels. *Comp. Biochem. Physiol.* **79,** 521–524.

Flik, G., van Rijs, J. H., and Wendelaar Bonga, S. E. (1985a). Evidence for high-affinity Ca^{2+}-ATPase and ATP-driven Ca^{2+}-transport in membrane preparations of the gill epithelium of the cichlid fish *Oreochromis mossambicus. J. Exp. Biol.* **119,** 335–347.

Flik, G., Fenwick, J. C., Kolar, Z. I., Mayer-Gostan, N., and Wendelaar Bonga, S. E. (1985b). Whole-body calcium flux rates in cichlid teleost fish *Oreochromis mossambicus* adapted to fresh water. *Am. J. Physiol.* **249,** R432–R437.

Flik, G., Wendelaar Bonga, S. E., and Fenwick, J. C. (1985c). Active Ca^{2+} transport in plasma membranes of branchial epithelium of the North-American eel, *Anguilla rostrata* LeSueur. *Biol. Cell.* **55,** 265–272.

Flik, G., Fenwick, J. C., Kolar, Z. I., Mayer-Gostan, N., and Wendelaar Bonga, S. E. (1986a). Effects of ovine prolactin on calcium uptake and distribution in *Oreochromis mossambicus. Am. J. Physiol.* **250,** R161–R166.

Flik, G., Fenwick, J. C., Kolar, Z., Mayer-Gostan, N., and Wendelaar Bonga, S. E. (1986b). Effects of low ambient calcium levels on whole body Ca^{2+} flux rates and internal calcium pools in the freshwater cichlid teleost, *Oreochromis mossambicus. J. Exp. Biol.* **120,** 249–264.

Flik, G., Fenwick, J. C., and Wendelaar Bonga, S. E. (1989). Calcitropic actions of prolactin in freshwater North American eel (*Anguilla rostrata* LeSueur). *Am. J. Physiol.* **257,** R74–R79.

Flik, G., Schoenmakers, T. J. M., Groot, J. A., van Os, C. H., and Wendelaar Bonga, S. E. (1990) Calcium absorption by fish intestine: The involvement of ATP- and sodium-dependent calcium extrusion mechanisms. *J. Membrane Biol.* **113,** 13–22.

Flik, G., van der Velden, J. A., Dechering, K. J., Verbost, P. M., Schoenmakers, T. J. M., Kolar, Z. I., and Wendelaar Bonga, S. E. (1993). Ca^{2+} and Mg^{2+} transport in gills and gut of tilapia, *Oreochromis mossambicus:* A review. *J. Exp. Zool.* **265,** 356–365.

Flik, G., Rentier-Delrue, F., and Wendelaar Bonga, S. E. (1994). Calcitropic effects of recombinant prolactins in *Oreochromis mossambicus. Am. J. Physiol.* **266,** R1302–R1308.

Foskett, J. K., and Scheffey, C. (1982). The chloride cell: Definitive identification as the salt-secretory cell in teleosts. *Science* **215,**, 164–166.

Fraser, R. A., Kaneko, T., Pang, P. K. T., and Harvey, S. (1991). Hypo- and hypercalcemic peptides in fish pituitary glands. *Am. J. Physiol.* **260,** R622–R626.

Furukawa, K. I., Ohshima, N., Tawada-Iwata, Y., and Shigekawa, M. (1991). Cyclic GMP stimulates Na^{+}/Ca^{2+} exchange in vascular smooth muscle cells in primary culture. *J. Biol. Chem.* **266,** 12337–12341.

Ghijsen, W. E. J. M., and Van Os, C. H. (1982). 1α,25-Dihydroxy vitamin D3 regulates ATP-dependent calcium transport in basolateral plasma membranes of rat enterocytes. *Biochim. Biophys. Acta* **689,** 170–172.

Ghijsen, W. E. J. M., De Jong, M. D., and Van Os, C. H. (1982). ATP-dependent calcium transport and its correlation with Ca^{2+} ATPase activity in basolateral membranes of rat duodenum. *Biochim. Biophys. Acta* **689,** 327–336.

Ghijsen, W. E. J. M., De Jong, M. D., and Van Os, C. H. (1983). Kinetic properties of Na^{+}/Ca^{2+}-exchange in basolateral plasma membranes of rat small intestine. *Biochim. Biophys. Acta* **730,** 85–94.

Ghosh, T. K., Bian, J., Short, A. D., Rybak, S. L., and Gill, D. L. (1991). Persistent intracellular calcium pool depletion by thapsigargin and its influence on cell growth. *J. Biol. Chem.* **266,** 24690–24697.

Grover, A. K., and Khan, I. (1992). Calcium pump isoforms: Diversity, selectivity and plasticity. *Cell Calcium* **13,** 9–17.

Hanssen, R. G. J. M., Lafeber, F. P. J. G., Flik, G., and Wendelaar Bonga, S. E. (1989). Ionic and total calcium levels in the blood of the European eel (*Anguilla anguilla*): Effects of stanniectomy and hypocalcin replacement therapy. *J. Exp. Biol.* **141,** 177–186.

Hennessen, U., Chabanis, S., Silva, S. L., Nabarra, B., Drueke, T., and Lacour, B. (1993). Increased brush-border membrane calcium transport in the intestine, but not the kidney tubule, of spontaneously hypertensive rats. *Am. J. Hypertens.* **6,** 593–601.

Hirano, T. (1986). The spectrum of prolactin actions in teleosts. *In* "Comparative Endocrinology: Developments and Directions" (C. L. Ralph, ed.). Alan R. Liss, New York.

Hirano, T. (1989). The corpuscles of Stannius. *In* "Vertebrate Endocrinology: Fundamentals and Biomedical Implications" (P. K. T. Pang and M. P. Schreibmann, eds.), Vol. 3, pp. 139–170. Academic Press, San Diego, CA.

Hwang, P.-P. (1989). Distribution of chloride cells in teleost larvae. *J. Morphol.* **200,** 1–8.

Ichii, T., and Mugiya, Y. (1983). Effects of a dietary deficiency in calcium on growth and calcium uptake from the aquatic environment in the goldfish, *Carassius auratus*. *Comp. Biochem. Physiol., A* **74,** 259–262.

Iger, Y., and Wendelaar Bonga, S. E. (1994). Cellular responses of the skin of carp (*Cyprinus carpio*) exposed to acidified water. *Cell Tissue Res.* **275,** 481–492.

Karnaky, K. J., Jr., Kinter, L. B., Kinter, W. B., and Stirling, C. E. (1976). Teleost chloride cell. II. Autoradiographic localization of gill Na/K-ATPase in killifish, *Fundulus heteroclitus*, adapted to low and high salinity environments. *J. Cell Biol.* **70,** 157–177.

Kaune, R., Kassianoff, I., Schroeder, B., and Harmeyer, J. (1992). The effects of 1,25-dihydroxyvitamin D-3 deficiency on calcium transport and calcium uptake into brush-border membrane vesicles from pig small intestine. *Biochim. Biophys. Acta* **1109,** 187–194.

Khoyi, M. A., Bjur, R. A., and Westfall, D. P. (1991). Norepinephrine increases Na–Ca exchange in rabbit abdominal aorta. *Am. J. Physiol.* **261,** C685–C690.

Kijima, Y., Ogunbunmi, E., and Fleischer, S. (1991). Drug action of thapsigargin on the Ca^{2+} pump protein of sarcoplasmic reticulum. *J. Biol. Chem.* **266,** 22912–22918.

Klaren, P. H. M., Flik, G., Lock, R. A. C., and Wendelaar Bonga, S. E. (1993). Ca^{2+} transport across intestinal brush border membranes of the cichlid teleost *Oreochromis mossambicus*. *J. Membr. Biol.* **132,** 157–166.

Lafeber, F. P. J. G., Flik, G., Wendelaar Bonga, S. E., and Perry S. F. (1988). Hypocalcin from Stannius corpuscles inhibits gill calcium uptake in trout. *Am. J. Physiol.* **254,** R891–R896.

Laurent, P., Höbe, H., and Dunel-Erb, S. (1985). The role of environmental sodium chloride relative to calcium in gill morphology of freshwater salmonid fish. *Cell Tissue Res.* **240,** 675–692.

Lin, S. H., and Russell, W. E. (1988). Two Ca^{2+}-dependent ATPases in rat liver plasma membrane. *J. Biol. Chem.* **263,** 12253–12258.

Lou, M., Garay, R., and Alda, J. O. (1991). Cadmium uptake through the anion exchanger in human red blood cells. *J. Physiol.* (*London*) **443,** 123–136.

Ma, S. W. Y., and Copp, D. H. (1978). Purification, properties and action of a glycopeptide from the corpuscles of Stannius which affects calcium metabolism in the teleost. *In* "Comparative Endocrinology" (P. J. Gaillard and H. H. Boer, eds.), pp. 283–286. Elsevier/North- Holland Biomed. Press, Amsterdam.

Maetz, J. (1974). Aspects of adaptation to hypo-osmotic and hyperosmotic environments. *In* "Biochemical and Biophysical Perspectives in Marine Biology" (D. C. Malins and J. R. Sargent, eds.), Vol. 1, pp. 1–167. Academic Press, New York.

Maetz, J., and Bornancin, M. (1975). Biochemical and biophysical aspects of salt excretion by chloride cells in teleost fish. *Fortschr. Zool.* **23,** 322–362.

Marshall, W. S., Bryson, S. E., Burghardt, J. S., and Verbost, P. M. (1995). Ca^{2+} transport by opercular epithelium of the freshwater-adapted euryhaline teleost, *Fundulus heteroclitus. J. Comp. Physiol. B* (in press).

Marshall, W. S., Bryson, S. E., and Wood, C. M. (1992). Calcium transport by isolated skin of rainbow trout. *J. Exp. Biol.* **166,** 297–316.

Mayer-Gostan, N., Bornancin, M., DeRenzis, G., Naon, R., Yee, J. A., Shew, R. L., and Pang, P. K. T. (1983). Extraintestinal calcium uptake in the killifish, *Fundulus heteroclitus. J. Exp. Zool.* **227,** 329–338.

McCormick, S. D., Hasegawa, S., and Hirano, T. (1992). Calcium uptake in the skin of a freshwater teleost. *Proc. Natl. Acad. Sci. U.S.A* **89,** 3635–3638.

McDonald, D. G., and Wood, C. M. (1993). Branchial mechanisms of acclimation to metals in freshwater fish. *In* "Fish Ecophysiology" (J. C. Rankin and F. B. Jensen, eds.), pp. 297–321. Chapman & Hall, London.

McWilliams, P. G., and Potts, W. T. W. (1978). The effects of pH and calcium concentrations on gill potentials in brown trout, *Salmo trutta. J. Comp. Physiol.* **126,** 277–286.

Ogino, C., and Takeda, H. (1978). Requirements of rainbow trout for dietary calcium and phosphorus. *Bull. Jpn. Soc. Sci. Fish.* **44,** 1019–1022.

Oh, Y., Smith, P. R., Bradford, A. L., Keeton, D., and Benos, D. J. (1993). Regulation by phosphorylation of purified epithelial sodium channels in planar lipid bylayers. *Am. J. Physiol.* **265,** C85–C91.

Pang, P. K. T. (1973). Endocrine control of calcium metabolism in teleosts. *Am. Zool.* **13,** 775–792.

Pang, P. K. T., Griffith, R. W., Maetz, J., and Pic, P. (1980). Calcium uptake in fishes. *In* "Epithelial Transport in the Lower Vertebrates" (B. Lahlou, ed.), pp. 122–132. Cambridge Univ. Press, Cambridge, UK.

Pang, P. K. T., Pang, R. K., and Wendelaar Bonga, S. E. (1989). Hormones and calcium regulation in vertebrates: An evolutionary and overall consideration. *In* "Vertebrate Endocrinology: Fundamentals and Biomedical Implications" (P. K. T. Pang, and M. P. Schreibmann, eds.), Vol. 3, pp. 343–352. Academic Press, San Diego, CA.

Pärt, P., and Svanberg, O. (1981). Uptake of cadmium in perfused rainbow trout (*Salmo gairdneri*) gills. *Can. J. Fish. Aquat. Sci.* **38,** 917–924.

Perry, S. F., and Flik, G. (1988). Characterization of branchial transepithelial calcium fluxes in freshwater trout, *Salmo gairdneri. Am. J. Physiol.* **254,** R491–R498.

Perry, S. F., and Wood, C. M. (1985). Kinetics of branchial calcium uptake in the rainbow trout: Effects of acclimation to various external calcium levels. *J. Exp. Biol.* **116,** 411–433.

Perry, S. F., Payan, P., and Girard, J. P. (1984). Adrenergic control of branchial chloride transport in the isolated perfused head of the freshwater trout (*Salmo gairdneri*). *J. Comp. Physiol., B* **154,** 269–274.

Perry, S. F., Goss, G. G., and Fenwick, J. C. (1992). Interrelationships between gill chloride cell morphology and calcium uptake in freshwater teleosts. *Fish Physiol. Biochem.* **10,** 327–337.

Pisam, M., Caroff, A., and Rambourg, A. (1987). Two types of chloride cells in the gill epithelium of a freshwater adapted euryhaline fish: *Lebistes reticulatus;* Their modifications during adaptation to saltwater. *Am. J. Anat.* **179,** 40–50.

Pisam, M., Prunet, P., Boeuf, G., and Rambourg, A. (1988). Ultrastructural features of chloride cells in the gill epithelium of the Atlantic salmon, *Salmo salar*, and their modifications during smoltification. *Am. J. Anat.* **183,** 235–244.

Poncet, V., Merot, J., and Poujeol, P. (1992). A calcium-permeable channel in the apical membrane of primary cultures of the rabbit distal bright convoluted tubule. *Pfluegers Arch. Eur. J. Physiol.* **422,** 112–119.

Pratap, H. B., and Wendelaar Bonga, S. E. (1993). Effect of ambient and dietary cadmium on pavement cells, chloride cells, and Na^+/K^+-ATPase activity in the gills of the freshwater teleost *Oreochromis mossambicus* at normal and high calcium levels in the ambient water. *Aquat. Toxicol.* **26,** 133–150.

Putney, J. W., and Bird, G. S. (1993). The inositol phosphate–calcium signalling system in nonexcitable cells. *Endocr. Rev.* **14,** ME241–ME631.

Rasmussen, H., and Barret, P. (1984). Calcium messenger system: An integrated view. *Physiol. Rev.* **64,** 938–984.

Schoenmakers, T. J. M., Klaren, P. H. M., Flik, G., Lock, R. A. C., Pang, P. K. T., and Wendelaar Bonga, S. E. (1992a). Actions of cadmium on basolateral plasma membrane proteins involved in calcium uptake by fish intestine. *J. Membr. Biol.* **127,** 161–172.

Schoenmakers, T. J. M., Visser, G. J., Flik, G., and Theuvenet, A. P. R. (1992b). Chelator: An improved method for computing metal ion concentrations in physiological solutions. *BioTechniques* **12,** 870–879.

Schoenmakers, T. J. M., Verbost, P. M., Flik, G., and Wendelaar Bonga, S. E. (1993). Transcellular intestinal calcium transport in freshwater and seawater fish and its dependence on sodium/calcium exchange. *J. Exp. Biol.* **176,** 195–206.

Strehler, E. E. (1991). Recent advances in the molecular characterization of plasma membrane Ca^{2+} pumps. *J. Membr. Biol.* **120,** 1–15.

Sundell, K., and Björnsson, B. T. (1988). Kinetics of calcium fluxes across the intestinal mucosa of the marine teleost, *Gadus morhua*, measured using an *in vitro* perfusion method. *J. Exp. Biol.* **140,** 170–186.

Sundell, K., and Björnsson, B. T. (1990). Effects of vitamin D_3, 25(OH) vitamin D_3, 24,25$(OH)_2$ vitamin D_3 and 1,25$(OH)_2$ vitamin D_3 on the *in vitro* intestinal calcium absorption in the marine teleost, Atlantic cod (*Gadus morhua*). *Gen. Comp. Endocrinol.* **78,** 74–79.

Sundell, K., Björnsson, B. T., Itoh, H., and Kawauchi, H. (1992a). Chum salmon (*Oncorhynchus keta*) stanniocalcin inhibits *in vitro* intestinal calcium uptake in Atlantic cod (*Gadus morhua*). *J. Comp. Physiol., B* **162,** 489–495.

Sundell, K., Bishop, J. E., Björnsson, B. T., and Norman, A. W. (1992b). 1,25-Dihydroxyvitamin D3 in the Atlantic Cod: Plasma levels, a plasma binding component, and organ distribution of a high affinity receptor. *Endocrinology* (*Baltimore*) **131,** 2279–2286.

Sundell, K., Norman, A. W., and Björnsson, B. T. (1993). 1,25$(OH)_2$ vitamin D3 increases ionized plasma calcium concentrations in the immature Atlantic cod *Gadus morhua*. *Gen. Comp. Endocrinol.* **91,** 344–351.

Swarup, K., and Srivastav, S. P. (1982). Note: Vitamin D_3-induced hypercalcemia in male catfish, *Clarias batrachus*. *Gen. Comp. Endocrinol.* **46,** 271–274.

Swarup, K., Das, V. K., and Norman, A. W. (1991). Dose-dependent vitamin D_3 and 1,25-dihydroxyvitamin D_3-induced hypercalcemia and hyperphosphatemia in male cyprinoid *Cyprinus carpio*. *Comp. Biochem. Physiol., A* **100,** 445–447.

Takagi, Y., Nakamura, Y., and Yamada, J. (1985). Effects of the removal of corpuscles of Stannius on the transport of calcium across the intestine of rainbow trout. *Zool. Science* **2,** 523–530.

Taylor, A. (1989). The role of sodium–calcium exchange in sodium-transporting epithelia. *In* "Sodium–Calcium Exchange" (T. J. A. Allen, D. Noble, and H. Reuter, eds.), pp. 298–323. Oxford Univ. Press, Oxford.

Van Heeswijk, M. P. E., Geertsen, J. A. M., and Van Os, C. H. (1984). Kinetic properties of the ATP-dependent Ca^{2+} pump and the Na^{+}/Ca^{2+} exchange system in basolateral membranes from rat kidney cortex. *J. Membr. Biol.* **79,** 19–31.

Van Os, C. H., Van den Broek, L. A. M., Van Corven, E. J. J. M., Timmermans, J. A. H., and Dirven, H. (1988). Calcium homeostasis of epithelial cells. *Comp. Biochem. Physiol., A* **90,** 767–770.

Verbost, P. M., Flik, G., Lock, R. A. C., and Wendelaar Bonga, S. E. (1987a). Cadmium inhibition of Ca^{2+} uptake in rainbow trout gills. *Am. J. Physiol.* **253,** R216–R221.

Verbost, P. M., Senden, M. H. M. N., and Van Os, C. H. (1987b). Nanomolar concentrations of Cd^{2+} inhibit Ca^{2+} transport systems in plasma membranes and intracellular Ca^{2+} stores in intestinal epithelium. *Biochim. Biophys. Acta* **902,** 247–252.

Verbost, P. M., Flik, G., Lock, R. A. C., and Wendelaar Bonga, S. E. (1988). Cadmium inhibits plasma membrane calcium transport. *J. Membr. Biol.* **102,** 97–104.

Verbost, P. M., van Rooij, J., Flik, G., Lock, R. A. C., and Wendelaar Bonga, S. E. (1989). The movement of cadmium through freshwater trout branchial epithelium and its interference with calcium transport. *J. Exp. Biol.* **145,** 185–97.

Verbost, P. M., Butkus, A., Atsma, W., Willems, P., Flik, G., and Wendelaar Bonga, S. E. (1993a). Studies on stanniocalcin: Characterization of bioactive and antigenic domains of the hormone. *Mol. Cell. Endocrinol.* **93,** 11–16.

Verbost, P. M., Flik, G., Fenwick, J. C., Greco, A. M., Pang, P. K. T., and Wendelaar Bonga, S. E. (1993b). Branchial calcium uptake: Possible mechanisms of control by stanniocalcin. *Fish Physiol. Biochem.* **11,** 205–215.

Verbost, P. M., Schoenmakers, T. J. M., Flik, G., and Wendelaar Bonga, S. E. (1994). Kinetics of ATP- and Na^{+}-gradient driven Ca^{2+} transport in basolateral membranes from gills of freshwater and seawater adapted tilapia. *J. Exp. Biol.* **186,** 95–108.

Wagner, G., and Friesen, H. G. (1989). Studies on the structure and physiology of salmon teleocalcin. *Fish Physiol. Biochem.* **7,** 367–374.

Wakasugi, H., Kimura, T., Haase, W., Kribben, A., Kaufmann, R., and Schulz I. (1982). Calcium uptake into acini from rat pancreas: Evidence for intracellular ATP-dependent calcium sequestration. *J. Membr. Biol.* **65,** 205–220.

Wassermann, R. H., and Fullmer, C. S. (1983). Calcium transport proteins, calcium absortion, and vitamin D. *Annu. Rev. Physiol.* **45,** 375–390.

Wendelaar Bonga, S. E., and Pang, P. K. T. (1986). Stannius corpuscles. *In* "Vertebrate Endocrinology: Fundamentals and Biomedical Implications" (P. K. T. Pang and M. P. Schreibman, eds.), Vol. 1, pp. 439–464. Academic Press, Orlando, FL.

Wendelaar Bonga, S. E., and Pang, P. K. T. (1992). Control of calcium regulating hormones in the vertebrates: Parathyroid hormone, calcitonin, prolactin, and stanniocalcin. *Int. Rev. Cytol.* **128,** 139–213.

Wendelaar Bonga, S. E., Löwick, C. J. M. and Van der Meij, J. C. A. (1983). Effects of external Mg^{2+} and Ca^{2+} on branchial osmotic water permeability and prolactin secretion in the teleost fish *Sarotherodon mossambicus. Gen. Comp. Endocrinol.* **52,** 222–231.

Wendelaar Bonga, S. E., Flik, G., Balm, P. H. M., and van der Meij, J. C. A. (1990). The ultrastructure of chloride cells in the gills of the teleost *Oreochromis mossambicus* during exposure to acidified water. *Cell Tissue Res.* **259,** 575–585.

Young, P. S., McCormick, S. D., Demarest, J. R., Lin, R. J., Nishioka, R. S., and Bern, H. A. (1988). Effects of salinity, hypophysectomy, and prolactin on whole-animal transepithelial potential in the tilapia *Oreochromis mossambicus. Gen. Comp. Endocrinol.* **71,** 389–397.

INDEX

G

H

I

J

K

M

N

O

P

R

U

OTHER VOLUMES IN THE FISH PHYSIOLOGY SERIES